Lecture Notes in Computer Science 1172

Edited by G. Goos, J. Hartmanis and J. van Leeuwen

Springer
Berlin
Heidelberg
New York
Barcelona
Budapest
Hong Kong
London
Milan
Paris
Santa Clara
Singapore
Tokyo

Josef Pieprzyk   Jennifer Seberry   (Eds.)

Information Security and Privacy

First Australasian Conference, ACISP'96
Wollongong, NSW, Australia, June 24-26, 1996
Proceedings

 Springer

Series Editors

Gerhard Goos, Karlsruhe University, Germany

Juris Hartmanis, Cornell University, NY, USA

Jan van Leeuwen, Utrecht University, The Netherlands

Volume Editors

Josef Pieprzyk
Jennifer Seberry
The University of Wollongong, Department of Computer Science
Wollongong, NSW 2522, Australia
E-mail: josef@cs.uow.edu.au

Cataloging-in-Publication data applied for

Die Deutsche Bibliothek - CIP-Einheitsaufnahme

Information security and privacy : first Australasian conference
; proceedings / ACISP '96, Wollongong, NSW, Australia, June
24 - 26, 1996. Josef Pieprzyk ; Jennifer Seberry (ed.).
[Sponsored by Australasian Society for Electronic Security and
Center for Computer Security Research, University of
Wollongong, NSW. In Co-operation with The International
Association for Cryptologic Research (IACR)]. - Berlin ;
Heidelberg ; New York ; Barcelona ; Budapest ; Hong Kong ;
London ; Milan ; Paris ; Santa Clara ; Singapore ; Tokyo :
Springer, 1996
  (Lecture notes in computer science ; Vol. 1172)
  ISBN 3-540-61991-7
NE: Pieprzyk, Josef [Hrsg.]; ACISP <1, 1996, Wollongong, New South
Wales>; GT

CR Subject Classification (1991): E.3, K.6.5, D.4.6, C.2, E.4,F.2.1-2, K.4.1,
K.5.1

1991 Mathematics Subject Classification: 94A60, 68P20, 68Q20, 68Q25

ISSN 0302-9743
ISBN 3-540-61991-7 Springer-Verlag Berlin Heidelberg New York

Typesetting: Camera-ready by author
SPIN 10548709   06/3142 – 5 4 3 2 1 0   Printed on acid-free paper

PREFACE

ACISP'96 was the inaugural conference on Information Security and Privacy. The conference was sponsored by the Australasian Society for Electronic Security (ASES) and the Center for Computer Security Research, University of Wollongong. The conference was organized in co-operation with the International Association for Cryptologic Research (IACR). The aim of the conference was to provide an opportunity for researchers from the region to meet, to discuss ongoing projects in the areas of Information Security and Privacy, and to plan and co-ordinate professional activity in the region.

The Program Committee invited three distinguished speakers: Mr Steve Orlowski, Mr Eugene Troy, and Dr Ravi Sandhu. Mr Orlowski from Attorney General's Office talked about Information Technology Security. Mr Troy from National Institute of Standards and Technology, USA gave an up-to-date account of the current activity in IT security criteria and product evaluations. Dr Sandhu from George Mason University, USA emphasized the importance of access control in information security designs.

A public forum entitled *The Internet and Privacy* which was addressed by the invited speakers, William Caelli from Queensland University of Technology, Vijay Varadharajan from University of Western Sydney, Jennifer Seberry from University of Wollongong, John Kane from Stanton and Partners, and Lyal Collins from the Commonwealth Bank of Australia provided a lively evening discussion.

The proceedings contains 26 refereed papers. Papers were presented in ten sessions covering access control, security models, intrusion detection, authentication, hashing, secret sharing, and encryption.

We are pleased to thank all the members of the Program Committee: Mark Anderson, Lawrie Brown, William Caelli, Chris Charnes, Nick Demytko, Ed Dawson, Jovan Golić, Andrzej Gościński, Thomas Hardjono, Svein Knapskog, Keith Martin, Luke O'Connor, Rei Safavi-Naini, Vijay Varadharajan, XianMo Zhang, and Yuliang Zheng.

We wish to thank all the authors for sending their submissions (successful or otherwise), the speakers, and all the participants of ACISP'96 conference.

Wollongong, New South Wales, Australia
September 1996

Josef Pieprzyk
Jennifer Seberry

AUSTRALASIAN CONFERENCE ON
INFORMATION SECURITY AND PRIVACY
ACISP'96

Sponsored by

Australasian Society for Electronic Security
and

Center for Computer Security Research
University of Wollongong, NSW

In co-operation with

The International Association for Cryptologic Research
(IACR)

General Chair:
Jennifer Seberry   *University of Wollongong*

Program Chair:
Josef Pieprzyk   *University of Wollongong*

Program Committee:

Mark Anderson	*DSTO*
Lawrie Brown	*University of NSW*
William Caelli	*Queensland University of Technology*
Chris Charnes	*University of Wollongong*
Nick Demytko	*Telstra Research Laboratories*
Ed Dawson	*Queensland University of Technology*
Jovan Golić	*Queensland University of Technology*
Andrzej Gościński	*Deakin University*
Thomas Hardjono	*University of Western Sydney*
Svein Knapskog	*University of Trondheim, Norway*
Keith Martin	*University of Adelaide*
Luke O'Connor	*Queensland University of Technology*
Rei Safavi-Naini	*University of Wollongong*
Vijay Varadharajan	*University of Western Sydney*
XianMo Zhang	*University of Wollongong*
Yuliang Zheng	*Monash University*

Organizing Committee (all from University of Wollongong):
Margot Hall, Mitra Heydari, Hossein Ghodosi, Mansour Esmaili.

CONTENTS

Session 8: THRESHOLD CRYPTOGRAPHY
Chair: E. Dawson

Session 9: HASHING
Chair: J. Golić

Session 10: ODDS AND ENDS
Chair: L. Brown

The Changing Face of Information Technology Security

Steve Orlowski
Special Adviser
IT Security Policy
Security Division
Australian Attorney-Generals Department

The views in this paper are those of the author and do not necessarily represent the views of the Australian Government

The Government in its pre-election policy statement *Australia Online* [1] said:

> *The development of private electronic commerce requires the construction of a regulatory and institutional framework which will convert the communications network from a chaotic electronic frontier into a secure marketplace where transactions can be made with the highest level of consumer confidence.*

> *The coalition recognises the scope of new technologies to dramatically improve business competitiveness and the critical importance of Australia achieving a first mover advantage in our geographical region.*

> *The combination of a stable regulatory framework and cheap, world-class telecommunications will be the platform upon which Australians will develop and market online services to Australia and the rest of the world.*

> *Four key areas of regulation requiring urgent reform are copyright law, privacy and security law, financial trading systems, and online content standards.*

The OECD in its 1992 *Guidelines for the Security of Information Systems* [2] said:

> *Recent years have witnessed ... growth of computer use to the point that, in many countries, every individual is an actual or potential user of computer and communication networks.*

Over the past two years, the OECD has embarked on a series of meetings on Global Information Infrastructures. The OECD interest in the Global Information Infrastructure relates not only to the direct impact of the infrastructures on national economies, but also on the economic impact of investment failures if the infrastructure is misused or not used to its expected capacity. User confidence is seen as a key factor in infrastructures reaching their full potential. It is from this position that the OECD is examining issues of security, privacy, the protection of intellectual property and the use of cryptography.

Turning again to the OECD Guidelines [2], they stated when addressing the question of building confidence:

> *Users must have confidence that information systems will operate as intended without unanticipated failures or problems. Otherwise, the systems and their underlying technologies may not be exploited to the extent possible and further growth and innovation may be inhibited.*

With the rapid expansion of technology, individuals and criminals now have access to sophisticated computer and communications equipment and software. Some of this equipment/software allows data and messages to be encrypted. Additionally this technology transcends national borders. Indeed, messages between parties within a country may, in fact, be routed through another country. How we control the use of such networks for criminal purposes while still respecting civil liberties will need to be resolved. This presents a problem for law enforcement authorities.

Cryptographic techniques have also led to the creation of anonymous electronic cash. Steps will need to be developed to ensure that such techniques do not facilitate money laundering or transfer of proceeds of crime. The Australian Government has established an Inquiry under the Chair of the Australian Cash Transactions Reporting Agency, AUSTRAC, to investigate the implications of the new technologies on law enforcement.

However, while much of the debate has focussed on the use of cryptography to protect data in transit, it has been at the cost of general IT security and good housekeeping. The majority of computer crimes such as the capture of credit card details do not occur while data is travelling over communications lines. They occur because someone has hacked into a system and recovered them, or placed a sniffer, or from the activities of insiders. It is unfortunate that the public debate on cryptography has moved attention away from the need to use traditional security techniques to ensure that the equipment itself is secure.

With increasing national dependency on computer networks, computer crimes may have an impact on the national interest. It would be possible to undermine the financial system for example by attacking the computers on which it is based. Moreover, what may appear to be a criminal attack could, feasibly, be an attack by another country with the object of undermining the national economy. The area of information warfare is starting to become widely debated.

We are also starting to see the emergence of requirements to extensively test mission critical computer systems before they are put into use. Australian Standard AS4400-1995 requires that:

> *...where systems are critical for the health and wellbeing of the patient the systems and software should be independently certified as being suitable for that use.*

Much has been discussed about the type of material available on networks. In particular, child pornography and material on bomb making has been of concern. A number of prosecutions have occurred as a result of investigations into child pornography rings operating over the Internet. Similarly there have been a significant number of instances of children harming themselves with homemade bombs based on instructions down loaded from Internet. These issues have been addressed by the Task Force into the Regulation of Online Services and are now being examined in a study by the Australian Broadcasting Authority.

Another emerging area of concern is the use of Internet for harassing or stalking individuals. Flaming is a minor form of this but increasingly there have been instances of people receiving death threats or sexually offensive messages by email. The obscene or threatening phone call has now moved to Internet. Persons have also been tracked through their use of Internet.

We are now also starting to see attacks on home pages; either by defacing or by changing information presented on them. In the case of business this could defeat the purpose for which an organisation had established its page. Techniques such as check sums on home page contents can allow such changes to be detected but in the meantime some damage could have been done to an organisations reputation.

Intellectual property protection is also a major issue. There are groups that advocate that the Internet should be completely open. I dont subscribe to that view. For Internet to reach its full potential there should be a wide variety of information available. However, authors of the material should be paid for their work if they so desire. If not, they will not make the work available. Similarly an authors reputation can suffer if their works are displayed in an unacceptable manner by someone copying them and displaying them in a different context. Techniques such as steganography where electronic watermarks can be incorporated in material, and access control techniques, can reduce this problem.

In 1980, the OECD issued *Guidelines Governing the Protection of Privacy and Transborder Flows of Personal Data* [3]. One of the Principles set out in the Guidelines was:

Security Safeguards Principle

> *Personal data should be protected by reasonable security safeguards against such risks as loss or unauthorised access, destruction, use modification or disclosure of data.*

In other words the focus of security measures was on the confidentiality and integrity of the data. The Guidelines also contained a principle which limited the use of data to the purpose for which it was collected. They did not, however, address the issue of the data being available for the purpose for which is was collected.

In 1992, the OECD issued *Guidelines for the Security of Information Systems* [2]. The Guidelines defined the security objective as:

> *The objective of security of information systems is the protection of the interests of those relying on information systems from harm resulting from failures of availability , confidentiality, and integrity.*

The following definitions were included in the guidelines:

> *availability means the characteristic of data, information and information systems being accessible and useable on a timely basis in the required manner;*
>
> *confidentiality means the characteristic of data and information being disclosed only to authorised persons, entities and processes at authorised times and in the authorised manner;*
>
> *integrity means the characteristic of data and information being accurate and complete and the preservation of accuracy and completeness.*

This took the approach beyond the protection of data to include the information systems themselves. We were looking at data and equipment.

In a footnote to its description of an information system, the Guidelines [2] recognised that the technology would change. They said:

> *The dynamism of information and communication technology dictates that this description of information systems may serve only to give an indication of the present situation and that new*

technological developments will arise to augment the potentialities of information systems.

Not only were these words prophetic, but the development and uptake of the new technologies has been more rapid than anyone envisaged at that time. These new technologies have not only changed the definition of information systems, but they have bought us to the stage where we have to reconsider our approach to IT security itself.

As mentioned before, the focus of security up until now has been on the data and equipment. Implicit in this was the assumption that there was a direct relationship between the owner of the data and the user. Access was based on the owner knowing who the user was and deciding whether to grant access to the data or equipment. This decision was quite often based on information concerning authorised users which was contained in the system itself. The traditional view of the hacker was someone who operated outside these parameters.

We are now entering a new world where we dont necessarily know who our user is. With the growth of the World Wide Web, data is becoming freely available for browsing and consequently issues of confidentiality do not always apply. Furthermore, the increasing availability of material offered by content providers on a fee for access basis has changed the traditional concept of confidentiality. In these cases the content providers want users to access their data. In many cases their business viability depends on it. In these cases it is not the confidentiality of the data which is at issue. Rather it is ensuring payment for access to the data. Having said that, there may still be a confidentiality issue concerning the re transmission of the data once it has been downloaded.

Additionally we may not be the owner of some of the data on our systems or alternatively our data may be on systems owned by others. Service providers now allow subscribers to generate their own home pages on their servers. In other cases they are merely providing links through which users and data travel. This has already raised questions of responsibility and liability for the material involved. Who owns the data on a server; the operator of the server or the originator of the data? Who then makes the rules regarding access to the data? Who then is responsible for the security of the data?

We are also moving to a situation where people within the system we control have access to systems outside our control. How do we control their activities outside our system to ensure they do not export data, import maleficient code, or simply incur charges for our organisation because of their outside activities?

These are just some of the issues which are starting to arise.

The role of the user and the owner of data and equipment has acquired greater significance as we have moved to the global information infrastructure. We now need to reexamine the approach to security to take users into account rather than just the data and the equipment alone.

The OECD Guidelines for the Security of Information Systems [2] are useful as they establish not so much the measures which should be taken to secure information systems, but rather the framework within which security should be applied. It addresses such issues as accountability, ethics, democracy, and awareness which are not traditionally seen as security matters. However, they are an integral part of a holistic approach to information security management.

Access control is a traditional element of information security and is now gaining increasing significance. Access will increasingly be granted on the basis of information provided by the user rather than information under the owners control. Owners are therefore seeking to accumulate more data on who has accessed there systems and what has been accessed.

There is a potential for misuse of this information to generate profiles of individual users. The aggregation of this data from a number of systems can create a powerful database for use by direct marketers. However, owners of information, particularly with intellectual property rights attached, are unlikely to make it available if they cannot ascertain who has accessed it and, where appropriate establish that the appropriate payment has been made. On the other hand, users may not wish to make use of the full range of material available on the Internet if they feel that such access is likely to result in personal profiles being built and used against them. There is, therefore a need to find a balance between the two requirements.

The development of anonymous cash techniques can allow a user to anonymously access material while still making the required payment for such access. However, this involves a trade off in security as, in the event of misuse of the system, the system owner has no way of knowing who has had access to the system. If anonymous access is permitted, the owner will need to ensure that adequate steps are taken to protect the system and information. While audit trails can still be used to ascertain what might have happened, they are of no assistance in determining who might have been responsible.

If a non anonymous access approach is either not appropriate, for example where warranties or licenses might attach to sale of goods, or not implemented, protection must be provided for users of the system. This could take the form of regulation or codes of conduct designed to protect the users of systems

Finally, as we become more dependent on systems, we need to be reassured that they do what they are supposed to do. For example as systems are increasingly used to collect, as well as store, personal data, we need to be able to rely on those systems to properly collect the data and transfer it to the appropriate storage area. We need to be able to trust the applications as well as protect the areas where the data is stored.

Authentication

Authentication techniques is an area where confidence needs to be engendered to ensure acceptance of new technologies. There is a need for a structure through which keys can be obtained and digital signatures authenticated in a trusted manner.

Much of the debate on cryptography has centred around encryption and the problems it poses for law enforcement and national security. This approach has largely failed to differentiate between cryptography for authentication,integrity and non repudiation; and cryptography for confidentiality. A general examination of the cryptography debate shows as much concern for items such as credit card information as it does for the message content.

I would like to go further and suggest that when faced with a choice between protecting personal data and protecting information likely to result in financial disadvantage, most people will prefer to protect the latter. The readiness of individuals to surrender data on buying patterns in return for incentives under loyalty schemes supports this assertion.

A significant part of the cryptography debate centres around the misuse of credit card information. This information is useless where systems are put into place which require authentication along with the credit card number. Similarly, while financial institutions are concerned about the confidentiality of their transactions, they are more concerned about people misusing their systems for financial advantage. In a significant proportion of activities, authentication takes precedence over confidentiality.

The distinction between encryption and authentication has been recognised by the US Government which does not impose export controls on cryptography for authentication purposes only. The problem is that most cryptographic techniques currently available do not differentiate between the two applications.

The time has come for academics and the computer industry to make this distinction. Resources need to be applied to developing systems which limit the use of cryptography to authentication. In parallel, key generation techniques need to be developed which ensure that keys generated for authentication can only be used in authentication applications. Once this has been achieved, the debate on law enforcement and national security access to encrypted communications can be seen in its proper perspective.

Public Key Infrastructures

A number of countries are developing public key infrastructures to facilitate the national development of electronic commerce. All recognise that such schemes need to interoperate for electronic commerce to develop internationally. Unfortunately some of these schemes are directed towards business applications and do not pay sufficient attention to individual users.

While business might account for the high value transactions, as the global information infrastructure develops, individual users are going to make up the largest volume of transactions. Many governments are developing national infrastructures to allow individual users access to a greater range of services. It is important that these planners recognise that customers are the base level of the business tree and provide facilities for users to transact their business with confidence.

Many of the schemes emerging combine the confidentiality and authentication functions in their key management schemes. They also provide for government access to keys held within such schemes. This will tend to destroy confidence in these key management infrastructures. People have already raised the possibility of law enforcement using an authentication key to fabricate electronic evidence tied to an individual. While the likelihood of such an event occurring is low, the possibility may reduce confidence in the use of such schemes, resulting in their failure.

Key Management Infrastructure - United States

In May this year, the United States Office of Management and Budget, released a draft paper *Enabling Privacy, Commerce, Security and Public Safety in the Global Information Infrastructure* [4] which proposed a Key Management Infrastructure.

The proposed scheme was voluntary and involved the establishment of government licensed certificate authorities. These authorities would hold both private and public keys and offer certification services when they are used. The scheme does not differentiate between keys used for authentication purposes and those used for confidentiality. The scheme also allows law enforcement access to the keys upon lawful authority. Self escrow is also allowable under the scheme provided performance requirements for law enforcement access are met.

The paper states that law enforcement access would only be to keys used for confidentiality purposes and not to signing keys. However, the fact that the signing key may be the same as the confidentiality key, which is accessible, can reduce confidence in the scheme. Additionally, by implication, persons choosing not to participate in the scheme cannot obtain the certification benefits available under the scheme.

The paper also proposes relaxation of export controls where keys are held in the country of destination and a bilateral agreement exists which will allow access to those keys upon lawful authority. As this provides overseas users with the choice of stronger encryption if they choose to agree to the keys being held, it is an advance over the existing situation where such choice is not available.

Trusted Third Parties - United Kingdom

In June this year the United Kingdom released a *Paper on Regulatory Intent Concerning Use of Encryption on Public Networks* [5]. The paper proposes a scheme of trusted third parties licensed by government. The scheme does not differentiate between confidentiality and authentication keys. It just states that private keys will be held in escrow. Again the scheme is voluntary however there is a stronger implication that persons who are not licensed under the scheme will not be able to participate in electronic commerce. The paper states that if the scheme were adopted, the UK would consider the relaxation of export controls.

Trusted Third Parties - France

France has also indicated that it is considering a trusted third party scheme. However, participation in the scheme would be mandatory to obtain a licence to obtain cryptography products.

Public Key Authentication Framework - Australia

Within Australia a Government Group developed a proposal for a Public Key Authentication Framework. The group's work was primarily focused on the needs of electronic commerce. In an unpublished paper [6] the group stated:

> There needs to be a wide scale informed debate about this issue before any decisions are taken as to choice of technology, the appropriate administrative structure, privacy issues, legal effect, method of implementation and the like. After such a debate the system will need to be introduced in a planned way with appropriate public education, legislation and the like in order that the use of the PKAF system will have the same standing and validity in the eyes of the community as a paper based signature

Since the project commenced, it has been recognised that the procedures involved could be applied to electronic documents in records management systems and access control, as well as to transactions in electronic commerce.

The proposal calls for a management structure to verify various key generation systems, supervise the issue of key pairs and maintain a directory of the public keys.

This proposal was referred to the Standards Association of Australia which established a task force to examine the establishment of an Australian Public Key Authentication Framework. The Task Force released a Draft Standard DR 96078 for public comment. The Task Force has redrafted its report on the basis of those comments and the final version is expected to be released in the next few weeks.

The basic philosophy behind the scheme is to develop a structure based on a series of minimum standards, both technological and managerial, which, when supplemented by legislation, would give an electronic document, signed with a digital signature, the same legal status as a paper document with a handwritten signature.

The key pairs will be either issued or certified at the bottom level of the chain but may be generated at a higher level depending on the structure of the organisation acting as certifying authority. For example a national organisation may wish to generate keys at a central facility and then distribute them at the lower level A smaller regional organisation would issue keys on the spot. Keys will not be generated by the national body which will manage the scheme. An applicant will need to establish their identity through the equivalent of the 100 points scheme used for passports or bank accounts before a key pair is linked to that identity.

It is proposed that the national body would be a company limited by guarantee with representatives of government, business and user groups on its board. It is not proposed that it be a government body.

The scheme would not prevent people generating and using their own key pairs. However, as the legislative support would be based on minimum standards established under the scheme, self generated key pairs would not have the legal standing of key pairs generated under the scheme unless it can be established that they were generated in accordance with the minimum standards set out in the scheme. It is proposed that a digital signature generated under the scheme would be presumed to be genuine unless it could be proved otherwise.

The scheme is designed purely for authentication and not for the issue of key pairs for encryption. This differs from the Key Management Infrastructure proposed by the United States and the Trusted Third Party schemes proposed by the United Kingdom and France.

It is recognised, however, that current technology will not necessarily prevent individuals from utilising keys obtained through this scheme for encryption purposes. The Task Force felt that this issue should be discussed in a different forum. It supports the need for technology to be developed which is capable of separating the two functions.

Australia has raised in the OECD the need to establish an international framework to ensure the effective use of public key cryptography as a tool for both international electronic commerce and individual use of the global information infrastructure.

While this proposal is driven, primarily, by commercial needs, there is scope for it to be extended to meet the needs of individuals who will also be using the information infrastructure. Any scheme such as this has to be better than the current process of passing credit card information over the network.

In the Australian context, any legislation required to support the use of cryptography in electronic commerce is likely to be written in broad terms rather than endorsing particular technology or algorithms. It would then be left to groups such as Standards Australia to specify the standards which at that particular point in time would meet the legislative requirement.

Chip Based Authenticators

Public acceptance of the new technology will depend, in part, on confidence that the privacy of personal information provided will be protected and that access will be restricted to the individual or persons entitled to access the information for official purposes.

With the advent of public access to information through government and community networks, users will need to be able to establish their entitlement to access information. Similarly the advent of electronic commerce will require the use of digital signatures to conduct the transaction. Electronic documents, for records purposes, will also need to be digitally signed. We are, therefore, moving to the stage where individuals will require a unique electronic authenticator to transact business on the superhighway. Such authenticators are likely to take the form of a digital signature on chip based devices such as smart cards or wrist bands.

Such devices could be issued within the public key authentication framework referred to earlier. An individual could apply to a certifying authority for a device which would be loaded with the secret key component of their digital signature. The device would be issued upon satisfaction of the one hundred point criteria currently used by banks to open accounts.

Technology also exists for blind signatures and anonymous cash transactions. It may be possible for both digital signature and anonymous cash techniques to be embedded in the one device, which the user could opt to use in either mode. Obtaining a device would, of course, be optional although there would be circumstances where it may be necessary for individuals to authenticate their identity to obtain access to a particular service. Regulations or Codes Of Practice issued by the Privacy Commissioner could limit the circumstances in which use of the authenticator is mandatory, in much the same way as use of tax file numbers is controlled.

The devices could then perform a number of activities. For example one concern has been access to restricted material over the network. It may be possible to convert date of birth information in the identified section of the chip into an anonymous age field which could be added when required to demonstrate an entitlement to restricted classification material. Access to restricted information could be limited to those who could demonstrate their age in this way.

As mentioned earlier, obtaining such devices would be optional, the user would determine what information other than basic identifying information would be held on the chip in addition to the authentication technique. The user would control which of this information would be released through a PIN pad or biometric input on the device.

The digital signature would be used in much the same way as a written signature is used to authenticate a person signing a document. As mentioned earlier, the Standards Australia Discussion Paper [7] raises the question of the legislative requirements to give legal effect to digital signatures.

OECD Activities

In 1995 the OECD established a Group of Experts on Security, Privacy and Intellectual Property Protection in the Global Information Infrastructure [7]. The Group met for the first time in Canberra in February this year. As a result of that meeting, the Group established an Ad Hoc Group of Experts on Cryptography Policy Guidelines to try to develop an international approach to the use of cryptography which would provide security and protect privacy, while recognising the needs of government to ensure public safety. The Ad Hoc Group is tasked to report by February 1997.

Other issues the Group of Experts is likely to address include a review of the existing guidelines for security and privacy, regulation of content, protection of intellectual property rights and securing electronic commerce.

As we move towards the 21st century, the issues of IT security will continue to change with the technology. We therefore need to focus more on the principles of what we are trying to achieve, rather than specific approaches. We need to be flexible enough to be able to respond to changes both in technology and the threat. This will require a stronger emphasis on risk assessment and the development of organisational IT security policies which focus on objectives rather than mechanics of implementing specific security regimes. It also means that both information owners and users need to have a better understanding of the issues and principles involved and how balances might be achieved.

Above all we need to ensure that security regimes generate public confidence in infomation infrastructures. With out that confidence, the infrastructures will not reach their full potential. In short we need more education and informed debate if we are to achieve our objective of an information superhighway which everyone feels they can use with confidence

References

[1]   *Australia Online* Liberal Party National Party Pre Election Policy Statement Canberra February 1996

[2]   OECD *Guidelines for the Security of Information Systems* Paris 1992 Document No OCDE/GD(92)190

[3]   OECD *Guidelines Governing the Protection of Privacy and Transborder Flows of Personal Data* Paris 1980

[4]   *Enabling Privacy, Commerce, Security and Public Safety in the Global Information Infrastructure* Draft Paper Office of Management and Budget Washington May 1996

[5]   *Paper on Regulatory Intent Concerning Use of Encryption on Public Networks* Department of Trade and Industry London June 1996

[6]   Unpublished Working Paper, *Public Key Authentication Framework (PKAF)* Australian Attorney-Generals Department, Canberra July 1993

[7]   Papers presented at the conference held in conjunction with the first meeting are available at   http://www.nla.gov.au/gii/oecdconf.html

Replicating the *Kuperee* Authentication Server for Increased Security and Reliability

(Extended Abstract)

Thomas Hardjono[1,2] and Jennifer Seberry[1]

[1] Centre for Computer Security Research, University of Wollongong,
Wollongong, NSW 2522, Australia
[2] Department of Computing and Information Systems,
University of Western Sydney - Macarthur, NSW 2560, Australia

The current work proposes a new scheme for the replication of authentication services in Kuperee based on a public key cryptosystem, in response to the two main shortcomings of the traditional single server solutions, namely those of low availability and high security risks. The work represents further developments in the Kuperee authentication system. The Kuperee server is presented in its simplified design to aid the presentation of the replication scheme. The replication approach is based on the sharing of session public keys, together with a threshold or secret sharing scheme. However, unlike previous approaches, in the current work the object to be shared-out is instead a session secret key which is not directly available to the (untrusted) Client. The scheme gains advantages deriving from the use of public key cryptology, as well as from the manner in which the secret is shared-out. A comparison with the notable work of Gong (1993) is also presented.

1 INTRODUCTION

Developments of networks and nodes in terms of size and number in the world today has made the issue of security is a pressing one. Expansion of these networks, such as the Internet, has brought various possibilities for electronic commerce. In such electronic-based activities a certain level of assurance must be provided, both for the users and the service-providers. This, in turn, has brought into the foreground the security issues related to such networks. These issues include the authenticity of users, the non-repudiation of a user's transactions, integrity and privacy of information traveling within the network, and a host of other issues.

Authentication of users and service-providers within a network is an important foundation stone for other security-related services, independent of whether these are provided by the network or by specific service-providers in the network. The variety of authentication systems and protocols, together with the fact that the encipherment of data must often accompany authentication events, point to the higher layers (eg. in the OSI/ISO stack or the TCP/IP stack) as being the best location to provide such security-related services.

The current work proposes a new scheme for the replication of authentication services based on a public key cryptosystem, in response to the two main shortcomings of the traditional single server solutions, namely those of low availability and high security risks. The work represents further developments in the *Kuperee* authentication system which was earlier presented in [1].

The choice of a public-key cryptosystem in Kuperee is largely motivated by conviction that as the needs of authentication increases with the expansion of networks, public-key techniques may offer better solutions to the problem of authentication in large-scale networks, compared to the approaches based exclusively on shared-key (symmetric) cryptosystems. This use of a public key, specifically that of [2], distinguishes Kuperee from other systems and provides it with more flexibility in tailoring the protocol steps following the particular demands. The use of the cryptosystems of [2] allows Kuperee to act either as an Authenticating Authority or as a Certifying Authority, or as a simultaneous combination of both. This ability is important in situations in which a special session key must be provided by an authority trusted by two communicating parties, eventhough both parties are already in possession of each other's (publicly available) public-key which have been generated and proclaimed by the parties respectively.

The earlier design of Kuperee suffered from a number of deficiencies which were noticeable soon after the design was published in [1]. The design was also unreasonably complicated. These problems prompted for Kuperee's simplification, reported in [3]. Some known deficiencies of the original design of [1] have again been confirmed recently in [4].

Kuperee, like other authentication systems – such as Kerberos [5, 6, 7] – was originally based on the use of a single authentication server to cater for clients in a given domain [1]. Although this traditional single-server approach has practical advantages, two of its more important shortcomings concerns the availability and performance of the server and the security of the server itself [8]. First, since entities in the distributed system rely on the server for their authentication operations, the server easily becomes a source of contention or bottleneck. The temporary unavailability of the server can lead to a degradation in the overall performance of the distributed system. Secondly, the fact that the server may hold cryptographic (secret) keys belonging to entities in the distributed system makes it the best point of attack for intruders wishing to compromise the distributed system. An attacker that successfully compromises the server has access to the (secret) keys of the entities under the server's jurisdiction. The attacker can then carry-out various active and passive attacks on the unsuspecting entities.

The use of a threshold scheme [9, 10] among a collection of security servers necessitates a client to consult a minimum of t out of M ($t \leq M$) honest servers before the authentication process succeeds or services are granted. This approach, besides increasing the overall security of the distributed system, also reduces the bottleneck present in the previous single-server solutions. An attacker wishing to compromise the cryptographic keys employed within the distributed system must compromise at least t of the M servers, something considerably harder than a single server. The problem of bottlenecks is eased by the existence of multiple servers, only a subset of which is required by the clients for an instance of authentication.

In the next section the background of Kuperee for the sharing of session keys will be presented. This is followed by the new scheme for the replication of authentication services in Kuperee in Section 3. Section 4 offers some comparisons with an existing approach, while Section 5 closes the paper with some remarks and conclusions. Readers wishing more specific details on the public key cryptosystem employed in Kuperee are directed to the Appendix.

2 AUTHENTICATION IN KUPEREE

In the authentication system the entities that interact with the Authentication Server are called *principals* [5]. The term can be used for Users, Clients or Servers. Commonly, a user directs a Client (eg. a program) on a machine to request a service provided by a another server (or another Client) on a remote machine.

There are a number of ways that the Kuperee server can be employed, either as an authentication server or as a certification server, or both [3]. In this current work we first employ Kuperee specifically as an authentication server, mimicking the actions of the Needham-Schroeder protocol [11, 12] within the Kerberos authentication system [5, 7]. That is, Kuperee will be used to deliver a pair of session keys to the principals that require secure and authentic communications. This is then followed by the description of a protocol in which the public-keys within Kuperee are used directly by the principals, thereby making Kuperee rather as a certification authority in the system.

In brief, the interactions within the authentication service consists of the Client requesting the Key Distribution Center (KDC) for a *ticket-granting ticket* to be submitted by the Client to the Ticket Granting Server (TGS). The TGS then authenticates the Client and issues the Client with a *service ticket* which has a shorter lifetime compared to the ticket-granting ticket. On presentation by the Client, the service ticket is used by the service-provider to authenticate the Client, since the service-provider places trust on the TGS (Figure 1). These two stages are also referred to as the credential-initialization and client-server authentication protocols respectively in [13].

2.1 Kuperee: Notations

In the public key cryptosystem of [2] a secret key is chosen randomly and uniformly from the integers in $[1, p - 1]$, where the prime p is public and is used to generate the multiplicative group $GF(p)^*$ of the finite field $GF(p)$. The generator of this multiplicative group is denoted as g and it is a publicly known value. This usage of g and p is inspired by notable works of [14] and [15]. The corresponding public key is then produced by using the secret key as the exponent of g modulo p. (In the remainder of this paper all exponentiations are assumed to be done over the underlying groups).

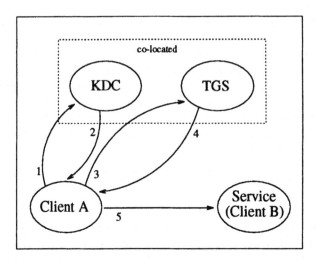

Fig. 1. Authentication in Kuperee following Kerberos

The KDC, the Client, the TGS and the Service provider (ie. destination server) have the secret-public key pairs $(X_{kdc}, Y_{kdc} \equiv g^{X_{kdc}})$, $(X_c, Y_c \equiv g^{X_c})$ $(X_{tgs}, Y_{tgs} \equiv g^{X_{tgs}})$, and $(X_s, Y_s \equiv g^{X_s})$ respectively.

The operation "$\{\}_K$" means that the contents within the braces "$\{\}$" are enciphered or deciphered using the key K. Thus, assuming that Y_c is the public-key of the Client, the operation "$\{\}_{Y_c}$" signifies encryption using the publicly known key Y_c of the Client using the modified encryption algorithm of [2]. The operation provides both confidentiality and integrity. All encrypted messages implicitly includes a nonce.

An important aspect of the protocol is that although a ticket is shown to be encrypted using the public-key of the targeted receiver, the sender's secret-key participates in the encryption. That is, there is *always* sender authentication built into every instance of encryption. In Kuperee, sender authentication and message secrecy are treated as inseparable in a single step. Hence, the receiver knows the identity of the sender and can verify the cryptogram as coming from the (alleged) sender, due to the fact that the receiver must use the sender's public-key in order to decipher the cryptogram.

2.2 Kuperee: Sharing of a Session Secret-Public Key Pair

Kuperee mimics Kerberos in that a session secret-public key pair is shared among the KDC-TGS-Client, while another is shared among the TGS-Client-Server. The session key pair is discarded after one instance of authentication.

Here, we assume that the KDC knows the public-key of all principals in the domain. The TGS knows the public key of the KDC and all Clients, while the Client knows the public-key of the TGS and the KDC. The KDC is assumed to have the highest level of trust equivalent to one holding private keys of the principals. This is necessary as the KDC will also function as a certification authority for the public keys of the entities in the domain. A Client wishing to communicate to another Client may request from the KDC a certified copy of the public key belonging to the second client (and vice versa).

In the following, each session secret-public key pair is denoted by $(k, K \equiv g^k)$, and their subscripts indicates which principals employ the key pair. For clarity, the session keys are shown next to the ticket, rather than within the ticket.

1. Client $\rightarrow$ KDC: c, tgs, N_c
2. KDC $\rightarrow$ Client: $\{K_{c,tgs}, tgs, N_c, N_{kdc}\}_{Y_c}, \{k_{c,tgs}, N_{kdc}, c, lifetime\}_{Y_{tgs}}$

    The KDC first generates the session key pair $(k_{c,tgs}, K_{c,tgs})$ to be used between the Client and the TGS.

    The session "public" key $K_{c,tgs}$ is delivered to the Client, enciphered under the Client's public-key. The session "secret" key $k_{c,tgs}$ is given to the TGS (via the client) encrypted under the TGS's public-key.

3. Client $\rightarrow$ TGS: $\{A_c, s, N_{kdc}\}_{K_{c,tgs}}, \{k_{c,tgs}, N_{kdc}, c, lifetime\}_{Y_{tgs}}, N'_c$

    The Client uses the session key $K_{c,tgs}$ obtained from the KDC to encipher the authenticator $A_c = (c, timestamp)$ [5, 7] destined for the TGS.

4. TGS $\rightarrow$ Client: $\{K_{c,s}, s, N'_c, N_{tgs}\}_{k_{c,tgs}}, \{k_{c,s}, N_{tgs}, c, lifetime\}_{Y_s}$

    Here the key $k_{c,tgs}$ is not used directly in the manner of public keys. The TGS uses the session key $k_{c,tgs}$ to compute another key $r_{tgs,c}$ as:

$$r_{tgs,c} \equiv (Y_c)^{X_{tgs}+k_{c,tgs}}$$

The Client can recover the key as

$$r_{tgs,c} \equiv (Y_{tgs} K_{c,tgs})^{X_c}$$

since Y_{tgs} is public.

Note that the Client cannot fabricate $\{k_{c,s}, N_{tgs}, c, lifetime\}_{Y_s}$ without being detected by the recipient Server. This is because in deciphering it, the Server will use keys X_s and Y_{tgs} simultaneously. Only these two keys can succeed. In order to fabricate it, the Client would need to know X_{tgs} which is known only to the TGS.

5. Client $\rightarrow$ Server: $\{A_c, N''_c, N_{tgs}\}_{K_{c,s}}, \{k_{c,s}, N_{tgs}, c, lifetime\}_{Y_s}$

    The Client uses the session key $K_{c,s}$ to encipher the authenticator A_c. The Server deciphers it using the matching secret session key $k_{c,s}$ obtained from within the ticket.

6. Server $\rightarrow$ Client: $\{c, s, N''_c\}_{k_{c,s}}$

If required, the Server may respond to the Client's request of proving the Server's identity. This can be done by the Server using the session key $k_{c,s}$ to create $r_{s,c}$ as:

$$r_{s,c} \equiv (Y_c)^{X_s + k_{c,s}}$$

which is recoverable by the Client as:

$$r_{s,c} \equiv (K_{c,s}\, Y_s)^{X_c}$$

since only the Client knows X_c.

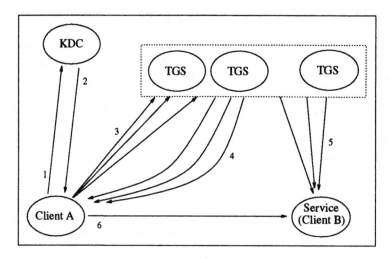

Fig. 2. Authentication in Kuperee using several TGSs

3   REPLICATING KUPEREE

In this section we extend the proposed protocol of the previous section based on a session secret-public key pair for the purpose of authentication using replicated TGSs. The replication approach is based on the use of secret-sharing schemes (or threshold schemes) [9, 10] to achieve a minimum consensus of honest TGSs. In the current work we do not present any secret-sharing scheme, although the practical scheme of [16] is currently under proposal for Kuperee.

Previous solutions – such as that in [8] – employ a thresholding function to split a parameter (eg. symmetric key) into a number of *shares*, t out of M of which is required at least to recover the origin.l parameter.

In the current work we also employ a thresholding function to achieve the same basic effect. However, in this case the parameter to be shared-out consists of a session *secret* key which is not directly available to the Client. This conforms with the underlying motivation that there should be a hierarchy of trust assignment in the systems, with the KDC being the most trusted and the Client the least. It is therefore preferable for the Client to hold the public half of the session key-pair, with the Server keeping the secret half.

The current approach is divided into two basic phases. The first is carried-out by the KDC in a continuous manner, where the KDC generates and distributes the keys and the shares to the multiple TGSs. The second phase is invoked whenever a Client requires access to the Server.

3.1 Share Generation and Distribution Phase

In this phase the KDC generates a set of session key pairs (secret and public) and associates an identity number with each pair. The secret session key is then broken into a number of shares. These shares, together with the share identity and the session public-key are then distributed to the TGSs in the system.

More specifically, given the session key pair $(k_{c,s}, K_{c,s})$ and given M of TGSs in the system, the KDC generates an identity $ID_{c,s}$ associated with the session key pair and splits the session secret-key into M pieces $P_{c,s_1}, P_{c,s_2}, \ldots, P_{c,s_M}$, where $P_{c,s_i} = thresh_{t.M}(k_{c,s}, i)$ for some threshold function $thresh_{t.M}()$ [8].

The KDC then distributes the parameters

$$(ID_{c,s}, K_{c,s}, P_{c,s_1})$$
$$(ID_{c,s}, K_{c,s}, P_{c,s_2})$$
$$\vdots$$
$$(ID_{c,s}, K_{c,s}, P_{c,s_M})$$

to the $TGS_1, TGS_2, \ldots TGS_M$ respectively. Each TGS securely holds a database containing these parameters, each entry of which is used for one authentication instance only.

This process is a continuous one, with the KDC ensuring that a ready supply of session key pairs are available for the TGSs.

3.2 Authentication Phase

In this phase the authentication over several TGSs occurs, initiated by the Client. The procedure is similar to the traditional single TGS approach, with the addition of one step between the TGSs and the Server (see Figure 2).

1. Client $\rightarrow$ KDC: c, tgs, N_c

   Here the Client approaches the KDC in the usual manner with a request to access the TGS.

2. KDC $\rightarrow$ Client:
   $\{ID_{c,s}, K_{c,tgs}, tgs, N_c, N_{kdc}\}_{Y_c}, \{ID_{c,s}, k_{c,tgs}, N_{kdc}, c, lifetime\}_{Y_{tgs_i}}$

The KDC provides the Client with a session public key $K_{c,tgs}$ in the ordinary manner together with an identity $ID_{c,s}$ to be presented later by the Client to the TGSs. The KDC also provides a ticket destined for the TGSs which contains the session secret key $k_{c,tgs}$ with its corresponding identity $ID_{c,s}$.

3. Client $\rightarrow$ TGS_i:
$$\{A_c, s, N_{kdc}\}_{K_{c,tgs}}, \{ID_{c,s}, k_{c,tgs}, N_{kdc}, c, lifetime\}_{Y_{tgs_i}}, N'_{c_i}$$

Here the Client obtains the session public key $K_{c,tgs}$ from the previous step and uses it to deliver an authenticator to at least t out of the M TGSs. The Client also forwards the ticket it received from the KDC to each of the TGSs.

4a. $TGS_i \rightarrow$ Client: $\{K_{c,s}, s, N'_{c_i}, N_{tgs}\}_{k_{c,tgs}}$

At least one of the TGSs must respond to the Client by delivering the session public key $K_{c,s}$ to the Client. Note that in fact the key used is $r_{tgs_i,c}$ computed as:
$$r_{tgs_i,c} \equiv (Y_c)^{X_{tgs_i}+k_{c,tgs}}$$
If the TGS happens to be dishonest (or has been subverted by an attacker) and delivers a false key, then the authentication process will fail in the ensuing steps.

4b. $TGS_i \rightarrow$ Server: $\{P_{c,s_i}, N_{tgs}, c, lifetime\}_{Y_s}$

Each of the t TGSs also deliver to the Server a ticket with the share P_{c,s_i} required by the Server to recover the session secret key $k_{c,s}$.

5. Client $\rightarrow$ Server: $\{A_c, N''_c, N_{tgs}\}_{K_{c,s}}$

The Client then uses the session public key $K_{c,s}$ to send the Client's authenticator to the Server.

6. Server $\rightarrow$ Client: $\{c, s, N''_c\}_{k_{c,s}}$

As in the single TGS case, the Server uses key $k_{c,s}$ which it now holds (from merging the shares) to compute the actual enciphering key $r_{s,c}$ as:
$$r_{s,c} \equiv (Y_c)^{X_s+k_{c,s}}$$

4   COMPARISON WITH OTHER APPROACHES

The notion of secret sharing schemes have been in use for sometime in a number of different application [17]. However, in the context of authentication in distributed systems and network security the first notable effort was given by Gong in [8]. It is therefore useful to compare the current proposed scheme for Kuperee with that given in [8].

One of the main underlying differences between the proposed scheme and Gong's scheme in [8] is the use of symmetric (shared-key) cryptosystem in [8]. This has impact on the system on a number of points:

– *Key management.* In [8] the Client and the TGSs (ie. Authentication Servers in [8]) must share a key on a one-to-one basis. This is established by computing the key K_{a_i} – shared between the Client A and the TGS_i – using a hash function h as $K_{a_i} = h(K_a, TGS_i)$ where K_a is the master key known only to the Client A and TGS_i is the unique identity of the i-th TGS.

In our case the overhead in the key management of these shared keys does not exist as the public keys of the Client and the TGSs are readily available in the public domain.

– *Role of the TGSs.* A more pronounced difference lies in the role of the TGSs in Kuperee and in the scheme of [8]. In the replication scheme of Kuperee the TGSs act as a *storage point* for the shares derived from the session secret key. The session secret key is chosen by the most trusted entity, namely the KDC, and it is only ever directly available to one other entity, which is the Server, at the successful completion of the authentication process. It is the KDC that performs the share generation and distribution.

In the scheme of [8] the TGSs (ie. Authentication Servers) act more as an intermediary in the exchange of two secret parameters x and y between the Client A and the Client B (or the Server B). That is, the TGSs collectively take the role of an *exchange point.* These two parameters x and y are then used by the two parties to compute a common key K_{AB} via some secure one-way function g (ie. Client A sends x to B via the TGSs; B sends y to A via the TGSs; each computes $K_{AB} = g(x, y)$). Thus, the TGSs (ie. Authentication Servers in [8]) do not actually store any shares for longer than the completion of the key exchange period.

– *Selection of participating TGSs.* In the scheme of [8] it is essential that the communicating parties Client A and Client B (or Server B) choose the *same* set of t TGSs. This stems from the fact that the t number of TGSs act as a common exchange point. Thus, if Client B chooses a slightly different set of t TGSs, with some TGS not being in Client A's chosen set, then these non-intersecting set of TGSs will not have the parameter x from Client A. Similarly, the TGSs selected by Client A which are not in the set chosen by Client B will not carry the parameter y from Client B.

The solution to this dilemma is for both Client A and Client B (or Server B) to select *all* the TGSs when delivering the parameters x and y initially. That is, all the TGSs become the exchange point, receiving copies of parameters x and y. After this has occurred, each of the parties can proceed to select any t of the TGSs.

In the replication scheme of Kuperee the above problem does not exist since all the TGSs are already acting as storage points and are in possession of all the shares respectively. The parties can select any t of the TGSs with no impact on the scheme.

Other differences between the two schemes derive largely from the use of a public key cryptosystem in Kuperee versus a private key (shared key) cryptosystem in [8]. The use of a public key cryptosystem simplifies key management in the domain since any new public-key can be broadcasted by the certification authority. Principal-to-principal secure communications can also be established much more readily without the aid of any trusted third party, albeit with the risk being fully burdened by the two communicating parties (eg. in the case that one of the two turns-out to be a masquerading attacker).

5 REMARKS AND CONCLUSION

In this work we have proposed a new scheme for the replication of authentication services based on a public key cryptosystem, in response to the two main shortcomings of the traditional single server solutions. First, since entities in the distributed system rely on the server for their authentication operations, the server easily becomes a source of contention or bottleneck. The temporary unavailability of the server can lead to a degradation in the overall performance of the distributed system. Secondly, the fact that the server may hold cryptographic (secret) keys belonging to entities in the distributed system makes it the best point of attack for intruders wishing to compromise the distributed system.

A comparison with the notable work of Gong [8] has been presented, focusing on the issues of key management, the role of the TGSs and the selection of the TGSs by the communicating parties in the system. Although the two approaches differ in their underlying use of a public key cryptosystem in Kuperee and a shared key (private key) cryptosystem in [8], some of the observations are useful to illustrate the various abilities and usefulness of the two approaches in differing environments.

Acknowledgements

We thank Anish Mathuria for indepth comments on Kuperee. This work has been supported in part by the Australian Research Council (ARC) under the reference number A49232172, A49130102 and A49131885, and by the University of Wollongong *Computer Security: Technical and Social Issues* research program.

References

1. T. Hardjono and J. Seberry, "Authentication via multi-service tickets in the *kuperee* server," in *Computer Security - ESORICS'94: Proceedings of the Third European Symposium on Research in Computer Security* (D. Gollmann, ed.), vol. 875 of *LNCS*, pp. 143–160, Springer-Verlag, 1994.
2. Y. Zheng and J. Seberry, "Immunizing public key cryptosystems against chosen ciphertext attacks," *IEEE Journal on Selected Areas in Communications*, vol. 11, no. 5, pp. 715–724, 1993.
3. T. Hardjono, "Kuperee simplified," Technical Report Preprint 95-5, Centre for Computer Security Research, Computer Science Department, University of Wollongong, December 1994.

4. Y. Ding and P. Horster, "Why the kuperee authentication system fails," *Operating Systems Review*, vol. 30, no. 2, pp. 42–51, 1996.

5. J. G. Steiner, C. Neuman, and J. I. Schiller, *"Kerberos:* an authentication service for open network systems," in *Proceedings of the 1988 USENIX Winter Conference*, (Dallas, TX), pp. 191–202, 1988.

6. S. M. Bellovin and M. Merritt, "Limitations of the Kerberos authentication system," *Computer Communications Review*, vol. 20, no. 5, pp. 119–132, 1990.

7. J. T. Kohl, "The evolution of the *kerberos* authentication service," in *Proceedings of the Spring 1991 EurOpen Conference*, (Tromsø, Norway), 1991.

8. L. Gong, "Increasing availability and security of an authentication service," *IEEE Journal on Selected Areas in Communications*, vol. 11, no. 5, pp. 657–662, 1993.

9. A. Shamir, "How to share a secret," *Communications of the ACM*, vol. 22, no. 11, pp. 612–613, 1979.

10. G. R. Blakley, "Safeguarding cryptographic keys," in *Proceedings of the National Computer Conference*, AFIPS Conference Proceedings, Vol.48, pp. 313–317, 1979.

11. R. M. Needham and M. D. Schroeder, "Using encryption for authentication in a large network of computers," *Communications of the ACM*, vol. 21, no. 12, pp. 993–999, 1978.

12. R. M. Needham and M. D. Schroeder, "Authentication revisited," *Operating Systems Review*, vol. 21, no. 1, p. 7, 1987.

13. T. Y. C. Woo and S. S. Lam, "Authentication for distributed systems," *IEEE Computer*, vol. 25, pp. 39–52, January 1992.

14. W. Diffie and M. E. Hellman, "New directions in cryptography," *IEEE Transactions on Information Theory*, vol. IT-22, no. 6, pp. 644–654, 1976.

15. T. El Gamal, "A public key cryptosystem and a signature scheme based on discrete logarithms," *IEEE Transactions on Information Theory*, vol. IT-31, no. 4, pp. 469–472, 1985.

16. Y. Zheng, T. Hardjono, and J. Seberry, "Reusing shares in secret sharing schemes," *The Computer Journal*, vol. 17, pp. 199–205, March 1994.

17. G. J. Simmons, "An introduction to shared secret and/or shared control schemes and their application," in *Contemporary Cryptology* (G. J. Simmons, ed.), pp. 441–497, IEEE Press, 1992.

APPENDIX: KUPEREE ALGORITHMS

The approach in Kuperee is based on the public key cryptosystem of [2]. Here we provide further notations for the cryptosystem and present the algorithm for the encipherment and decipherment of tickets based on a modified version of the original cryptosystem of [2]. The algorithms expresses only the encipherment (decipherment) of the plaintext (ciphertext) tickets, and do not incorporate the steps taken by the KDC, Client, TGS and the Server.

The following notation is taken directly from [2]. The cryptosystem of [2] employs a n-bit prime p (public) and a generator g (public) of the multiplicative group $GF(p)^*$ of the finite field $GF(p)$. Here n is a security parameter which is greater that 512 bits, while the prime p must be chosen such that $p - 1$ has a large prime factor. Concatenation of string are denoted using the "$\|$" symbol and the bit-wise XOR operations of two strings is symbolized using "$\oplus$". The notation $w_{[i \cdots j]}$ $(i \leq j)$ is used to indicate the substring obtained by taking the bits of string w from the i-th bit (w_i) to the j-th bit (w_j).

The action of choosing an element x randomly and uniformly from set S is denoted by $x \in_R S$. G is a cryptographically strong pseudo-random string generator based on the difficulty of computing discrete logarithms in finite fields [2]. G stretches an n-bit input string into an output string whose length can be an arbitrary polynomial in n. This generator produces $O(\log n)$ bits output at each exponentiation. All messages to be encrypted are chosen from the set $\Sigma^{P(n)}$, where $P(n)$ is an arbitrary polynomial with $P(n) \geq n$ and where padding can be used for messages of length less than n bits. The polynomial $\ell = \ell(n)$ specifies the length of tags. The function h is a one-way hash function compressing input strings into ℓ-bit output strings.

In the process of getting an initial ticket the Clients asks the KDC to prepare the ticket to be submitted by the Client to the TGS. The KDC generates a session key pair by first calculating $k_{c,tgs} \in_R [1, p-1]$ followed by the calculation $K_{c,tgs} \equiv g^{k_{c,tgs}}$.

The KDC then enciphers the session key $K_{c,tgs}$ intended for the Client who owns the public-key Y_c by invoking *Encipher* (Algorithm 1) with the input parameters $(p, g, r_c, K_{c,tgs})$ where $r_c \equiv (Y_c)^{X_{kdc}+x_c}$ for some random $x_c \in_R [1, p-1]$. The output of *Encipher* (Algorithm 1) that is sent to the Client is in the form $C_c = \{C \| y_c\}$ where $y_c \equiv g^{x_c}$.

Algorithm 1 *Encipher*(p, g, r, T)
    1. $z = G(r)_{[1 \cdots (P(n)+\ell(n))]}$.
    2. $t = h(T \oplus r)$.
    3. $m = (T \| t)$.
    4. $C = z \oplus m$.
    5. output (C).

end

Upon receiving the ciphertext $C_c = \{C \| y_c\}$ the Client attempts to decipher the ciphertext by first computing $r' \equiv (Y_{tgs} y_c)^{X_c}$ and using this as input to *Decipher* (Algorithm 2). More specifically, the Client inputs That is, the TGS inputs (p, g, r', C) resulting in the output $K_{c,tgs}$.

Algorithm 2 *Decipher(p, g, r', C)*
1. $z' = G(r')_{[1\cdots(P(n)+\ell(n))]}$.
2. $m = z' \oplus C$.
3. $T' = m_{[1\cdots P(n)]}$.
4. $t' = m_{[(P(n)+1)\cdots(P(n)+\ell(n))]}$.
5. if $h(T' \oplus r') = t'$ then
    output (T')
  else
    output $(\emptyset)$.

end

In general, the same procedure is followed by each principal who must compute the parameter r (either directly or by selecting a random number x) and use it as input to either *Encipher* (Algorithm 1) or *Decipher* (Algorithm 2).

Readers interested in the security of Kuperee are directed to [2] which discusses the security of the public key cryptosystem upon which Kuperee is directly built.

Non-repudiation Without Public-Key

R. Taylor

Defence Science and Technology Organisation
PO Box 783
Jamison ACT Australia 2614
(Extended Abstract)

Abstract. An unconditionally secure non-repudiation scheme is optimised to maximise the number of messages that may be sent with a given amount of shared key data. By dropping the unconditional security property a related scheme with less memory storage requirements is constructed. In relation to the options for providing non repudiation some discussion of the complexity of the cryptanalysis of public and private key cryptosystems is provided.

1   Introduction

Non-repudiation techniques prevent the sender of a message from subsequently denying that they sent the message. Digital Signatures using public-key cryptography and hash functions are the generally accepted means of providing non-repudiation of communications (see the general reference [13]). However other means exist of providing non-repudiation, and some of these methods are the subject of this paper.

Public-key methods depend for their security on the difficulty of solving particular mathematical problems. Currently the most popular public-key methods relate to one of two problems – The Integer Factoring Problem and The Discrete Logarithm Problem. Furthermore all these methods belong to a special class (**NP ∩ coNP**) with respect to their theoretical complexity (see [17], [18] and the general reference [7] on complexity theory). Note that the proof in [18] that The Discrete Logarithm Problem is in (**NP ∩ coNP**) applies also to the very general form of the discrete log problem as described in [1]. This includes all variations of the Discrete Logarithm Problem that the author is aware of.

This means that the level of confidence that these problems will remain difficult is not as high as that of many other mathematical problems (**NP-complete** problems). Unfortunately no public-key method has been found the security of which corresponds to an **NP-complete** problem. In fact any public-key system presents an attack problem that is essentially in **NP ∩ coNP** (see [4], [6]). In Section 6 we note that similar complexity results in a probabilistic sense hold also for symmetric cryptosystems.

In this context it is useful to consider other methods of providing non-repudiation, particularly those that are unconditionally secure – meaning that they cannot be cryptographically broken with any amount of time or effort from information available on public channels. We describe a thread of research developing such methods, and build on these techniques in this paper.

In [14] (see full paper [15]) an authentication scheme is presented in which an independent party or arbiter participates to prevent misuse by the sender or receiver. Systems of this type have received some attention in the literature recently and are often described as A^2 codes (see [8], [9] and 11]). However, as pointed out in [14], [15], the arbiter can impersonate the sender in a way that the receiver will not detect. This problem is addressed in Desmedt and Yung [5] where a scheme in which the arbiter cannot cheat is presented. The basic idea is used in Taylor [19] where the efficiency and practicality of [5] is improved. Note that a critique of [19] was presented in [8], and a scheme that improves the efficiency of that given in [19]. However the scheme [8] requires that all the key material be brought together and processed. It is unclear how this can be done without having the undesirable effect of placing complete trust in some single party.

We briefly describe the system of Taylor [19] and show how to tune the method to maximise the number of messages that may be sent with a given amount of shared key data. Examples are presented that indicate the practicality of the scheme. By dropping the unconditional security property a related scheme with less memory storage requirement is also constructed and discussed.

2   The Basic Scheme

The basic scheme which forms the starting point of the remainder of this work is described. We refer the interested reader to [19] for a more detailed presentation and analysis of the system.

Let a sender, receiver, arbiter and some hostile outsider be denoted by Sally, Ray, Alice, and Oliver, respectively. Sally wishes to send messages to Ray and for this communication to be unconditionally secure against the following attacks. Note that these properties amount to sender non-repudiation. Thus prevention of Attack 4 must in turn prevent Attacks 1-3. Also note that it is assumed that Alice will arbitrate correctly, even though she may attempt to generate bogus messages.

Attacks
1/   Oliver generates a message, or alters one in transit, and attempts to send this to Ray as if it came from Sally. In this attack Ray or Sally do not defer to Alice.

2/   Alice generates a message, or alters one in transit, and attempts to send this to Ray as if it came from Sally. In this attack Ray or Sally do not defer to Alice.

3/   Ray generates a message, or alters one received from Sally, and attempts to claim that it was sent by Sally. In this attack Alice is deferred to in an attempt to detect this attack.

4/   Sally sends a message that is accepted by Ray as coming from Sally, and later attempts to deny that the message was sent by her. As in attack 2, Alice is deferred to in an attempt to detect this attack.

A message M is represented in terms of blocks of n w-bit words $m_1, m_2, ..., m_n$. A prime number $p > 2^w$ and another number t is chosen. The choice of p determines the level of security offered (see Theorem 3), while the scheme provides sufficient key for t message blocks to be sent from Sally to Ray. Hyperplanes in $n+t+3$ space are used with arithmetic over the field modulo p.

Key sharing

- Alice randomly generates hyperplanes P and Q and secretly shares $P \cap Q$ with Sally. Note that this can be done without revealing either P or Q to Sally, by sending other hyperplanes P' and Q' in which $P \cap Q = P' \cap Q'$. Alice also shares Q secretly with Ray.

- Ray randomly generates a hyperplane R and secretly shares this with Sally.

This key sharing process is represented in Figure 1.

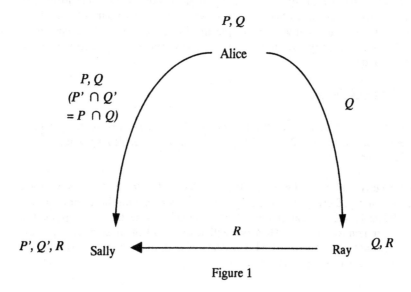

Figure 1

Transmission

- To send a message block M Sally generates a point p_M in the subspace $P' \cap Q' \cap R$ and sends this to Ray. This point is formed in such a way that the initial block of n coordinates correspond to the message block, and the construction of each of the t message blocks that Sally may send takes into account the position of each message block in the message sequence (from 1 to t). The collection of t points generated by Sally is such that they correspond to a linearly independent collection of vectors. This prevents anyone but Sally from generating an acceptable message block from knowledge of the previous message blocks sent.

- Ray verifies that the message block is in order by checking that p_M lies on $Q \cap R$.

Arbitration

- In case of a disputed message block M, Ray takes the corresponding point p_M to Alice.

- Alice checks that p_M lies on $P \cap Q$. If so p_M is deemed to have been sent by Sally.

Theorem 1. The probabilities of successful attacks of Types 1 - 4 are all bounded by $1/(p-1)$.

Proof: See [19].

Of course in practice each message that needs to be sent will not be of exactly nw bits in length. Shorter messages will need to be padded with zeros, and longer messages sent in blocks of nw bits each, with the last block padded as necessary.

3 Message Block Size Optimisation

A question arises of how to choose n, the message block size, for a given distribution of message lengths and limited capacity tokens for key distribution, so as to maximise the number of messages that may be sent. In the following we shall consider this question in the context of a system of u users, all of whom may wish to communicate with non-repudiation, and provide examples that indicate the practicality of the scheme.

Define the following terms and notation:

u - the number of users in the system
mem - the memory capacity of tokens for storage of keys (in bits).
num - the number of messages that may be sent from a user to any other user
av - the average message length (in bits)
$|m|$ - the length of a message m (in bits)
$\lceil x \rceil$ - the least integer greater than or equal to x

Recall the notation

p - the prime modulus
w - the word size used (in bits)
n - the message block size (in w-bit words)
t - the number of message blocks that may be used by each user

Alice generates u^2 planes P_{ij}, Q_{ij} $i,j=1,...,u$, and sends user k, $k=1,...,u$, the data Q_{jk}, P'_{kj}, Q'_{kj}, $j=1,..., k-1,k+1,...,u$, where $P_{ij} \cap Q_{ij} = P'_{ij} \cap Q'_{ij}$. Similarly each user k generates R_{jk}, $j=1,..., k-1,k+1,...,u$, and sends user l R_{lk}, $l=1,..., k-1,k+1,...,u$. This enables each user to send and receive messages from all other users in the system. The problem considered is to maximise num, subject to fixed values for u, mem, av, p, and w.

From the amount of key data that must be sent from Alice to each user

$$mem \geq 3w(u-1)(n+t+3). \tag{1}$$

Also the number of message blocks t required in terms of a sum over the messages m sent may be calculated as

$$t = \sum_{|m| \le nw} 1 + \sum_{|m| > nw} \left\lceil \frac{|m|}{nw} \right\rceil. \tag{2}$$

Manipulating the right hand side of (2) we have

$$\sum_{|m| \le nw} 1 + \sum_{|m| > nw} \left\lceil \frac{|m|}{nw} \right\rceil$$

$$\le \sum_{|m| \le nw} 1 + \sum_{|m| > nw} \left(\frac{|m|}{nw} + 1 \right)$$

$$\le \sum_{|m| \le nw} 1 + \sum_{|m| > nw} 1 + \sum_{|m| > nw} \left(\frac{|m|}{nw} \right) \tag{3}$$

$$\le \sum_{m} 1 + \sum_{|m| > nw} \left(\frac{|m|}{nw} \right)$$

$$\le num + \frac{av.num}{nw}.$$

Thus from (1), (2) and (3) we may derive a lower bound num^* for num where num^* is defined as the solution to

$$mem = 3w(u-1)(n+t+3) \tag{4}$$

$$t = num^* + \frac{av.num^*}{nw}. \tag{5}$$

Combining (4) and (5)

$$\frac{mem}{3w(u-1)} - n - 3 = num^* + \frac{av.num^*}{nw}. \tag{6}$$

From (6) num^* is maximised by choosing

$$n = \sqrt{\frac{av.num^*}{w}}. \tag{7}$$

This gives a value of num^*

$$num^* = \left[-\sqrt{\frac{av}{w}} + \sqrt{\frac{av}{w} + \frac{mem}{3w(u-1)} - 3} \right]^2 \tag{8}$$

The following examples indicate the practicality of this scheme. We shall assume that $w = 30$ and $p = 2^{31}-1$. This allows easy multiplication modulo p (see [19], [10]) and provides a probability of attack bounded by $1/p-1$ or about one chance in $2,000,000,000$ (see Theorem 1).

Example 3.1. Let the memory token be a standard floppy disk of 1.45×10^6 bytes. Also let the average message be of about two pages (say 2400 bytes) of text, and consider a system of 100 users. Thus $u=100$, $mem =1.16 \times 10^7$ and $av=1.92 \times 10^4$. Choosing $n=474$ then from (8) the number of messages that may be sent from a user to any other user is at least $num^*=351$.

Example 3.2. Let the memory token be a compact disk of 600×10^6 bytes. Allow messages to average 100 pages of text, and consider a system of 1000 users. Thus $mem =4.8 \times 10^9$, $av=9.6 \times 10^5$ and $u=1000$. Choosing $n=20,271$ gives $num^*=12,841$.

4   PRNG Scheme

The system described in Section 2 suffers from a number of deficiencies. The amount of key data retained by the arbiter increases as the square of the number of users and so may be unmanageable if the number of users is large. This problem is compounded if the arbiter needs to be prepared to arbitrate for years into the future, and so needs to retain the key data from many key distributions. Also the requirement for each user to distribute data to all other users may not be reasonable, and could more practically be managed centrally for a large number of users. These problems may be mitigated by combining the basic framework of the above scheme with the use of a cryptographically secure pseudorandom number generator (*PRNG*).

Consider the non repudiation scheme of Section 2 with the difference that the hyperplanes generated by Alice are generated by a *PRNG* rather than a true random number generator. Thus Alice needs to store only seeds for the *PRNG*, from which hyperplanes can be re-generated for arbitration purposes at some future time. Similarly let the planes R_{ij} be generated centrally using a *PRNG* by a fourth party independent of Alice, Ray and Sally, say Alan. Thus Alan generates R_{ij}, $i,j=1,...,u$, $i \neq j$, and sends user k R_{kj}, R_{jk}, $j=1,...,k-1,k+1,...,u$. At the end of the life of this key distribution phase when the key data has been used up, the received message data is securely stored. At this point Sally and Ray may safely discard the stored key data that was distributed to them by Alice and Alan, and are ready for a new key distribution. In future Ray can retrieve stored message data that he knows can be verified by Alice in case of a dispute. This scheme is illustrated in Figure 2.

Security
Since the non repudiation scheme is based on an unconditionally secure scheme it is easy to show that the strength of the non repudiation scheme against attacks 1 - 4 corresponds to the strength of the *PRNGs* used by Alice and Alan.

Furthermore it is worth noting that the *PRNGs* used are not distributed, unlike symmetric crypto-algorithms for example, thus the risk of the design of the *PRNGs* leaking out is reduced, and/or the effort made to conceal the *PRNG* designs (through tamperproof hardware for example) may be reduced.

Of course Ray and Sally depend upon Alice and Alan not colluding. If they do they can together impersonate Sally, and Ray will not be able detect this. Clearly however the function that Alan carries out can be duplicated by several similar parties, thus reducing the risks of effective collusion.

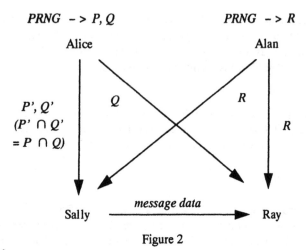

Figure 2

Secret Sharing

In order to reduce the amount of key material that needs to be secretly shared in each key distribution phase of this scheme a threshold based key agreement scheme may be used (see for example [3]). This provides an efficient mechanism whereby secret seeds may be shared among each pair of participants. This efficiency is achieved however at the cost of allowing any subset of sufficiently many (the threshold) users to combine their key data and obtain the key data of the other users. The seeds distributed by the threshold key agreement scheme may then be used as encryption keys enabling the relatively large amount of key material of the non-repudiation scheme to be encrypted and distributed over public channels.

5   Storage Integrity

It is natural to ask how the receiver Ray is to store received messages, possibly for long periods of time, while knowing that the message data is intact when it is needed. The following method provides a highly secure way of checking the integrity of large volumes of stored data.

In this application some small portable storage device such as a floppy disk is used. The information is partitioned into subsets (these may be files or directories for example). Sufficient random data is generated and stored on the floppy disk to enable a type of secret check-sum to be calculated for any one data subset. Check-sums (CS) for each data subset are then stored on the floppy disk (see Figure 3). For this purpose the secret check-sum function may be reused for every subset. The floppy disk is then secured for the required duration (for example by locking in a safe). Subsequently the integrity of each data subset may be checked by recalculating the checksums of the stored data using the secret check-sum function on the floppy disk. These checksums are then compared with the checksums previously stored on the floppy disk.

Note that there is a trade off involved here. The larger the subsets the fewer the number of check-sums that need to be stored on the floppy, but if data becomes corrupted the position of the corruption is harder to pinpoint. In the following we use check sums in the form of universal hashing functions due to Wegman and Carter [20], but modified slightly in [16] and [19]. A survey reference to these types of functions may be found in [2].

Example 5.1. If the modified Wegman-Carter method (see [20], [16] and [19]) is used with $p=2^{31}-1$, $w=30$ and $n=10^{10}$, this is sufficient to calculate check-sums for any data subset of up to 30×10^{10} bits or about 40 gigabytes. The probability of successful attack is bounded by $10log_2(10)/p$ which is about one chance in 65,000,000. Each check-sum requires 31 bits of storage, and so a 1.44×10^6 byte floppy can support a database of about 370,000 such data subsets. A compact disk of 600 megabytes can store around 155,000,000 data subsets.

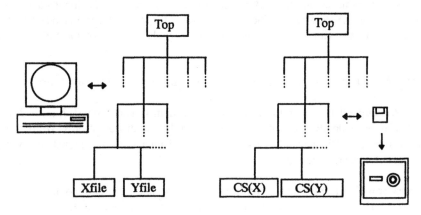

Figure 3

6   Complexity

Following the use of cryptographically strong PRNGs or stream ciphers suggested in Section 4 we shall present some results on the complexity of attacking such systems. Specifically we show that most instances of the cryptanalysis problem for stream ciphers are in NP ∩ coNP. Similar observations have been made in relation to asymmetric systems (see [4], and [6]).

Thus the construction of secure cryptosystems that are not one-time pads, whether symmetric and asymmetric, may be considered as attempts to find hard problems in NP ∩ coNP. Since no NP-complete problem is known to be in coNP (and there are reasons to suppose that no such problems exist, see [7]) this may explain why no cryptosystem has been found that is provably as secure as solving an NP-complete problem.

Let $s(l, k, i)$ denote the ith bit of a stream cipher s with initial state $(i=0)$ determined by a key k of length l. Assume that $s(l, k, i)$ may be evaluated efficiently, that is in a number of steps polynomial in l. We model a cryptanalysis decision problem CA. Note that this serves only as a representative model of a typical cryptanalysis problem, but does not attempt to capture every possible variation of a meaningful cryptanalysis problem.

CA   Input: a key length l;

                  a bound b greater than l;

                  an integer j with $0 \le j \le b-1$;

                  for $j>0$, cipher outputs $o_1 = s(l, k, 1)$, $o_2 = s(l, k, 2)$, ..., $o_j = s(l, k, j)$;

          Question: Is $o_{j+1} = s(l, k, j+1)$ equal to 1?

Note that we assume in the problem **CA** that the cipher outputs o_i are legitimate in that they correspond to the outputs of s under some key k. Thus the correctness of the input cipher data to **CA** is assumed. Also we assume that the attacker knows the cipher design s. Attempts to avoid assumptions of correctness or legitimacy of the input data in the formulation of the cryptanalysis problem has motivated the study of a non-standard branch of complexity theory (for results in this area and a guide to some of the literature see [12]).

Theorem 2.   An instance of **CA** is in **NP $\cap$ coNP** with probability at least $1-l/b$.

Proof: We assume that j is a uniformly distributed random variable from $0, 1, ..., b-1$, and that k is chosen randomly and uniformly from the collection of l bit strings. We describe certificates for the YES and NO answer to **CA**, and then show the probability of the correctness of those certificates.

If the answer to **CA** is YES then a certificate that shows this consists of some key k^* for which $s(l, k^*, i) = o_i$, $i = 1, ..., j$, and $s(l, k^*, j+1) = 1$. If the answer to **CA** is NO then a certificate that shows this consists of some key k^* for which $s(l, k^*, i) = o_i$, $i = 1, ..., j$, and $s(l, k^*, j+1) = 0$. Clearly these certificates may be checked in a number of steps polynomial in the input length $log_2(l)+log_2(b)+ log_2(j)+j$.

For the certificates above to be valid the outputs o_i, $i = 1, ..., j$, need to uniquely determine the next bit of output o_{j+1}. To measure the probability of this define

        n_i = the number of legitimate (corresponds to some key) output strings,
        $o_1, o_2, ..., o_i$, of length i.

        p_i = the probability that a legitimate string of length i uniquely determines the
        next bit (the legitimate string of length i extends to a legitimate string of length $i+1$
        in only one way).

Thus by counting strings that bifurcate and those that do not

$$n_{i+1} = 2n_i(1 - p_i) + p_i n_i = (2 - p_i)n_i .\tag{9}$$

Defining $n_0=1$ it follows that

$$n_b = \prod_{i=0}^{b-1}(2 - p_i).\tag{10}$$

Since the number of legitimate strings must be bounded by the number of keys,

$$\prod_{i=0}^{b-1}(2-p_i) \le 2^l.$$ (11)

If E denotes the expectation that the next bit of output is uniquely determined in an instance of **CA** we have

$$E = \frac{1}{b}\sum_{i=0}^{b-1}p_i.$$ (12)

This sum is minimised subject to the product inequality (11) when l of the p_i are 0 and the others 1. Thus

$$E \ge \frac{1}{b}(b-l) = 1-\frac{l}{b}. //$$ (13)

Remarks. Similar results may be obtained straightforwardly for a number of variations of **CA**. These include the case where the attacker does not know the design of the stream cipher s, but only knows bounds on the length of s; ciphertext only attacks where some redundant language is being used; and also block cipher systems.

References

[1]　E. Bach, Intractable Problems in Number Theory, Advances in Cryptology - CRYPTO '88, proceedings, Springer-Verlag 1989, pp. 77-93.

[2]　J. Bierbrauer, T. Johansson, G. Kabatianskii and B. Smeets, On families of Hash functions via Geometric Codes and Concatenation, Advances in Cryptology - CRYPTO '93, proceedings, Springer-Verlag 1994, pp. 331-342.

[3]　R. Blom, An Optimal class of Symmetric Key Generation Systems, Advances in Cryptology - EUROCRYPT '84, proceedings, Springer-Verlag 1985, pp. 335-338.

[4]　G. Brassard. A note on the complexity of cryptography. IEEE Transactions on Information Theory , 25, 1979, pp. 232-233.

[5]　Y. Desmedt and M. Yung, Arbitrated Unconditionally Secure Authentication Can Be Unconditionally Protected against Arbiters Attacks, Advances in Cryptology - CRYPTO '90, proceedings, Springer-Verlag 1991, pp. 177-188.

[6]　S. Even and Y. Yacobi, Cryptocomplexity and NP-completeness. In Proc. of 8th Colloq. on Automata, Languages, and Programming, Lecture Notes in Computer Science, Springer-Verlag 1980, pp. 195-207.

[7]　M. Garey and D. Johnson, *Computers and Intractability: A Guide to the Theory of NP-Completeness*, W. H. Freedman, San Francisco, 1979.

[8]   T. Johansson and B. Smeets, On A^2-codes Including Arbiters Attacks, Advances in Cryptology - EUROCRYPT '94, proceedings (rump session), Springer-Verlag 1995.

[9]   T. Johansson, On the construction of Perfect Authentication Codes that permit Arbitration, Advances in Cryptology - CRYPTO '93, proceedings, Springer-Verlag 1993, pp 341-354.

[10]   H. J. Knobloch, A Smart Card Implementation of the Fiat-Shamir Identification Scheme, Advances in Cryptology - EUROCRYPT '88, proceedings, Springer-Verlag 1989, pp. 87-96.

[11]   K. Kurosawa, New bounds on Authentication Codes with Arbitration, Advances in Cryptology - CRYPTO '94, proceedings, Springer-Verlag 1994, pp 140-149.

[12]   A. L. Selman, Complexity Issues in Cryptography, Proceedings of Symposia in Applied Mathematics, 38, 1989, pp. 92-107.

[13]   B. Schneier, *Applied Cryptography*, 2nd Edition, John Wiley and Sons, ISBN 0-471-12845-7, 1995.

[14]   G. J. Simmons, Message Authentication with Arbitration of Transmitter/Receiver disputes, Advances in Cryptology - EUROCRYPT '87, proceedings, Springer-Verlag 1988, pp. 151-165.

[15]   G. J. Simmons, A Cartesian Product Construction for Unconditionally Secure Authentication Codes that permit Arbitration, Journal of Cryptology, 2(2), pp. 77-104, 1990.

[16]   D. Stinson, Universal Hashing and Authentication codes, Advances in Cryptology - CRYPTO '91, proceedings, Springer-Verlag 1992, pp 74-85.

[17]   R. Taylor, Factoring and Cryptography, Telecom Australia, Research Laboratories Report 8048, July 1991.

[18]   R. Taylor, Discrete Logs, Factoring and Cryptography, Unpublished work completed in 1991.

[19]   R. Taylor, Near Optimal Unconditionally Secure Authentication, Advances in Cryptology - EUROCRYPT '94, proceedings, Springer-Verlag 1995, pp. 244-253.

[20]   M. N. Wegman and J. L. Carter, New Hash Functions and their use in Authentication and set equality, Journal of Computer and System Sciences 22, 1981, pp. 265-279.

Investigation of Non-repudiation Protocols

Yongfei Han
Information Security Group
Institute of Systems Science
Heng Mui Keng Terrace
Kent Ridge
Singapore 119597

Abstract. The paper surveys the state of non-repudiation protocols. A fair non-repudiation protocol should provide an equal protection to the sender and the recipient. A number of current non-repudiation protocols expect the protection from or partly from a Trusted Third Party (TTP). In practice, the sender and the recipient that do not trust each other, do not expect or are not able to find an TTP or a strong TTP in some circumstances. A simultaneous secret exchange protocol seems to be one of efficient solutions without an TTP to prevent entities from denying the transferring (sending or receiving) of certain messages. The secret exchange bit by bit, however, is neither very efficient nor convenient to the sender and the recipient in some cases. We introduce a model and a fair non-repudiation protocol without an TTP. In the protocol, the transferring of the message is split into three parts, a commitment C, a key K and an address for the key. Therefore, without bit by bit exchange, the protocol is more efficient than the existing non-repudiation protocols.

1 Introduction

In both the Internet and mobile telecommunication environments, security threats are often divided into three categories: breach of confidentiality, failure of authenticity, and unauthorized denial of service [3, 17, 19]. One of services in these systems is to prevent entities from denying the transferring of certain messages. Non-repudiation service with a Trusted Third Party (TTP), giving protection against attacks from outsiders and denial of insiders, have been studied [1, 2, 22]. For this type of non-repudiation protocols, we assume that the TTP performs evidence generation, evidence transmission, evidence recording or evidence verification, and adjudicator could get enough evidences from the TTP to reason and judge the dispute between the sender and the recipient, although some evidences could be created by a non-repudiation initiator. In practice, however, this is not always applicable, for example, some countries or companies can not find an TTP in between them and the sender and the recipient can not trust each other.

In ISO/IEC CD 13888-3 and ISO/IEC DIS 10181-4 [1, 2], non-repudiation mechanisms without an TTP participation allow for generation of evidences for both non-repudiation of origin (NRO) and non-repudiation of delivery (NRD). With assuming that each of the signature keys of the entity A and the entity B

is in place at each entity, and that the corresponding verification keys are known to all the entities concerned. The entity A wishes to send a message to the entity B and thus will be the originator of the non-repudiation sender. The entity B will be the recipient. The mechanisms can not give a reasonable judgement if B had received the message from A but B denied having received the message.

Zhou and Gollmann [22] proposed a fair non-repudiation protocol that requires a trusted third party but attempts to minimize its involvement in the execution of the protocol. The main idea of their protocol is to spilt the definition of the message into two parts, a commitment C and a key K. The commitment is exchanged between A and B. B can not see the message before B got the key from the TTP and the key is submitted to the TTP by A. The protocol is efficient in exchanging commitment C between A and B. Without the TTP, however, the protocol can not deliver the key, and the adjudicator can not collect enough evidences to judge a dispute. Hence the TTP is still a critical party in the protocol. Moreover, the transmission among the TTP, A and B seems not secret if the TTP has no protection for the key since the key is stored in a public access area.

In a simultaneous secret exchange protocol, entity A produces a as entity B generates b, and then they exchange $f(a)$ and $g(b)$ where A can not obtain b from $g(b)$ as B can not get a from $f(a)$. Further, A and B open a and b bit by bit [6, 18]. The protocol without an TTP seems to be superior to other protocols with an TTP applied in non-repudiation. The checking of validity of each bit, however, is time-consuming. The remaining i-bits in n-bit secret a or b become more and more easily to attack as the i gets smaller and smaller. Moreover, The entity with more computing power has advantage since it can stop the secret exchange when $n - i$ bits has been released, then compute the remaining i bits with its advantage computing power in the practical protocols.

The purpose of this paper is to introduce a new model and protocol, which is based on assumption of no trusted third party existing. The criterion in non-repudiation protocols is how to force one entity to submit a receipt to another entity when the entity has sent to or received from another entity. Non-repudiation protocols have to provide an equal protection to two entities since. This means that the entities are equally trusted and thus we want them to have equal protection for the different attacks and denial. The protocol which we present here successfully forces A and B to leave evidence while they are sending or receiving certain messages.

The paper is organized as follows: In Section 2 we briefly describe a model of non-repudiation with an arbitration and introduce the basic notation. In Section 3, we give a non-repudiation protocol without an TTP, which has equal protection to the sender and the recipient if a communication channel is not reliable and the communicating parties do not play fair. In Section 4, we discuss and analyse some aspects of the protocol such as a pub and encryption. In Section 5, an adjudicator judges the dispute between two participants using those evidences provided. We investigate formal analysis of the protocol with SVO logic in Section 6. The conclusion and further research are in Section 7.

2 The Model of Non-Repudiation with an Adjudicator

The model assumes a communication situation where there are three participants: *the sender A, the recipient B and the adjudicator*, and the communication
channel is not reliable and A and B neither trust each other nor play fair. The
sender intends to send some messages, called a source state, to the recipient in
such a way that the recipient could both recover the transmitted message state
and verify that the transmitted messages originated from the legal sender [15].
The adjudicator only has to be active in the event of a dispute, and then settle
disputes.

The sender has a *pub* that is a public access system.

Definition 1. pub: A public access system, where every operation such as modifications, accesses, fetchings and inputs has been automatically recorded in a
recorder, is a pub.

The *pub* is different from an TTP. It is not an independent party. A can
modify, write to all of the *pub* except for the recorder, and there is no reason for
B to trust the *pub*. the adjudicator and A only can read the recorder in the *pub*.
The *pub* can be implemented by current technology.

Definition 2. The recorder is a black box which can not be written and modified
by a user.

Definition 3. evidence: Some piece of information can be used to prove an
event happened.

The adjudicator is also different from an TTP. It is neither active and on-line
nor transferring any message for A and B.

The adjudicator is the supervisory person or a communication management
that can check the *pub* and read all information in the recorder. He does not
take part in any communication activities on the channel and his only task is to
solve disputes between the sender and the receiver by checking the recorder in
pub and evidences provided by A and B whenever they occur. The adjudicator
is defined to be justice.

In the model, we define denial as follows.

Definition 4. denial is:

1. An entity denies having sent a certain message though it has sent.
2. An entity denies having received a certain message though it has received

There are different attacks, cheat and denial from outsiders and insiders.
They are, at least, the following:

1. The opponent sends a message to A or B and succeeds if A or B accepted
   the message.

2. The opponent observes a message that is transmitted and replaces this message with another message, and succeeds if the message is accepted.

3. The opponent gets a message and a key from A, and succeeds if the opponent sees the message.

4. The sender transfers a message to the receiver, with the intention to later deny the transferring. The sender succeeds if the receiver accepts the message although the message could not be generated by encryption of A or B does not hold any evidence.

5. The receiver claims having received a message from the sender that was never actually sent. The receiver succeeds if the message could be generated by the sender due to his encryption rule and B holds an evidence.

6. The receiver claims not receiving a message from the sender that has been sent. The receiver succeeds if the sender can not provide reasonable evidence.

7. The receiver received a message from the sender A but claims to have received another message, containing a different source state. The receiver succeeds if this other message could have been generated by the sender due to his encoding rule and A can not provides the evidence for the difference between two messages.

8. A or B claims having sending or receiving a message though the message has never been sending or receiving, it succeeds if it can provides reasonable evidence.

The non-repudiation protocol should have two basic objectives, the first is to make violations and denial difficult; the second is to make them to the adjudicator, and the adjudicator easily makes a judgement when the violations and the denial happen.

We give the basic notation presented in the paper as follows.

- A is the distinguishing identifier of an originator A.
- B is the distinguishing identifier of a recipient B.
- M is the message which is sent from entity A to entity B with protection through the use of non-repudiation service.
- $X\|Y$ denotes the concatenation of the data element X and Y.
- S_X is a digital signature of message M with private key of entity X.
- $V_X(Z)$ is verification result of signed data Z obtained by applying verification algorithm V and verification key of entity X.
- f is data field indicating the non-repudiation in effect.
- X_p is the encryption with the public key of principal X.
- C is a commitment (ciphertext) for message M, e.g. M encrypted under a key K.
- T_{X_i} is the i-th time stamp of entity X.
- D_k is decryption with k.

3 The Protocol

The denial of service is easily to happen but difficult to judge in current internet and mobile telecommunications, and bank commercial transferrings. The fair non-repudiation intends to provide a service which the denial of the service is difficult to happen but easily to judge. We assume that the authentication of entity identities has been done and the integrity mechanism ensure that subsequent messages are coming from the entity.

We propose that the transferring of the message is split into three parts, a commitment C (encrypted M), a key K and an address of the key. In each step of the transferring, the entity has to leave evidence. For example, A sends the commitment at first, and sends the address with evidence after A received the certification of the receipt from B.

We have the following basic definitions for the protocol.

A non-repudiation of origin (NRO) is generated by the originator. The originator A creates a message M which it will send to the recipient B. The originator signs the message and then the recipient B can ensure that this message is from the claimed originator A by verifying the associated NRO.

$$NRO = S_A(text\|C)$$

A non-repudiation of receipt (NRR) is generated by the recipient. Recipient B receives and verifies the NRO. If $V_B(NRO)$ is positive, an NRR can be generated and sent to acknowledge A that the NRO has been received.

$$NRR = S_B(text\|C)$$

A non-repudiation of address submission (NRA) is generated by the originator. The originator receives and verifies the NRR. If $V_A(NRR)$ is positive, then an NRA can be generated and sent to B acknowledge that the NRR has been received, and an address of the key has been forwarded.

$$NRA = S_A(text\|B_p(address))$$

A non-repudiation of fetching the key (NRF) is generated by the recipient. Recipient B receives and verifies the NRA. If $V_B(NRA)$ is positive, then an NRF can be generated and sent to A acknowledge that the NRA has been received, and fetching key has gone ahead.

$$NRF = S_B(text\|ftp(address))$$

A non-repudiation of key submission (NRS) is generated by the originator. The originator received and verifies the NRF. If V_A is positive, then an NRS can be generated and sent to B.

$$NRS = S_A(text\|B_p(key))$$

We propose *time stamp chain* method in the protocol which is defined as follows.

Definition 5. Time Stamp Chains are a sequence of A's and B's time stamps.

In the protocol, we assume that A, B each holds their own private signature key and the relevant public verification keys. A and B obtain time stamp by synchronizing clock.

The communication between A and B is as follows where $a = address$, $k = KEY$, and C is an encryption of M.

1. $A \rightarrow B$: $f_{NRO}, B, S_A(f_{NRO}, B, T_{A_1}, C)$
2. $B \rightarrow A$: $f_{NRR}, A, S_B(f_{NRR}, A, T_{A_1}, T_{B_1}, C)$
3. $A \rightarrow B$: $f_{NRA}, B, S_A(f_{NRA}, B, T_{B_1}, T_{A_2}, B_p(a))$
4. $B \rightarrow A$: $f_{NRF}, A, S_B(f_{NRF}, A, T_{A_2}, T_{B_2}, FTP(A_p(a)))$
5. $A \rightarrow B$: $f_{NRS}, B, S_B(f_{NRS}, B, T_{A_2}, T_{B_2}, T_{A_3}, B_p(k))$

We observe the procedure of the protocol execution, and examine the effect of each step.

1. $A \rightarrow B$: If A has sent NRO and C to B, but they fail to reach B, then the protocol stops.
2. $B \rightarrow A$: After receiving NRO and C from A, B has to submit the receipt to A if B wants to see the message since C is only a ciphertext. B has to verify the signature and check the time stamp chain before $S_B(text||C)$ has been sent to A as a receipt.
3. $A \rightarrow B$: The protocol stops if A receives the receipt from B then give up the communication to B. A has to verify the signature and check the time stamp chain before $S_A(text||NRA)$ has been sent to B to provide an address of the key and an evidence of the sending.
4. $B \rightarrow A$: The protocol stops if B receives the $S_A(text||NRA)$ then ends the communication to A. B has to verify the signature and address, and check the the time stamp chain before $S_B(text||NRF)$ has been sent to A to fetch the key and provide an evidence.
5. $A \rightarrow B$: The protocol stops if A received $S_B(text||NRF)$, then ends the communication. A has to verify the signature and the address then let the ftp to the *pub*. The recorder in the *pub* records B's access and ftp operation.

In the protocol, both the address and the key are much shorter than the full messages. Both signature and public key encryption can be accelerated by schemes in [4, 7, 11, 13, 12, 16]. Time stamp chains makes sure the entity respond the message which is sent by another entity just before the reply, and provides evidence for delayed transferring.

4 Discussion of the Protocol

We discuss several aspects of the protocol in the section.

4.1 Commitment

In the protocol, a message M is defined a commitment C and a key K in which the key defined by an address and a key. Like other non-repudiation protocols, the commitment need not restrict the content of the message in any way. In practice, the computation of $C = D_k(M)$ and $M = D_k(C)$ should not be time-consuming so that one of possible resolution is that K is much shorter than M. Further, we suggest the key is selected from symmetric algorithms so that the key is short and $C = E_k(M)$ and $M = D_k(C)$ is fast.

4.2 Address and Fetching

A provides B an address encrypted by B's public key to protect others to intercept the address. B sends A a ftp with the address encrypted by A's public key to prevent from intercepting. Ftp will fetch the stored message in the address after A verifies the signature of B and time stamps.

4.3 Pub

The pub consists of a public access area and a recorder. The recorder records every access with user name, address in the public access area and what time the access happens. The pub owns a number of public access areas which store a number of keys. A must place a key for decryption of C which is forwarded to B in the address since the recorder keeps the access trace and B can provide NRA as an evidence.

If A places a forged key in the address, A will be fail in later dispute.

We assume that each entity is supplied with a pub during the installation of the protocol. A monitor of the pub can be in ROM and the recorder can be special EPROM. The system must hand over to the monitor when any access to the pub is required.

5 Resolution for Dispute

We observe the procedure of the protocol running, and examine the effect of each step. Disputes can arise over the origin and receipt of a message M and K. In the first case, A claims having sent M to B while B denies having received M. In second case, B claims to have received M from A while A denies having sent M to B. These dispute can be resolved by the adjudicator who evaluates the evidences held by the participants and judges whether the claim is right or wrong.

5.1 Repudiation of Receipt of M

If A claims that B had received M, the adjudicator will require A provide M, C, K and the time stamp chains, and the non-repudiation evidence NRR and NRF as well as the record of the sending. If A cannot provide these evidences or one of the following check fails, A's claim will be judged invalid.

1. NRR is with B's signature and the time stamp chain is right.

2. NRF is with B's signature and A' public key encrypted address and right time stamp chains.

3. $M = D_k(C)$.

4. Record of sending.

Once the first and second check are positive, the adjudicator will assume A has sent commitment C and an address in pub to B. Then the adjudicator check the record in the pub after checking the key is positive. The adjudicator will declare that A's claim is right if the recorder shows B has fetched stored message in the address and the message is the key. The adjudicator can know whether A has changed the stored message in the address after A sends NRA because the recorder keeps the record of access and modification.

5.2 Repudiation of Origin of M

If B claims having received M from A, the adjudicator will require B provide M, C, K, and the non-repudiation evidences NRO and NRA and NRS including the time stamp chains. If B can not provide these evidence or one of the following checks fails, B's claim will be judged invalid.

1. NRO is with A's signature and the time stamp chain is right.
2. NRA is with A's signature and the time stamp chain is right.
3. NRS is with A's signature and the time stamp chain is right.
4. $M = D_k(C)$.

If the first check is positive, the adjudicator will assume A has sent the commitment to B. Then the adjudicator will judge the address of the key has been sent to B if the second check is positive. It will claim that B has got the stored message in the address if the third check is positive. If the final check is positive, the adjudicator will say that B is right.

6 Formal Analysis

The protocol can be analysed by the SVO logic in [20] as beliefs held by the adjudicator (AJD).

1. Non-repudiation of origin:

   AJD believes $(A$ said $C) \wedge AJD$ believes $(A$ said $K) \supset AJD$ believes $(A$ said $M)$.

2. Non-repudiation of receipt:

   AJD believes $(B$ sees $C) \wedge AJD$ believes $(B$ sees $K) \supset AJD$ believes $(B$ sees $M)$.

The formal proof of the protocol would be implemented in terms of existing axioms referring to digital signature [20, 21]. However, there are a few specific issues which may be need to extend the existing logics for analysing cryptographic protocols [8, 9, 10]

7 Conclusion

Fair non-repudiation protocols without a trusted third party are superior to other non-repudiation protocols with an TTP. The protocol presented in the paper has the following properties.

1. Without an TTP.
2. Provided an equal protection to entities thus it is fair.
3. Without the assumption that fareness depends on communicating parties have equal computing power.
4. Against attacks from outsiders and denial from insiders.
5. The protocol does not require the reliability of the communication channel, and the communicating parties to play fair.
6. The transferring of the key is not time-consuming.

The fair non-repudiation protocol provides the equal protection to entities if one of them denies having sent or received a certain message and the entity in fact has sent or received the message.

Open Question:

1. An entity forges a certain message, then claim it is from another entity,
2. An entity has not sent a certain message to another entity, but he claims having sent.

In [14], Han presents a resolution to the open question.

8 Acknowledgements

The author was with Information Security Group, Dept. of Computer Science, Royal Holloway, Univ. of London. He would like to thank Prof. Chris J. Mitchell, Dr. Dieter Gollmann and Mr. Zhou Jianying for their discussins on a fair non-repudiation protocol with a TTP.

References

1. ISO/IEC DIS 10181-4. Information technology - open systems interconnection - security frameworks in open systems, part 4: Non-repudiation. *ISO/IEC JTC1*, 1995-04-21.
2. ISO/IEC CD 13888-3. Information technology - security techniques - non-repudiation - part 3: Using asymmetric techniques. *ISO/IEC JTC1/SC27 N1107*, 1995-09-21.
3. Ross J. Anderson. Why cryptosystems fail. *Communications of the ACM*, **37**, No. 11:32–40, 1994.
4. E.F. Brickell, D.M. Gordon, K.S. McCurley, and D.B. Wilson. Fast exponentiation with precomputation. In *Advances in Cryptology: Proceedings of Eurocrypt '92, LNCS 658*, pages 200–207. Springer-Verlag, New York, 1993.

5. Liqun Chen, Dieter Gollmann, Yongfei Han, and Chris Mitchell. Identification protocol. *IEEE Transaction on Computer*, 1996.

6. Ivan Bjerre Damgård. Practical and provably secure release of a secret and exchange of signatures. *J. of Cryptology*, **8, No.** 4:201–222, 1995.

7. D. Gollmann, Y. Han, and C.J. Mitchell. Redundant integer representations and fast exponentiation. *Designs, Codes and Cryptography*, **7**:135–151, 1996.

8. Y. Han and D. J. Evans. Parallel inference on systolic arrays and neural networks. *Parallel Algorithms and Application*, **10, No.** 1&2:169–175, 1996.

9. Y. Han and D.J. Evans. The simulation of EDC with OCCAM on multitransputer system. In *Modelling and Simulation*, pages 277–279, York, U. K., 1992. Simulation Councile.

10. Y. Han and D.J. Evans. Parallel inference algorithms for the connection method on systolic arrays. *International Journal of Computer Mathematics*, **53**:177–188, 1994.

11. Y. Han, D. Gollmann, and C.J. Mitchell. Minimal k-SR representations. In *Proceedings of fifth IMA Conference on Cryptography and Coding, LNCS 1025*, pages 34–43, Cirencester, U.K, 1995. Springer-Verlag, Berlin.

12. Y. Han, D. Gollmann, and C.J. Mitchell. Fast modular exponentiation for RSA on systolic arrays. *International Journal of Computer Mathematics*, **61**, 1996.

13. Yongfei Han. *Fast Algorithms for Public Key Cryptography*. Ph.D Thesis, University of London, 1995.

14. Yongfei Han. An assurance protocol. In *IEEE Information Theory and Its application*, 1996.

15. T. Johansson. Authentication codes for nontrusting parties obtained from rank metric codes. *J. of Cryptology*, **6**:205–218, 1995.

16. C.H. Lim and P.J. Lee. More flexible exponentiation with precomputation. In *Advances in Cryptology: Proceedings of CRYPTO '94*, pages 95–105, Santa Barbara, Ca., 1994. Springer-Verlag, New York.

17. Roger M. Needham. Denial of service. *Communication of the ACM*, **37, No.** 11:42–46, 1994.

18. T. Okamoto and K. Ohta. How to simultaneously exchange secret by general assumptions. In *Proceedings of 1994 IEEE Symposium on Research in Security and Privacy*, pages 14–28, Fairfax, Virginia, November, 1994.

19. A. Shamir. How to share a secret. *Communications of the ACM*, **22**:612–613, 1979.

20. Paul F. Syverson and Paul C. von Oorschot. On unifying some cryptographic protocol logics. In *Security and Pravicy*. IEEE, 1995.

21. Paul C. van Oorschot. Extending cryptographic logics of belief to key agreement protocols. In *Proceedings of 1st ACM Conference on Computer and Communications Security*, pages 232–243, Fairfax, Virginia, 1993. ACM press.

22. J. Zhou and D. Gollmann. A fair non-repudiation protocol. In *Proceedings of the IEEE Symposium on Research in Security and Privacy to appear*. IEEE, 1996.

A Dynamic Secret Sharing Scheme with Cheater Detection

Shin-Jia Hwang; and Chin-Chen Chang*

Department of Information Management, Chaoyang Institute of Technology,
WuFeng,Taichung Country, Taiwan 143, R. O. C.
E-mail: hwangsj@winston.cis.nctu.edu.tw
*Institute of Computer Science and Information Engineering, National Chung
Cheng University, Chiayi, Taiwan 621, R.O.C.,E-mail: ccc@cs.ccu.edu.tw

Abstract. We propose an efficient dynamic threshold scheme with cheater detection. By our scheme, without collecting and changing any secret shadows, the secret shadows can be reused after recovering or renewing the shared secret. Thus the new scheme is efficient and practical. In addition, the new scheme can detect the cheaters. Furthermore, the amount of public data is still proportional to the number of shadowholders.

Keywords: Secret sharing, dynamic threshold scheme, cheater detection.

1   Introduction

In our real world, some secrets are so critical that it cannot be protected by only one person. For example, the secret is used to set up a launch program for a nuclear missile. If this critical secret is held only by one person, it is easy to be lost, destroyed, and modified. Therefore, Shamir [1979] and Blakely [1979] independently introduced the threshold scheme (secret sharing scheme) to maintain critical secrets.

In a (t, n) threshold scheme, the critical secret is shared by n different shadowholders at the same time, where t is the threshold value and n is the number of holders. Each holder holds one secret shadow. Then the shared secret can be recovered by any t or more true shadows while the shared secret is undetermined by any t-1 or less true shadows, where the true shadow is derived from the corresponding secret shadow. Since 1979, many (t, n) threshold schemes have been proposed [Shamir 1979, Blakley 1979, Karn et al. 1983, Asmuth and Bloom 1983, Xian 1988].

However, these threshold schemes are inefficient and impractical for their two common disadvantages. The first one is the renew problem, which is defined as the shared secret cannot be renewed without modifying the secret shadows [Laih et al. 1989]. The second one is the reuse problem, which is defined as all of the secret shadows cannot be reused for the new shared secret after recovering the old shared secret [Zheng et al. 1994].

To eliminate the renew problem, Laih et al. [1989] proposed the concept of the dynamic secret sharing. Some dynamic threshold schemes providing perfect security were proposed [Sun 1990, Chang and Hwang 1992]. In these schemes, the renew times are limited. Recently, Sun and Shieh [1994a] showed that no dynamic threshold scheme providing perfect security can change the shared secret unlimited times. Moreover, the bit length of a secret shadow is proportional to the modifiable times of shared secrets. To conquer the reuse problem, some dynamic threshold schemes providing computational security had been proposed [Charnes et al. 1994, He and Dawson 1994, Zheng et al. 1994, Sun and Shieh 1994b, He and Dawson 1995]. In these schemes, the modifiable times of shared secrets are allowed to be reasonable large while the length of each shadow is as same as the length of shared secrets. Therefore, these schemes providing computational security are more practical than the schemes providing perfect security.

The renew and reuse problems cannot be eliminated by collecting the secret shadows in the dynamic scheme, because the collection is inefficient and will causes the revealing damage of secret shadows. An efficient and practical dynamic threshold scheme must have the feature that changing the shared secret will not lead to modify and collect any secret shadow.

In the dynamic threshold scheme, the shadowholders are assumed to be honest; but this assumption is not practical in the real world. If the holder offers false true shadow to recover the shared secret, he is a cheater. Thus the cheater detection is necessary in the dynamic threshold scheme. Among the proposed dynamic threshold

schemes, Sun and Shieh's scheme was the only one that can detect cheaters. Unfortunately, their cheater detection mechanism may cause the number of members who can cooperatively solve the shared secret will be less than the predetermined threshold value. In addition, the random integers cannot be reused, so the times of

changing shared secret in Sun and Shieh's scheme are limited.

Being inspired by the dynamic threshold scheme in [Hwang et al. 1995], we will propose an efficient dynamic threshold scheme with cheater detection. The

review and the weakness of Sun and Shieh's scheme will be described in Section 2. In Section 3, the new scheme will be presented. The security analysis will be given in Section 4. Finally, Section 5 gives the conclusions.

2   The Weakness of Sun and Shieh's Scheme

Sun and Shieh's (t, n) dynamic threshold scheme [1994] is described as below. The trustworthy dealer firstly publishes a large prime number P and the generator α of GF(P). Each shadowholder H_i randomly selects his secret shadow s_i and computes his public parameter $p_i = \alpha^{s_i} \mod P$. Next, the dealer randomly constructs a polynomial $f_0(X)$ of degree t-1 over GF(P) such that the constant item of $f_0(X)$ is the shared secret K_0. Then the dealer chooses a random integer k_0 in the range [1, P-1]. Finally, the dealer publishes $R_0 = \alpha^{k_0} \bmod P$, $C_{0i} = f_0(i) \times (p_i)^{k_0} \bmod P$ and the verification values $V_{0i} = \alpha^{f_0(i)} \bmod P$ for i=1, 2, ..., n. To reconstruct the threshold system after recovering or renewing the shared secret, the dealer repeats the above process to publish R_T, $C_{Ti} = f_T(i) \times (p_i)^{k_0} \bmod P$ and $V_{Ti} = \alpha^{f_T(i)} \bmod P$ for i=1, 2, ..., n, where T denotes the renew and recovery times of the shared secret.

Suppose H_1, H_2, ..., H_t want to recover the shared secret K_T cooperatively. Each holder H_i derives his true shadow $f_T(i)$ by $f_T(i) = C_{Ti} \times (R_T)^{-S_i} \bmod P$. Then each holder H_i shows his $f_T(i)$ to the other t-1 holders. Each holder H_i verifies all $f_T(j)$'s by $V_{Tj} \equiv \alpha^{f_T(j)} \pmod P$. If all of the $f_T(j)$'s are correct, then these t holders use the t points $(1, f_T(1))$, $(2, f_T(2))$, ..., $(t, f_T(t))$ to reconstruct the polynomial $f_T(X)$ by the technique of Lagrange interpolating polynomials as follows:

$$f_T(X) = \sum_{k=1}^{t} f_T(x_k) \prod_{\substack{j=1 \\ j \neq k}}^{t} (X - x_j)/(x_k - x_j) \pmod P, \text{ where } x_i = i \text{ for } i = 1, 2, ..., t.$$

Finally, they get the current shared secret K_T.

The weakness of Sun and Shieh's scheme is that their cheater detection mechansim may decrease the threshold value t. Moreover, this weakness cannot be avoided by the dealer. If a holder H_i finds that $V_{Ti} = V_{Tj}$ for i≠ j and T'≤ T, then H_i has two points $(i, f_T(i))$ and $(j, f_T(j)) = (j, f_T(i))$ on $f_T(X)$. To recover the current secret $f_T(0) = K_T$, H_i needs to cooperate with only t-2 holders. That is the threshold value is decreased by one. On the other hand, the dealer cannot construct a polynomial $f_T(X)$ such that all of the values of $f_T(i)$'s are unused because the degree of $f_T(X)$ is t-1. Therefore, this weakness cannot be removed. A threshold scheme is used to protect the critical secret, so the decrease the threshold value t should be forbade.

Though ElGamal public key cryptosystem and digital signature scheme [ElGamal 1985] also has the weakness caused by reusing the same random integer, the reuse problem of random integers can be overcome completely. In ElGamal public key cryptosystem and digital signature scheme, the sender can store the value of $\alpha^k \bmod P$ in a table for the used random integer k. When he selects a new integer k', he first computes $\alpha^{k'} \bmod P$. Then he check whether the value of $\alpha^{k'} \bmod P$ appears in the table. Consequently, he is sure that he uses the same integer k just once. In contrast, Sun and Shieh's scheme cannot adopt the above method to overcome the reuse problem of random integers. The reuse problem of random integers in Sun and Shieh's scheme is more serious than that in ElGamal public key cryptosystem and signature scheme.

Finally, let us consider whether Sun and Shieh's scheme can work unlimited times. Consider that t holders have recovered the shared secret K_T. Then anyone of the t holders knew all of $f_T(i)$'s, consequently. They also obtained all of β_i's, where $\beta_i = (p_i)^{k_T} \bmod P$. If k_T is reused as k_T, then the t holders can find any $f_T(i)$ at T. Therefore, k_T cannot be reused. That is, the number of reconstruction times of this dynamic system must be limited, which is equal to $(P-1-n)$.

3   Oor Scheme with Cheater Detection

Being inspired by the dynamic threshold scheme in [Hwang et al. 1995], we propose a new (t, n) dynamic threshold scheme with cheater detection. The new scheme is divided into three parts: the construction part, the reconstruction part, and the recovery part.

Part 1: Construction

The trustworthy dealer first publishes a large prime number P such that $P-1$ contains at least one large prime factor and the generator α of $GF(P)$. Each holder H_i randomly selects his secret shadow s_i and computes his public value $p_i = \alpha^{s_i} \bmod P$ for $i = 1, 2, ..., n$. Then each holder H_i gives his identity ID_i and the public value p_i to the dealer. Assume that all the public values of the holders are distinct.

After collecting all the public values of the holders, the dealer performs the following steps to construct the whole threshold system.

Step 1: Randomly choose a polynomial $f_0(X)$ of degree $t-1$ over $GF(P)$ such that the constant item is the shared secret K_0.

Step 2: Select an unused random integer z_0 in $[1, P-1]$ and compute the system parameter $Z_0 = \alpha^{z_0} \bmod P$. Then compute the difference values $d_{0,i}$'s, where $d_{0,i} = f_0(ID_i) - (p_i)^{z_0} \bmod P$ for $i = 1, 2, ..., n$.

Step 3: For each $f_0(ID_i)$, select n unused and distinct random integers in $[0, P-1]$ as $k_{0,1}, k_{0,2}, ...,$ and $k_{0,n}$. Then compute $R_{0,i} = \alpha^{k_{0,i}} \bmod P$ and construct $S_{0,i}$ such that $z_0 \times R_{0,i} \equiv S_{0,i} + k_{0,i} \times f_0(ID_i) \pmod{P-1}$ for $i = 1, 2, ..., n$.

Step 4: Publish a read-only table as Table 1 and the system parameter Z_0.

Table 1: A read-only public table

Identity	Verification	Difference
ID_1	$(R_{0,1}, S_{0,1})$	$d_{0,1}$
ID_2	$(R_{0,2}, S_{0,2})$	$d_{0,2}$
.		.
.		.
.		.
ID_n	$(R_{0,n}, S_{0,n})$	$d_{0,n}$

Part 2: Reconstruction

Suppose that the current shared secret is K_{T-1} and the current system parameter is Z_{T-1}. There are two cases that the dealer needs to reconstruct the threshold system. One is that the shared secret K_{T-1} had been recovered, and the other is that the dealer renews the shared secret for some security reasons. The dealer randomly selects K_T as the newly shared secret, and then repeats Steps 1-3 in the construction part. Finally, the dealer modifies the public table and publishes the new system parameter $Z_T = \alpha^{Z_T} \bmod P$.

We should point out here that the new secret random integer z_T must be unused. An unused random integer means that an integer has not been used as any secret shadow z_T, $k_{T,1}$, $k_{T,2}$, ..., and $k_{T,n}$ for $T' < T$. The secret integers $k_{T,i}$'s should also be unused.

Part 3: Recovery

Without losing the generality, suppose that the t holders H_1, H_2, .., H_t want to recover the shared secret K_T. Each holder H_i computes his true shadow $f_T(ID_i) = d_{m,i} + (Z_T)^{S_i} \bmod P$, and then sends $f_T(ID_i)$ with the other $t-1$ holders. Each holder H_i verifies the true shadows by the verification $(Z_T)^{R_{T,j}} \equiv \alpha^{S_{T,j}} \times (R_{T,j})^{f_T(ID_j)} \pmod{P}$ for $j = 1, 2, ..., t$ and $j \neq i$. If someone provides incorrect true shadow, the incorrect one will be found by this verification. That is, this verification can detect cheaters. If all these true shadows are correct, the t true shadows $(ID_1, f_T(ID_1))$, $(ID_2, f_T(ID_2))$, ..., $(ID_t, f_T(ID_t))$ are used to recover $f_T(X)$ and K_T through the technique of Lagrange interpolating polynomials.

4   Security Analysis

We first consider the security of the secret data: secret shadows and secret random integers. To derive s_i from p_i, z_T from Z_T, and $k_{T,i}$ from $R_{T,i}$ are equivalent to solve the discrete logarithm problem, so s_i, z_T, and $k_{T,i}$ are secure. Because s_i and z_T are secure, only the holder H_i has the ability to use his own secret shadow s_i to compute $(Z_T)^{s_i}$ mod P. That is, only the holder H_i has $f_T(ID_i) = d_{0,i} + (Z_T)^{s_i}$ mod P. Therefore, the true shadow $f_T(ID_i)$'s are secret.

Next, we discuss the threshold value of the new scheme at some T. Without losing the generality, consider the worst situation that t-1 holders, H_1, H_2, ..., H_{t-1}, wants to derive the shared secret K_T. Because the true shadows are secret, there are at most t-1 true shadows $f_T(ID_1)$, $f_T(ID_2)$, ..., $f_T(ID_{t-1})$. Since each shared secret K_T is protected by Shamir's threshold scheme, the t-1 true shadows cannot be used to recover K_T [Shamir 1979, Blakely and Measows 1984]. According to Theorem 1 in [He and Dawson 1994], the public difference values $d_{T,i}$'s do not reveal any information about the shared secret K_T. Therefore, the threshold value of our scheme at some T is t.

Now, let us consider whether or not the threshold value t will be decreased after reconstructing the whole threshold system. Because there is no relation among the polynomials used to protect the shared secrets, the threshold value will not be decreased by changing the polynomials. Since $k_{T,i}$'s and z_T's are secure and unused, nobody knows any information about the current shared secret K_T from $k_{T,i}$'s and z_T's for T'≤ T. Consequently, the threshold value is not decreased by reconstructing the threshold system.

In the following, we give the security analysis of the cheater detection. If the holder H_i wants to derive a false $f_T(ID_i)'$ to pass the verification $(Z_T)^{R_{T,i}} \equiv \alpha^{S_{T,i}} \times (R_{T,i})^{f_T(ID_i)'}$ (mod P). he should solve the discrete logarithm to get $f_T(ID_i)'$ from $(Z_T)^{R_{T,i}} \equiv \alpha^{S_{T,i}} \times (R_{T,i})^{f_T(ID_i)'}$ (mod P). If H_i wants to forge $(R'_{T,i}, S'_{T,i})$ for the false $f_T(ID_i)'$, he should solve the discrete logarithm problem whatever he determines $R'_{T,i}$ ($S'_{T,i}$) before $S'_{T,i}$ ($R'_{T,i}$). Thus this verification can detect false true shadows. Hence the cheater who provides the false true shadow can be found by the other honest holders.

The reuse problem of the secret random integers is discussed below. All T's, $k_{T,i}$'s and z_T's must be distinct; otherwise, the secret data $k_{T,i}$'s and z_T's are derived by three holders whose $k_{T,i}$'s are the same. The reason is that the number of unknown variables is less than the number of equations. Because the secret random integers cannot be reused, the new scheme can be used at most $(P-1-n)/(n+1)$ times. Because $P \gg n$, $(P-1-n)/(n+1)$ is large, the new scheme can work practically.

Finally, we give the reason why the weakness of Sun and Shieh's scheme does not exist in our dynamic threshold scheme. In our scheme, anyone can compute $R_{T,i}{}^{f_T(ID_i)} \bmod P$. Though anyone can obtain $R_{T,i}{}^{f_T(ID_i)} \equiv R_{T,i}{}^{f_T(ID_i)} \pmod P$ for $T \geq T'$ and $i \neq j$, he can not get $f_T(ID_j)$ because both $k_{T,i}$ and $k_{T,i}$ are secret, so he still has only one true shadow at T.

5   Conclusions

Sun and Shieh's dynamic threshold scheme was the first dynamic one to detect cheaters. Unfortunately, their cheater detection may cause the number of members who can cooperatively solve the shared secret will be less than the predetermined threshold value. Being inspired of the dynamic threshold scheme proposed by Hwang et al. [1995], we proposed a new efficient dynamic threshold scheme with

cheater detection to remove the weakness of Sun and Shieh's scheme. To change the shared secret, the dealer need not collect or change any secret shadow, so our new scheme is efficient and practical. The new scheme also provides the function to detect cheaters. Moreover, in the new scheme, the dealer needs not keep any secret data for the threshold system. The amount of public data is $O(n)$, where n is the number of the shadowholders

References

1. Asmuth, A. and Bloom, J. (1983): "A Modular Approach to Key Safeguarding," *IEEE Transactions on Information Theory*, Vol. IT-29, 1983, pp. 208-210.
2. Blakley, G. R. (1979): "Safeguarding Cryptographic Keys," *Proceedings of the National Computer Conference, 1979, American Federation of Information Processing Societies*, Vol.. 48, 1979, pp. 242- 268.
3. Blakley, G. R. and Meadows, C. (1984): "Security of Ramp Schemes," *Advances in Cryptology- Crypto '84*, Springer-Verlag, Berlin, Heidelberg, 1984, pp. 242- 268.
4. Chang, C. C. and Hwang, S. J. (1992): "Sharing a Dynamic Secret," *IEEE International Phoenix Conference on Computer and Communications*, Wyndham Paradise Vally Resort Scottsdale Arizona, U.S.A., April, 1992, pp. 3.4.4.1-3.4.4.4.
5. Charnes, Pieprzyk, and Safavi-Naini (1994): "Using the Discrete Logarithm for Dynamic (t, n) Secret Sharing Schemes", *2nd ACM Conference on Computer & Communication Security*, 1994, pp. 89-95.

6. Diffie, W. and Hellman, M. E.(1976): "New Directions in Cryptography," *IEEE Transactions on Information Theory*, Vol. IT-22, No. 6, 1976, pp. 644-654.

7. ElGamal, T. (1985): "A Public Key Cryptosystem and a Signature Scheme Based on Discrete Logarithms," *IEEE Transactions on Information Theory*, Vol. IT-31, No. 4, July 1985, pp. 469- 472.

8. He, J. and Dawson, E. (1994): "Multistage Secret Sharing Based on One-way Function," *Electronics Letters*, Vol. 30, No. 19, September, 1994, pp. 1591-1592.

9. He, J. and Dawson, E. (1995): "Multisecret-Sharing Scheme Based on One-Way Function," *Electronics Letters*, Vol. 31, No. 2, January, 1995, pp. 93- 95.

10. Hwang, S. J., Chang, C. C. and Yang, W. P. (1995): "An Efficient Dynamic Threshold Scheme," *IEICE TRANS. INF. & SYST.*, Vol. E79-D, No. 7, 1996, pp. 936-942.

11. Karnin, E. D., Greene, J. W. and Hellman, H. E. (1983): "On Secret Sharing Systems," *IEEE Transactions on Information Theory*, Vol. IT-29, 1983, pp. 35-41.

12. Knuth, D. (1981): *The Art of Computer Programming, Vol. 2: Seminumerical Algorithms*, Second Edition, Addison-Wesley, Reading, Mass, 1981.

13. Laih, C., Harn, L., Lee, J. and Hwang, T. (1989): "Dynamic Threshold Scheme Based on the Definition of Cross-Product in an N-Dimensional Linear Space," *Proceeding of Crypto '89*, Santa Barbara, California, U.S.A., August 1989, pp. 20-24.

14. Shamir, A. (1979): "How to Share a Secret," *Communications of the Association for Computing Machinery*, Vol. 22, No. 11, 1979, pp. 612-613.

15. Sun, H. M. (1990): "Key Management on Public-Key Cryptosystems and Threshold Schemes," *Master Thesis of Institute of Applied Mathematics, National Cheng Kung University*, Tainan, Taiwan, R.O.C., 1990, pp. 36-67.

16. Sun H. M. and Shieh, S. P. (1994a): "On Dynamic Threshold Schemes," *Information Processing Letters*, Vol. 52, 1994, pp. 201-206.

17. Sun H. M. and Shieh, S. P. (1994b): "Construction of Dynamic Threshold Schemes", *Electronics Letters*, Vol. 30, No. 24, November, 1995, pp. 2023-2024

18. Xain, Y. Y. (1988): "Relationship Between MDS Codes and Threshold Scheme," *Electronics Letters*, Vol. 24, No. 3, 1988, pp. 154- 156.

19. Zheng, Y., Hardjono, T. and Seberry, J. (1994): "Reusing Shares in Secret Sharing Schemes," *The Computer Journal*, Vol. 37, No. 3, 1994, pp. 200- 205.

A Nonlinear Secret Sharing Scheme

Ari Renvall and Cunsheng Ding

Department of Mathematics
University of Turku
Fin-20014 Turku, Finland

Abstract. In this paper, we have described a nonlinear secret-sharing scheme for n parties such that any set of $k-1$ or more shares can determine the secret, any set of less than $k-1$ shares might give information about the secret, but it is computationally hard to extract information about the secret. The scheme is based on quadratic forms and the computation of both the shares and the secret is easy.

1 Introduction

In a secret-sharing scheme a number of parties share a secret. The information about the secret a party has is called a share or shadow. The earliest contributions on this topic are linear ones [1, 9]. There are many later contributions on this topic (see [2, 3, 4, 8, 10, 11, 12] for examples). Some nonlinear secret-sharing schemes can be found in [2, 11]. The most studied secret-sharing schemes are the (k, n) threshold schemes in which the secret can be recovered with any set of k shares while any set of $k-1$ shares gives no information about the secret. By a proper formulation linear (k, n) threshold schemes are equivalent to maximum distance separable (MDS) error-correcting codes, but there is only a very limited class of MDS linear codes. The progress in coding theory shows that finding more MDS linear codes, and therefore more linear (m, n) threshold schemes, is not easy. This motivates the study of nonlinear secret-sharing schemes.

In secret-sharing schemes for n parties the shares t_i for $i = 1, 2, \cdots, n$ are often computed according to some arithmetic functions

$$t_i = f_i(s_1, s_2, \cdots, s_u),$$

where s_1 is the secret and $s_2, \cdots, s_u$ are randomly chosen elements of some fields. We call such a scheme *linear* if all f_i are linear with respect to the variable $\mathbf{s} = (s_1, \cdots, s_u)$, and *nonlinear* otherwise.

Most of the secret-sharing systems are designed so that when a subset of parties pools their shares together, they get either no information or all information about the secret. Thus, in such a scheme only authorized parties can get information about the secret. Such secret-sharing schemes are called *perfect*. An alternative could be that the unauthorized sets of shares are allowed to give a limited amount of information about the secret, but extracting the information by such a subset of parties is computationally hard. Thus, theoretically they can get some information about the secret, but computationally they cannot do so.

This is the case for many cipher systems in which a set of plaintext-ciphertext pairs usually gives information about the key, but finding the actual key by making use of the information provided by the set of plaintext-ciphertext pairs could be computationally infeasible. From practical point of view such a secret-sharing system with similar properties could be as good as a perfect scheme.

In this paper we describe a nonlinear secret sharing scheme for n parties such that

- any $k - 1$ or more shares are enough to recover the secret;
- any $k - 2$ or fewer shares might give information about the secret, but it is computationally hard to extract the information.

Our secret-sharing system has the following properties:

1. it is nonlinear;
2. both computing the shares and recovering the secret are efficient;
3. with decoding techniques for Reed-Solomon codes it can correct cheatings when enough parties pool their shares together.

The secret-sharing scheme, based on the theory of quadratic forms, is nonperfect since it is nonlinear. It can be proved that all linear secret-sharing schemes are perfect, but nonlinear schemes are often nonperfect.

2 Designing the parameters

Let p be a large prime of the form $p = 3 \bmod 4$. Our secret is a number between 0 and $(p-1)/2$. We shall use $GF(p)$ to denote the field consisting of the integers $\{0, 1, \cdots, p-1\}$ with addition and multiplication modulo p. We assume that each number between 0 and $(p-1)/2$ is equally likely to be the secret.

The secret-sharing scheme described in this paper is intended for key safeguarding, so we assume that the prime p has at least 60 bits. The other parameters of the system are a set of distinct nonzero integers $a_1, \cdots, a_n$.

Before specifying the parameters we need some notations. In the sequel for each set of indices $1 \leq i_1 < \cdots < i_w \leq n$, where $1 \leq w \leq n$, let

$$M(i_1, \cdots, i_w) = \begin{bmatrix} a_{i_1} & a_{i_1}^2 & \cdots & a_{i_1}^w \\ a_{i_2} & a_{i_2}^2 & \cdots & a_{i_2}^w \\ \vdots & \vdots & \vdots & \vdots \\ a_{i_w} & a_{i_w}^2 & \cdots & a_{i_w}^w \end{bmatrix}$$

and $N(i_1, \cdots, i_w) = [n(i_1, \cdots, i_w)_{u,v}]$ be the inverse of $M(i_1, \cdots, i_w)$. Furthermore, let

$$\beta_2(i_1, \cdots, i_{k-2}) = 1 + \sum_{u=2}^{k-1} \left[\sum_{v=1}^{k-2} n(i_1, \cdots, i_{k-2})_{u,v} \right]^2$$

$$\gamma_2(i_1,\cdots,i_{k-2}) = 1 + \sum_{u=2}^{k-1}\left[\sum_{v=1}^{k-2} n(i_1,\cdots,i_{k-2})_{u,v}a_{i_v}^{k-1}\right]^2$$

$$\delta(i_1,\cdots,i_{k-2}) = 2\sum_{u=2}^{k-2}\left[\sum_{v=1}^{k-2} n(i_1,\cdots,i_{k-2})_{u,v}\right]\left[\sum_{v=1}^{k-2} n(i_1,\cdots,i_{k-2})_{u,v}a_{i_v}^{k-1}\right].$$

The parameters a_i are chosen such that

C1: $a_1,\cdots,a_n$ are distinct nonzero elements of $GF(p)$;

C2: for any set of indices $1 \le i_1 < \cdots < i_{k-1} \le n$

$$1 + \sum_{u=2}^{k}\left(\sum_{v=1}^{k-1} n(i_1,\cdots,i_{k-1})_{u,v}\right)^2 \ne 0; \tag{1}$$

C3: for any set of indices $1 \le i_1 < \cdots < i_{k-2} \le n$ one of the following conditions holds:
  - $\delta(i_1,\cdots,i_{k-2}) \ne 0$ and $\delta(i_1,\cdots,i_{k-2})^2 - 4\beta_2(i_1,\cdots,i_{k-2})\gamma_2(i_1,\cdots,i_{k-2})$ is not a quadratic nonresidue;
  - $\delta(i_1,\cdots,i_{k-2}) = 0$, $\gamma_2(i_1,\cdots,i_{k-2}) \ne 0$, and $-\frac{\beta_2(i_1,\cdots,i_{k-2})}{\gamma_2(i_1,\cdots,i_{k-2})}$ is not a quadratic nonresidue.

Here we regard 0 as neither a quadratic residue nor a quadratic nonresidue. The above conditions are necessary in controlling the amount of information provided by sets of shares, as seen in the sequel.

In order to design the parameters a_i, we need to find the inverses of the matrices $M(i_1,\cdots,i_{k-1})$, and those of the matrices $M(i_1,\cdots,i_{k-2})$. Since they are Vandermonde matrices, it is easy to compute their inverse matrices by the Chinese remainder algorithm.

Thus, the procedure of designing the above parameters is the following. Choose first a large prime p of form $p = 3 \bmod 4$. Then take randomly n distinct elements a_i of $GF(p)$. Then compute the inverse of the matrices $M(i_1,\cdots,i_{k-1})$ and $M(i_1,\cdots,i_{k-2})$. Finally, check whether conditions C2 and C3 are satisfied. If not, choose another set of parameters and repeat the same procedure until the two conditions are satisfied. Note that the parameter p is usually large enough, while the parameters k and n are very small, the probability of satisfying conditions C2 and C3 with a randomly chosen set of parameters a_i is large. Because it is easy to check whether a number is a quadratic residue modulo p with an algorithm described in [5, pp.45-46], it is computationally easy to find a set of n distinct nonzero a_i satisfying conditions C2 and C3.

3　The nonlinear secret sharing system

With the parameters specified above we are ready to describe the secret-sharing system. The s_1 with $0 \le s_1 \le (p-1)/2$ is the secret to be shared among n parties. Then we choose randomly $k-1$ values $s_2, s_3, \cdots, s_k \in GF(p)$. Let

$\mathbf{s} = (s_1, s_2, \cdots, s_k)$. The shares are computed with the quadratic form $f(\mathbf{x}) = \mathbf{x}\mathbf{x}^T = x_1^2 + x_2^2 + \cdots + x_k^2$ over $GF(p)$.

Let

$$\alpha_i = (1, a_i, a_i^2, \cdots, a_i^{k-1}), \; i = 1, 2, \cdots, n.$$

It follows from properties of Vandermonde matrices that every k of these α_i are linearly independent. Let G be the matrix having α_i as its ith column vector, then G is a generator matrix of a Reed-Solomon code over $GF(p)$ [6].

The shares are computed as follows. The dealer calculates $t_0 = f(\mathbf{s})$ and

$$t_i = f(\mathbf{s} + \alpha_i), \;\; i = 1, 2, \cdots, n.$$

Then he distributes (t_0, t_i) to each party P_i. This is efficient since only $(2k - 1)n + k - 1$ additions and nk multiplications of $GF(p)$ are needed. It is noted that the parameters α_i are known to each party.

To describe the algorithm for computing the secret when $k - 1$ parties pool their shares together, we need the following conclusion.

Theorem 1. *The sets of equations*

$$\begin{cases} t_0 = f(\mathbf{s}), \\ t_i = f(\mathbf{s} + \alpha_i), \; i = 1, 2, \cdots, m \end{cases} \tag{2}$$

and

$$\begin{cases} t_0 \quad = f(\mathbf{s}), \\ t_i - t_0 = 2\alpha_i \mathbf{s}^T + \alpha_i \alpha_i^T, \; i = 1, 2, \cdots, m \end{cases} \tag{3}$$

have the same solution space.

Proof: By equivalence of two equations we understand that they have the same solution space. First we note that equation (2) is equivalent to

$$t_0 = f(\mathbf{s}), \; t_i - t_0 = f(\mathbf{s} + \alpha_i) - f(\mathbf{s}), i = 1, \cdots, m.$$

But $f(\mathbf{s} + \alpha_i) - f(\mathbf{s}) = 2\alpha_i \mathbf{s}^T + \alpha_i \alpha_i^T$. This proves the theorem. $\qquad \square$

When k parties pool their shares together the secret can be computed as follows. Assume parties $P_1, P_2, \cdots, P_k$ pool their shares together. The values $t_0, t_1, ..., t_k$ are known. Since $\alpha_1, \alpha_2, \cdots, \alpha_k$ are linearly independent, solving equation

$$2\alpha_i \mathbf{s}^T = t_i - t_0 - \alpha_i \alpha_i^T, i = 1, 2, \cdots, k \tag{4}$$

gives $\mathbf{s}^T$, and therefore the secret. There are efficient algorithms for solving such a set of linear equations such as the Gaussian elimination algorithm having complexity $O(k^3)$.

Now we deal with the case when $k - 1$ parties pool their shares together in which we need a special algorithm for solving the quadratic equation

$$y^2 = a \bmod p.$$

Since $p = 3 \bmod 4$, -1 is a quadratic nonresidue. Thus, an efficient algorithm for solving this equation exists, which has the worst-case complexity $O((\log p)^4)$ (see [5, pp.47-49] for details). We shall refer to this algorithm as Algorithm A in the sequel.

Let

$$b_i = t_i - t_0 - \alpha_i \alpha_i^T, \quad i = 1, 2, \cdots, n.$$

Suppose that $k - 1$ shares $(t_0, t_{i_1}), \cdots, (t_0, t_{i_{k-1}})$ are known. Then we want to solve

$$M(i_1, \cdots, i_{k-1}) \begin{bmatrix} s_2 \\ s_3 \\ \vdots \\ s_k \end{bmatrix} = \begin{bmatrix} b_{i_1} - s_1 \\ b_{i_2} - s_1 \\ \vdots \\ b_{i_{k-1}} - s_1 \end{bmatrix}.$$

By solving this equation we have

$$s_u = \sum_{v=1}^{k-1} n(i_1, \cdots, i_{k-1})_{u,v} + c_u, \quad u = 2, \cdots, n, \tag{5}$$

where

$$c_u = \sum_{v=1}^{k-1} n(i_1, \cdots, i_{k-1})_{u,v} b_{i_v}.$$

Combining equation (5) and $f(s) = \sum s_i^2$ yields

$$t_0 = \left\{ 1 + \sum_{u=2}^{k} \left(\sum_{v=1}^{k-1} n(i_1, \cdots, i_{k-1})_{u,v} \right)^2 \right\} s_1^2 \\ + 2 \sum_{u=2}^{k} c_u \sum_{v=1}^{k-1} n(i_1, \cdots, i_{k-1})_{u,v} s_1 + \sum_{u=2}^{k} c_u^2. \tag{6}$$

This is an equation of the form

$$a s_1^2 + b s_1 + c = 0,$$

where by Condition C2 the coefficient $a \neq 0$. It is easily written as

$$\left(s_1 + \frac{b}{2a} \right)^2 = -\frac{c}{a} + \frac{b^2}{4a^2}. \tag{7}$$

When choosing the parameters a_i, the inverses of the matrices

$$M(i_1, \cdots, i_{k-1}) \text{ and } M(i_1, \cdots, i_{k-2})$$

have been already computed, and they are computed once for all. Thus, the coefficients of equation (6) are easy to compute. Since $p = 3 \bmod 4$, -1 is a quadratic nonresidue modulo p. Thus, by Algorithm A for solving $y^2 = u$, we get two solutions for s_1. Only one of them is less than $(p + 1)/2$, and it is the secret.

In summary, when more than $k-1$ shares are known, the secret can be recovered by solving linear equations. When $k-1$ shares are known, the coefficients of (6) are first computed, and then (7) is solved with Algorithm A. The computation of the shares and that for the secret for both cases are as efficient as a $(k-1, n)$ threshold scheme based on a Reed-Solomon code including the Shamir scheme.

4 Theoretical analysis of the system

In this section we shall do theoretical analysis for the secret-sharing system described in the previous section. We first analyze whether it is possible to determine the secret computationally when $k-2$ shares are given. Then we prove that this secret-sharing system is able to correct cheatings when enough parties come together.

Consider now the following equation

$$\beta_2 x^2 + \beta_1 x + \delta x y + \gamma_2 y^2 + \gamma_1 y = d, \quad \beta_i, \gamma_i, \beta, \delta, d \in GF(p) \tag{8}$$

with $\delta \neq 0$. Let $\Delta(x, y)$ denote the number of solutions (x, y) of equation (8), and $\Delta(x)$ the number of distinct x such that (x, y) is a solution of (8) for some $y \in GF(p)$.

Theorem 2. *Assume that (8) has solutions. If $\delta^2 - 4\beta_2\gamma_2 = 0$ or $\delta^2 - 4\beta_2\gamma_2$ is a quadratic residue modulo p, then $\Delta(x) \geq (p-1)/2$.*

Proof: The proof is divided into several cases.
Case 1: Suppose $\gamma_2 = 0$. Then we have

$$d - \beta_2 x^2 - \beta_1 x = (\delta x + r_1)y.$$

Hence for any $x \neq \gamma_1/\delta$, there is a $y = (d - \beta_2 x^2 - \beta_1 x)/(\delta x + r_1)$ such that (x, y) is a solution of (8). Thus, $\Delta(x) \geq p - 1$.
Case 2: Suppose $\gamma_2 \neq 0$. We divide this case into several subcases. First suppose that $\beta_2 = 0$. We then have

$$x(\delta y + \beta_1) = d - \gamma_2 y^2 - \gamma_1 y.$$

It is then easily seen that $\Delta(x) \geq (p-1)/2$ since for each x the above equation has at most two solutions y. Then suppose $\beta_2 \neq 0$. If (x, y_1) and (x, y_2) are two solutions of (8), where $y_1 \neq y_2$, then

$$\delta x(y_1 - y_2) + \gamma_2(y_1^2 - y_2^2) + \gamma_1(y_1 - y_2) = 0$$

Hence

$$\Delta(x) \geq \Delta(x, y)/2. \tag{9}$$

Let $z = x + (2\beta_2)^{-1}\delta y$, then equation (8) becomes

$$\beta_2 \left(z + \frac{\beta_1}{2\beta_2} \right)^2 + \sigma_2 y^2 + \sigma_1 y = d + \frac{\beta_1^2}{4\beta_2}, \tag{10}$$

where

$$\sigma_2 = \gamma_2 - \frac{\delta^2}{4\beta_2}, \quad \sigma_1 = \gamma_1 - \frac{\beta_1 \delta}{2\beta_2}.$$

Note also that

$$\begin{bmatrix} z \\ y \end{bmatrix} = \begin{bmatrix} 1 & (2\beta)^{-1}\delta \\ 0 & 1 \end{bmatrix} \begin{bmatrix} x \\ y \end{bmatrix}$$

and that the matrix of the above equation is invertible, the number of solutions (x, y) of (8) is equal to that (z, y) of (10).

By assumption either $\sigma_2 = 0$ or $\delta^2 - 4\beta_2\gamma_2$ is a quadratic residue. Assume $\sigma_2 = 0$, then equation (10) is equivalent to

$$\beta_2 z^2 + \beta_1 z + \sigma_1(z - x)2\beta_2\delta^{-1} = d \tag{11}$$

If $\sigma_1 = 0$, it is easily seen that $\Delta(x) = p$. If $\sigma_1 \neq 0$, it is easy to see that $\Delta(x) = (p - 1)/2$.

Now assume that $\sigma_2 \neq 0$, then equation (10) can be written as

$$\beta_2 \left(z + \frac{\beta_1}{2\beta_2} \right)^2 + \sigma_2 \left(y + \frac{\sigma_1}{2\sigma_2} \right)^2 = d', \tag{12}$$

where $d' = d + (4\beta_2)^{-1}\beta_1^2 + (4\sigma_2)^{-1}\sigma_1^2$. Note that in this case $\delta^2 - 4\beta_2 r_2$ is a quadratic residue, and so is $-\sigma_2/\beta_2$. By Theorem 6.26 of [7] $\Delta(x, y)$ must equal one of $2p - 1$, $p + 1$, and $p - 1$. It then follows from (9) that $\Delta(x) \geq (p - 1)/2$.

Summarizing the above cases, we have proved this theorem. □

Similarly to the proof of Theorem 2, we can prove the following result.

Theorem 3. *Assume that the equation*

$$\beta_2 x^2 + \beta_1 x + \delta xy + \gamma_2 y^2 + \gamma_1 y = d, \quad \gamma_2 \neq 0 \tag{13}$$

has solutions, where $\beta_i, \gamma_i, d \in GF(p)$. If $-\beta_2/\gamma_2 = 0$ or it is a quadratic residue, then $\Delta(x) \geq (p - 1)/2$.

Now assume that $k - 2$ parties pool together their shares

$$(t_0, t_{i_1}), \cdots, (t_0, t_{i_{k-2}}).$$

We now solve this equation

$$M(i_1, \cdots, i_{k-2}) \begin{bmatrix} s_2 \\ s_3 \\ \vdots \\ s_{k-1} \end{bmatrix} = \begin{bmatrix} b_{i_1} - s_1 - a_{i_1}^{k-1} s_k \\ b_{i_2} - s_1 - a_{i_2}^{k-1} s_k \\ \vdots \\ b_{i_{k-2}} - s_1 - a_{i_{k-2}}^{k-1} s_k \end{bmatrix}.$$

By solving this equation we have

$$s_u = -s_1 \sum_{v=1}^{k-2} n(i_1, \cdots, i_{k-2})_{u,v} - s_k \sum_{v=1}^{k-2} r(i_1, \cdots, i_{k-2})_{u,v} a_{i_v}^{k-1} + d_u,$$

$$u = 2, \cdots, k-1,$$

where

$$d_u = \sum_{v=1}^{k-2} n(i_1, \cdots, i_{k-2})_{u,v} b_{i_v}, \quad u = 2, \cdots, k-1.$$

Combining the above equation and $f(\mathbf{s}) = \sum s_i^2$ yields

$$t_0 = \beta_2(i_1, \cdots, i_{k-2})s_1^2 + \gamma_2(i_1, \cdots, i_{k-2})s_k^2 + \delta(i_1, \cdots, i_{k-2})s_1 s_k$$
$$- 2\sum_{u=2}^{k-1} d_u \left[\sum_{v=1}^{k-2} n(i_1, \cdots, i_{k-2})_{u,v} a_{i_v}^{k-1} \right] s_k$$
$$- 2\sum_{u=2}^{k-1} d_u \left[\sum_{v=1}^{k-2} n(i_1, \cdots, i_{k-2})_{u,v} \right] s_1 + \sum_{u=2}^{k-1} d_u^2,$$

where $\beta_2(i_1, \cdots, i_{k-2}), \gamma_2(i_1, \cdots, i_{k-2})$ and $\delta(i_1, \cdots, i_{k-2})$ are defined as in Section 2.

By condition C3 this is an equation of form (8) or (13), and further by Theorems 2 and 3 it has at least $(p-1)/2$ solutions. Thus, if the secret s_1 is supposed to take any element of $GF(p)$, the conditional uncertainty of the secret is

$$H(s_1|(t_0, t_{i_1}), \cdots, (t_0, t_{i_{k-2}})) \geq \log_2(p-1)/2.$$

It follows that the mutual information

$$I(s_1|(t_0, t_{i_1}), \cdots, (t_0, t_{i_{k-2}})) \leq \log_2 p - \log_2(p-1)/2 = 1 + \log_2(1 + 1/(p-1)).$$

Thus, any set of $k-2$ shares does give information about the secret, but at most $1 + \log_2(1 + 1/(p-1))$ bits if we consider the secret to be any element of $GF(p)$.

It should be noted that the knowledge $0 \leq s_1 \leq (p-1)/2$ could be very useful. However, to make use of the knowledge to extract information about the secret the best one can do is to solve one of the several equations in the proof of Theorem 2. To find all the possible x such that (x, y) is a solution of the equation, one has to do at least $(p-1)/2$ multiplications or additions over $GF(p)$, and this is computationally infeasible when p is large enough. Thus, a set of $k-2$ shares might give information about the secret, but extracting the information is hard when p is large enough.

In secret sharing some parties may change their shares in order to cheat. Thus, secret-sharing systems with the ability to detect and/or correct cheatings may be required in some applications. This secret-sharing scheme is obviously nonlinear, and the share vector $(t_1, \cdots, t_n)$ does not correspond to the codeword of a linear code. However, by some transform this secret-sharing scheme is closely related to some linear codes. As a consequence it can correct cheatings with necessary shares.

Clearly, the detection and correction of t_0 are very easy, since every party has one copy of it. So it is natural to assume that there will be no cheating on t_0. Let

$$G = [\alpha_1^T, \cdots, \alpha_n^T].$$

As mentioned before, G is the generator matrix of a $[n, k, n-k+1]$ Reed-Solomon MDS code. By the definition of the shares t_i it is easily seen that

$$(s_1, \cdots, s_k)G = (w_1, \cdots, w_n),\qquad(14)$$

where $w_i = (t_i - t_0 - \alpha_i \alpha_i^T)/2$. Thus, the vector $(w_1, \cdots, w_n)$ is a codeword of the Reed-Solomon code. Note that the cheating from party P_i results in exactly an error entry in the vector $(w_1, \cdots, w_n)$, and that the Reed-Solomon code can correct $\lfloor(n - k)/2\rfloor$ errors [6]. The following result follows from the fact that a linear $[n, k, d]$ code can correct $\lfloor(d - 1)/2\rfloor$ errors.

Theorem 4. *This secret-sharing scheme can correct up to $\lfloor(n-k)/2\rfloor + k + j - n$ cheaters when $k + j$ parties come together for the secret.*

Since there are efficient decoding algorithms for Reed-Solomon codes, the correction of cheatings in this system is easy when enough parties pool their shares together. To summary, the secret sharing system described in the previous section has the properties outlined in Section 1.

5   An example

In this section we give an example of the nonlinear secret-sharing scheme described before, here we consider the case $k = 3$ and $n = 3$. As before, the prime p is of the form $p = 3 \bmod 4$.

Let $f(x) = x_1^2 + x_2^2 + x_3^2$ and

$$\alpha_i = (1, a_i, a_i^2), \ i = 1, 2, 3.$$

Also let

$$M = \begin{bmatrix} a_{i_1} & a_{i_1}^2 \\ a_{i_2} & a_{i_2}^2 \end{bmatrix}.$$

Then we have

$$M^{-1} = \begin{bmatrix} a_{i_2}^2/a & -a_{i_1}^2/a \\ -a_{i_2}/a & a_{i_1}/a \end{bmatrix},$$

where $a = a_{i_1} a_{i_2}(a_{i_2} - a_{i_1})$. It is easy to see that Condition C2 becomes that for each $1 \le i_1 < i_2 \le 3$

$$(a_{i_1} + a_{i_2})^2 + a_{i_1}^2 a_{i_2}^2 + 1 \ne 0.\qquad(15)$$

Assume that (t_0, t_i) is known. Combining $s_1 + a_i s_2 + a_i^2 s_3 = w_i$ and $s_1^2 + s_2^2 + s_3^2 = t_0$ yields

$$(1 + a_i^4)s_1^2 + (a_i^4 + a_i^2)s_2^2 + 2a_i s_1 s_2 - 2w_i s_1 - 2a_i w_i s_2 + w_i^2 - a_i^4 t_0 = 0.$$

Apparently $\delta = 2a_i \neq 0$, and

$$\delta^2 - 4\beta_2\gamma_2 = 4a_i^4[-(a_i^4 + a_i^2 + 1)],$$

where $\beta_2 = 1 + a_i^4$ and $\gamma_2 = a_i^4 + a_i^2$. Thus, we need only to choose a_i such that (15) is satisfied and that $-(a_i^4 + a_i^2 + 1)$ is not a quadratic nonresidue. This is a secret-sharing scheme in which any two shares are enough to recover the secret, one share could give information about the secret, but it is computationally hard to extract the information.

To illustrate this system better, we consider the special prime $p = 19$. It is easily seen that the set of parameters $a_1 = 1, a_2 = 2, a_3 = 3$ satisfies all the above conditions. Hence

$$\alpha_1 = (1,1,1), \quad \alpha_2 = (1,2,4), \quad \alpha_3 = (1,3,9).$$

The computation of the shares are carried out as follows:

$$t_0 = s_1^2 + s_2^2 + s_3^2, \qquad\qquad t_1 = (s_1 + 1)^2 + (s_2 + 1)^2 + (s_3 + 1)^2,$$
$$t_2 = (s_1 + 1)^2 + (s_2 + 2)^2 + (s_3 + 4)^2, \; t_3 = (s_1 + 1)^2 + (s_2 + 3)^2 + (s_3 + 9)^2,$$

where $0 \leq s_1 \leq 9$ is the secret and s_2, s_3 are randomly chosen independent elements of $GF(19)$. The pair (t_0, t_i) is the share of party P_i for $i = 1, 2$ and 3.

When all three shares are known, the secret s_1 is obtained by solving the set of linear equations

$$2s_1 + 2s_2 + 2s_3 = t_1 - t_0 - 3,$$
$$2s_1 + 4s_2 + 8s_3 = t_2 - t_0 - 2,$$
$$2s_1 + 6s_2 - s_3 = t_3 - t_0 + 4.$$

When (t_0, t_1) and (t_0, t_2) are known, the secret s_1 is obtained by solving the following equation

$$s_1^2 + (3u - 9v)s_1 + 3u^2 + 3v^2 - 3t_0 = 0,$$

where

$$u = 5t_2 + 9t_1 + 5t_0 + 1, \quad v = -5t_2 + t_1 + 4t_0 + 7.$$

This shows the computation of the shares and that of the secret.

References

1. G. R. Blakley, *Safeguarding cryptographic keys,* Proc. NCC AFIPS (1979), 313-317.
2. E. F. Brickell, *Some ideal secret sharing schemes,* in "Advances in Cryptology — Eurocrypt'89", LNCS **434** (1990), 468-475.
3. C. Charnes, J. Pieprzyk and R. Safavi-Naini *Conditionally secure secret sharing scheme with disenrollment capability,* in "2nd ACM Conference on Computer and Communications Security," ACM Press, 1994, 89-95.
4. E. Dawson, E. S. Mahmoodian and A. Rahilly, *Orthogonal arrays and ordered threshold schemes,* Australian Journal of Combinatorics **8** (1993), 27-44.
5. N. Koblitz, *A Course in Number Theory and Cryptography,* New York: Springer, 1985.
6. F. J. MacWiliams and N. J. A. Sloane, *The Theory of Error-Correcting Codes,* North-Holland, 1978.
7. R. Lidl and H. Niederreiter, *Finite Fields,* Cambridge University Press and Addison-Wesley, 1983.
8. K. M. Martin, *New secret sharing schemes from old,* Journal of Combinatorial Mathematics and Combinatorial Computing **14** (1993), 65-77.
9. A. Shamir, *How to share a secret,* Comm. ACM **22** (1979), 612-613.
10. G. J. Simmons, *How to (really) share a secret,* in "Advances in Cryptology — Crypto'88", Goldwasser, ed., LNCS **403** (1989), 390-448.
11. G. J. Simmons, *Geometric shared secret and/or shared control schemes,* in "Advances in Cryptology — Crypt'90", LNCS **537** (1991), 216-241.
12. Y. Zheng, T. Hardjono and J. Seberry, *Reusing shares in secret sharing schemes,* The Computer Journal **37** (1994), 199-205.

The Access Structure of Some Secret-Sharing Schemes

Ari Renvall and Cunsheng Ding

Department of Mathematics
University of Turku
Fin-20014 Turku, Finland

Abstract. In this paper, we determine the access structure of a number of secret-sharing schemes that are based on error-correcting linear codes with respect to two approaches. Some secret-sharing schemes based on linear codes are also constructed. The relation between the minimum distance of codes and the access structure of secret-sharing schemes is also investigated.

1 Introduction

Secret sharing is an important topic of cryptography. In a secret-sharing scheme, a dealer has a secret. The dealer gives each party in the scheme a share of the secret. Let $\mathbf{P}$ denote the set of parties involved in the secret sharing. There is a set $\Gamma \subseteq 2^{\mathbf{P}}$ such that any subset of parties that is in Γ can determine the secret. The set Γ is called the access structure of the secret-sharing scheme.

Since the first construction of secret-sharing schemes by Blakley [1] and Shamir [13], many other schemes have been proposed and studied [2, 4, 5, 9, 16]. So far the most studied secret-sharing system is the (k,n) threshold schemes. A (k,n) threshold scheme is a secret-sharing scheme such that

- the secret can be constructed from any k shares; and
- no subset of $k-1$ shares reveals any information about the secret.

Two kinds of approaches to the construction of secret-sharing schemes based on linear codes have been so far considered. Suppose that the secret is an element of $GF(q)$, and we wish to share the secret among n parties. The two kinds of approaches are described as follows.

The first one uses an $[n,k]$ linear code over $GF(q)$. Let G be a generator matrix of an $[n,k]$ linear code over $GF(q)$, i.e., G is a $k \times n$ matrix over $GF(q)$ with rank k whose row vectors generate the code. Let s_1 denote the secret, and let $s_2, \cdots, s_k$ be next chosen independently and uniformly at random in $GF(q)$. Taking $\mathbf{s} = (s_1, s_2, \cdots, s_k)$ as the information vector, we calculate the codeword corresponding to this information vector $\mathbf{s}$ as

$$\mathbf{t} = (t_1, t_2, \cdots, t_n) = \mathbf{s}G. \tag{1}$$

Then give t_i only to party P_i as the share for each i.

The secret-sharing schemes based on linear codes with respect to the first approach were considered by McEliece and Sarwate [12] where the Shamir scheme formulated in terms of polynomial interpolation was generalized in terms of Reed-Solomon codes. In terms of vector spaces, Brickell studied also this kind of secret-sharing schemes [3]. A generalization of this approach was given by Bertilsson according to [6] and by van Dijk [6]. The following lemma, which is frequently needed, is equivalent to Proposition 1 of [3].

Lemma 1. *Let G be a generator matrix of an $[n, k]$ linear code, and $1 \leq i_1 < \cdots < i_m \leq n$ be a set of indices, where $1 \leq m \leq n$. In the secret-sharing scheme based on G with respect to the first approach a set of shares $\{t_{i_1}, t_{i_2}, \cdots, t_{i_m}\}$ determines the secret s_1 if and only if the vector $\mathbf{e}_1 = (1, 0, \cdots, 0)^T$ is a linear combination of the $\mathbf{g}_{i_1}, \mathbf{g}_{i_2}, \cdots, \mathbf{g}_{i_m}$, where $\mathbf{g}_i$ is the ith column of the generator matrix G. Furthermore, the secret-sharing scheme is perfect.*

In this secret-sharing scheme the computation of the secret is carried out as follows. Given a set of shares $\{t_{i_j}\}$ which can determine the secret, let

$$\mathbf{e}_1 = \sum_{j=1}^{m} x_j \mathbf{g}_{i_j}. \tag{2}$$

It follows that

$$s_1 = s\mathbf{e}_1 = \sum_{j=1}^{m} x_j s\mathbf{g}_{i_j} = \sum_{j=1}^{m} x_j t_{i_j}. \tag{3}$$

Thus, the secret s_1 is determined, when the shares $t_{i_1}, \cdots, t_{i_m}$ are given. The computational complexity of computing the secret by solving linear equations is $O(m^3)$

The second one uses an $[n+1, k']$ linear codes over $GF(q)$. Let G' be a generator matrix of a $[n+1, k']$ linear code over $GF(q)$, i.e., G' is a $k' \times (n+1)$ matrix over $GF(q)$ with rank k'. Let s denote the secret, and $\mathbf{g}_0' = (g_{10}', g_{20}', \cdots, g_{k'0}')^T$ be the first column of the generator matrix G'. Then the information vector $\mathbf{s} = (s_0, \cdots, s_{k'-1})$ is chosen to be any vector of $GF(q)^{k'}$ such that

$$s = s\mathbf{g}_0' = \sum_{i=0}^{k'-1} s_i g_{i0}'.$$

Taking $\mathbf{s} = (s_0, s_1, \cdots, s_{k'-1})$ as the information vector, we calculate the codeword corresponding to this information vector $\mathbf{s}$ as

$$\mathbf{t} = (t_0, t_1, \cdots, t_n) = \mathbf{s}G'. \tag{4}$$

Then give t_i only to party P_i as the share for each $i \geq 1$. The first component $t_0 = s$ of the codeword $\mathbf{t}$ is the secret.

Secret-sharing schemes based on codes with respect to the second approach were considered by Karnin, Green and Hellman [7]. Massey considered the second approach and introduced the concept of minimal codewords. He proved that the access structure of a secret-sharing scheme based on an $[n, k]$ linear code with respect to the second approach is determined by the minimal codewords of the dual code of the linear code [10, 11]. The following Lemma 2 is equivalent to the characterization given in [11, Prop. 2], which we need in the sequel.

Lemma 2. Let $G' = [g'_0 g'_1 \cdots g'_n]$ be a generator matrix of an $[n+1, k']$ code over $GF(q)$ such that g'_0 is a linear combination of the other n columns $g'_1, \cdots, g'_n$. For the secret-sharing scheme based on the generator matrix G' with respect to the second approach the secret t_0 is determined by the set of shares $\{t_{i_1}, \cdots, t_{i_m}\}$ if and only if g'_0 is a linear combination of the vectors $g'_{i_1}, \cdots, g'_{i_m}$, where $1 \leq i_1 < \cdots < i_m \leq n$ and $m \leq n$. Furthermore, this secret-sharing scheme is perfect.

In the secret-sharing scheme based on codes with respect to the second approach the computation of the secret is carried out similarly to that in the first approach.

The two kinds of approaches are different but related. The differences between the two approaches are the following:

1. Given an $[n, k]$ linear code, the first one gives a secret-sharing scheme among n parties, while the second one gives a scheme among only $n - 1$ parties.
2. In the first approach all the shares form a *complete* codeword of the linear code, while in the second one all the shares form only part of a codeword. Thus, from the viewpoint of cheating correction and detection the first one could be better than the second.
3. The access structure of secret-sharing schemes with respect to the second approach is completely determined by the structure of the dual code and is independent of the choice of a generator matrix G' of the linear code, as shown clearly by Lemma 2, while that with respect to the first approach cannot be determined by the structure of the dual code and is heavily dependent on the choice of the generator matrix G. Thus, given a linear code, the second approach gives only one secret-sharing scheme, while the first gives possibly a large number of secret-sharing schemes.

Secret-sharing schemes based on codes with respect to linear codes usually have a multilevel access structure. Multilevel secret-sharing schemes were also studied in [3, 14, 15].

In this paper we determine the access structure of some secret-sharing schemes based on linear codes with respect to the two approaches. Some secret-sharing schemes with almost the same properties as (k, n) threshold schemes will be constructed. Finally, we shall investigate the relation between the access structure of secret-sharing schemes based on codes and the minimum distance of codes.

2 Secret-sharing schemes related to MDS codes

When G' is a generator matrix of an $[n + 1, k, n - k + 2]$ MDS code over $GF(q)$, the access structure of the secret-sharing scheme based on G' with respect to the second approach is quite clear by Lemma 2, i.e., any k shares determine the secret, but any set of $k-1$ or fewer shares give no information about the secret. Such secret-sharing schemes are called (k, n) threshold schemes.

When G is a generator matrix of an $[n, k, n-k+1]$ MDS code over $GF(q)$, the access structure of the secret-sharing scheme based on G with respect to the first approach is not so simple, we shall discuss the access structure from case to case.

Theorem 3. *Let G be a generator matrix of an $[n, k, n-k+1]$ MDS code over $GF(q)$. For the secret-sharing scheme based on G with respect to the first approach any k shares determine the secret.*

Proof: Since the code generated by G is MDS, every k columns of a generator matrix of the code are linearly independent, and thus the vector e_1 is a linear combination of the k column vectors of G. By Lemma 1 any k shares determine the secret. □

This theorem does not say anything about whether $k-1$ or fewer shares are enough to determine the secret. It is important to note that the access structure of secret-sharing schemes based on codes with respect to the first approach depends on the choice of the generator matrix, and thus it is more complicated than that with respect to the second approach. Given an MDS code, it is possible to have different secret sharing schemes with respect to the first approach by taking different generator matrices, while the access structure is independent of the choice of the generator matrix with respect to the second approach.

Theorem 4. *Let G be a generator matrix of an $[n, k, n-k+1]$ code over $GF(q)$ such that its ith column is a multiple of e_1. In the secret-sharing scheme based on G with respect to the first approach a set of shares determine the secret if and only if it contains the ith share or its cardinality is no less than k.*

Proof: By Theorem 3 a set of shares with cardinality no less than k determines the secret. We now consider $k-1$ or fewer shares.

Since e_1 is a multiple of the ith column of G, by Lemma 1 the ith share alone determines the secret. Thus, any set of shares containing t_i determines the secret.

We conclude that any $k-1$ shares without t_i cannot determine the secret. If not so, suppose that the shares $t_{i_1}, \cdots, t_{i_{k-1}}$ determine the secret, where $i_j \neq i$ for $j = 1, 2, \cdots, k-1$. Then by Lemma 1

$$e_1 = \sum_{j=1}^{k-1} a_j g_{i_j}.$$

By assumption let $e_1 = a g_i$. It follows that

$$a g_i = \sum_{j=1}^{k-1} a_j g_{i_j}.$$

Hence the k columns $g_i, g_{i_1}, \cdots, g_{i_{k-1}}$ of G are linearly dependent. This is impossible since G is a generator matrix of an $[n, k, n - k + 1]$ MDS code. □

The access structure of the secret-sharing schemes described by Theorem 4 could be interesting in applications where the president of some organization alone should know the secret, and a set of other parties can determine the secret if and only if the number of parties is no less than k.

A practical question concerning such a secret-sharing scheme is whether each $[n, k, n - k + 1]$ MDS code has such a generator matrix and how to find them if there are some. The answer is "yes" and such a generator matrix is found as follows. First, take any generator matrix G'. Then find a coordinate of the first row vector of G' and add multiples of this row to other rows to make the coordinate of other rows equal zero. The obtained matrix is the desired one.

Theorem 5. *Let G be a generator matrix of an $[n, k, n - k + 1]$ MDS code such that e_1 is a linear combination of g_1 and g_2, but not a multiple of any of them, where g_i denotes the ith column vector of G. The access structure of the secret-sharing scheme based on G with respect to the first approach is then described as follows:*

1. *any k shares determine the secret;*
2. *a set of shares with cardinality $k - 2$ or less determines the secret if and only if the set contains both t_1 and t_2;*
3. *a set of $k - 1$ shares $\{t_{i_1}, t_{i_2}, \cdots, t_{i_{k-1}}\}$*
   - *determines the secret when it contains both t_1 and t_2;*
   - *cannot determine the secret when it contains one and only one of t_1 and t_2;*
   - *determines the secret if and only if $\operatorname{rank}(e_1, g_{i_1}, \cdots, g_{i_{k-1}}) = k - 1$ when it contains none of t_1 and t_2.*

Proof: The first part is the conclusion of Theorem 3. By assumption and Lemma 1 any set of shares containing both t_1 and t_2 determines the secret. Suppose that a set of $k - 2$ shares $\{t_{i_1}, \cdots, t_{i_{k-2}}\}$ not containing t_1 and t_2 determines the secret. By Lemma 1 e_1 is a linear combination of $g_{i_1}, \cdots, g_{i_{k-2}}$, and by assumption e_1 is a linear combination of g_1 and g_2. Hence the k column vectors $g_1, g_2, g_{i_1}, \cdots, g_{i_{k-2}}$ of G are linearly dependent. This is impossible since G is a generator matrix of an $[n, k, n - k + 1]$ MDS code. Similarly, we can prove that any $k - 2$ shares containing one and only one of t_1 and t_2 cannot determine the secret. This completes the proof of the second part.

We now prove the conclusion of part three. As already proved, any $k - 1$ shares containing both t_1 and t_2 determines the secret. Assume that $\{t_{i_1}, \cdots, t_{i_{k-1}}\}$ containing t_1 but not t_2 determines the secret. By Lemma 1 e_1 is a linear combination of $g_{i_1}, \cdots, g_{i_{k-1}}$. Note that e_1 is a linear combination of g_1 and g_2, the k distinct columns $g_2, g_{i_1}, \cdots, g_{i_{k-1}}$ are linearly dependent. This is impossible since G is a generator matrix of an $[n, k, n - k + 1]$ MDS code. By symmetry any set of $k - 1$ shares containing t_2 but not t_1 cannot determine the secret.

The remaining conclusion of part three follows from Lemma 1 and the fact that any $k - 1$ columns of G are linearly independent.

$\square$

Secret-sharing schemes described by Theorem 5 could be practically interesting in applications where there is one group of parties that is the superpower in the group of parties involved in the secret-sharing.

Theorem 6. *Let G be a generator matrix of an $[n, k, n - k + 1]$ MDS code over $GF(q)$. In the secret-sharing scheme based on the matrix G with respect to the first approach if two sets of shares $\{t_{i_1}, \cdots, t_{i_l}\}$ and $\{t_{j_1}, \cdots, t_{j_m}\}$ determine the secret respectively, where $i_1, \cdots, i_l, j_1, \cdots, j_m$ are pairwise distinct, then*

$$l + m \geq k + 1.$$

Proof: By assumption e_1 is a linear combination of both $\{g_{i_1}, \cdots, g_{i_l}\}$ and $\{g_{j_1}, \cdots, g_{j_m}\}$. It follows that $g_{i_1}, \cdots, g_{i_l}, g_{j_1} \cdots, g_{i_m}$ are linearly dependent. Because every k columns of G are linearly independent and those i_s and j_t are different, we have $l + m \geq k + 1$. $\square$

Theorem 6 means that in such a secret-sharing scheme there can only be one superpower consisting of a few parties. For example, if parties P_1 and P_2 are able to determine the secret, there is no other group of two without P_1 and P_2 that can determine the secret if $k \geq 4$.

Let G be a generator matrix of an $[n, k, n - k + 1]$ code. It follows from Lemma 1 that the secret-sharing scheme based on G with respect to the first approach is a (k, n) threshold scheme if and only if e_1 is not a linear combination of any $k - 1$ column vectors of G.

Now we consider the access structure of some secret-sharing schemes based on some codes relating to MDS codes with respect to the second approach.

Let $G'' = [g_0' g_1' \cdots g_{n-1}']$ be a generator matrix of an $[n, k, n - k + 1]$ MDS code. For two nonzero elements a, b of $GF(q)$ define

$$G' = [g_0' g_1' \cdots g_{n-1}' a g_0' + b g_1']. \tag{5}$$

Clearly, G' does not generate an MDS code. Consider now the secret-sharing scheme based on G' with respect to the second approach. The access structure of the secret-sharing scheme is described by the following theorem.

Theorem 7. *The access structure of the secret-sharing scheme based on G' of (5) with respect to the second approach is described as follows:*

1. *a set of $k - 1$ or fewer shares determines the secret when and only when it contains both t_1 and t_n;*
2. *any set of $k + 1$ shares determines the secret:*
3. *a set of k shares $\{t_{i_1}, \cdots, t_{i_k}\}$ determines the secret*
   - *when it does not contain t_n;*
   - *when it contains both t_1 and t_n;*
   - *if and only if $a g_0' + b g_1'$ is not a linear combination of $g_{i_1}', \cdots, g_{i_{k-1}}'$ when the set of shares contains t_n but not t_1.*

Proof: By Lemma 2 a set of $k - 1$ or fewer shares containing both t_1 and t_n determines the secret. Suppose now a set of $k - 1$ shares $\{t_{i_1}, \cdots, t_{i_{k-1}}\}$ determines the secret, but does not contain both t_1 and t_n. Then by Lemma 2 g_0' is a linear combination of $g_{i_1}', \cdots, g_{i_{k-1}}'$. First, suppose that t_1 but not t_n is in the set of shares. Then it is not hard to see that the k distinct column vectors $g_0', g_{i_1}', \cdots, g_{i_{k-1}}'$ of G'' are linearly dependent. This is impossible since G'' is a generator matrix of an $[n, k, n - k + 1]$ MDS code. Similarly it is also impossible for the set to include t_1 but not t_n. For similar reason it is impossible for the set not to include both t_1 and t_n. Hence both t_1 and t_n belong to the set of shares. This proves the first part.

Since any $k + 1$ columns of G' has rank k, g_0' must be a linear combination of any $k + 1$ columns of G'. By Lemma 2 any set of $k + 1$ shares determines the secret. This proves part two.

The proof of part three is divided into several cases. When the set of shares does not contain t_n, the set of vectors $\{g_{i_1}', \cdots, g_{i_k}'\}$ has rank k. Hence g_0' is a linear combination of the k vectors. By Lemma 2 the k shares determine the secret.

When the set of shares includes both t_1 and t_n, it follows from Lemma 2 that the set determines the secret.

When the set of shares contains t_n but not t_1, we prove now that g_0' is a linear combination of $g_{i_1}', \cdots, g_{i_k}'$ if and only if $ag_0' + bg_1'$ is not a linear combination of $g_{i_1}', \cdots, g_{i_{k-1}}'$, where $i_k = n$. If g_0' is a linear combination of $g_{i_1}', \cdots, g_{i_k}'$, let

$$g_0' = y_1 g_{i_1}' + \cdots + y_{k-1} g_{i_{k-1}}' + y(ag_0' + bg_1'). \tag{6}$$

Since any k columns of G'' are linearly independent, all the coefficients $y_1, \cdots, y_{k-1}, y$ are nonzero. It follows from (6) that

$$g_0' = \frac{y_1}{1 - ay} g_{i_1}' + \cdots + \frac{y_{k-1}}{1 - ay} g_{i_{k-1}}' + \frac{by}{1 - ay} g_1'. \tag{7}$$

For the same reason the constant $1 - ay \neq 0$. Suppose that $ag_0' + bg_1'$ is a linear combination of $g_{i_1}', \cdots, g_{i_{k-1}}'$, let

$$ag_0' + bg_1' = z_1 g_{i_1}' + \cdots + z_{k-1} g_{i_{k-1}}'. \tag{8}$$

Since any k columns of G'' are linearly independent, all the coefficients $z_1, \cdots, z_{k-1}$ are nonzero. It follows from (8) that

$$g_0' = \frac{z_1}{a} g_{i_1}' + \cdots + \frac{z_{k-1}}{a} g_{i_{k-1}}' - \frac{b}{a} g_1'. \tag{9}$$

Comparing the coefficients of g_1' in (7) and (9) we have

$$\frac{by}{1 - ay} = -\frac{b}{a}.$$

It follows from this equation that $b = 0$. This is contrary to the assumption that $b \neq 0$. Thus, $ag_0' + g_1'$ cannot be a linear combination of $g_{i_1}', \cdots, g_{i_{k-1}}'$.

If $ag'_0 + g'_1$ is not a linear combination of $g'_{i_1}, \cdots, g'_{i_{k-1}}$, the set of vectors

$$\{ag'_0 + g'_1, g'_{i_1}, \cdots, g'_{i_{k-1}}\}$$

must have rank k since $g'_{i_1}, \cdots, g'_{i_{k-1}}$ are linearly independent. Hence g'_0 is a linear combination of $g'_{i_1}, \cdots, g'_{i_k}$. By Lemma 2 the last conclusion of part three is true. □

The access structure of the secret-sharing scheme described by Theorem 7 is similar to that described by Theorem 5 in that there is only one group consisting of a few parties that can determine the secret.

3 The access structure and minimum distance of codes

The minimum distance of linear codes is one of the most important parameters since it determines the error-correcting capacity of codes. In this section we first investigate the meaning of the minimum distance of linear codes in secret-sharing with respect to the two approaches in general, then we prove some general results about the access structure of secret-sharing schemes based on cyclic codes where the secret-sharing meaning of the minimum distance of codes is more clear.

Theorem 8. *Let G be a generator matrix of an $[n, k]$ code C over $GF(q)$ and $C^\perp$ be the dual code of C. In the secret-sharing scheme based on G with respect to the first approach if each of two sets of shares $\{t_{i_1}, \cdots, t_{i_l}\}$ and $\{t_{j_1}, \cdots, t_{j_m}\}$ determines the secret, but no subset of any of them can do so, then*

$$|\{t_{i_1}, \cdots, t_{i_l}\} \cup \{t_{j_1}, \cdots, t_{j_m}\}| - |\{t_{i_1}, \cdots, t_{i_l}\} \cap \{t_{j_1}, \cdots, t_{j_m}\}| \geq d^\perp, \quad (10)$$

where $d^\perp$ is the minimum distance of the dual code C.

Proof: By assumption no one of the two sets $\{i_1, \cdots, i_l\}$ and $\{j_1, \cdots, j_m\}$ includes the other, and by Lemma 1

$$e_1 = a_1 g_{i_1} + \cdots + a_l g_{i_l}, \quad e_1 = b_1 g_{j_1} + \cdots + b_m g_{j_m},$$

where all $a_i \neq 0$ and $b_i \neq 0$. It follows that

$$\sum_{s=1}^{l} a_s g_{i_s} = \sum_{s=1}^{m} b_s g_{j_s}.$$

Hence

$$|\{i_1, \cdots, i_l\} \cup \{j_1, \cdots, j_m\}| - |\{i_1, \cdots, i_l\} \cap \{j_1, \cdots, j_m\}|$$

column vectors of G are linearly dependent. Finally, the conclusion follows from the definition of the minimum distance and the dual code. □

This is a generalization of Theorem 6. If the two set of shares are disjoint, we have

$$l + m \geq d^\perp.$$

Thus, a general meaning of the minimum distance of codes in secret-sharing is clear.

Theorem 9. *Let G' be a generator matrix of an $[n+1, k]$ linear code over $GF(q)$. In the secret-sharing scheme based on G' with respect to the second approach any set of $d^\perp - 2$ or fewer shares cannot determine the secret, where $d^\perp$ is the minimum distance of the dual code of the code generated by G'.*

Proof: Assume that a set of shares $\{t_{i_1}, \cdots, t_{i_l}\}$ determines the secret, where $1 \leq i_1 < \cdots < i_l \leq n$. By Lemma 2 the first column g'_0 of G' is a linear combination of $g'_{i_1}, \cdots, g'_{i_l}$, where g'_i is the $(i+1)$th column of G'. Let

$$g'_0 + c_{i_1} g_{i_1} + \cdots + c_{i_l} g'_{i_l} = 0.$$

Then $(1, 0, \cdots, 0, c_{i_1}, 0, \cdots, 0, c_{i_l}, 0, \cdots, 0)$ is a codeword of the dual code. Hence $1 + l \geq d^\perp$. This proves the theorem. □

The secret-sharing meaning of the minimum distance of linear codes with respect to the second approach is more clear. $d^\perp$ is a threshold in the sense that less than $d^\perp - 1$ shares cannot determine the secret.

A code C is cyclic if it is linear and if any cyclic shift of a codeword is in C then so is $(c_{n-1}, c_0, \cdots, c_{n-2})$. The above Theorems 8 and 9 of course apply to secret-sharing schemes based on cyclic codes, but due to the cyclic property we have further results for secret-sharing schemes based on cyclic codes.

Theorem 10. *Let G' be a generator matrix of an $[n+1, k]$ linear code over $GF(q)$, and $d^\perp$ be the minimum distance of the dual code of the code generated by G'. In the secret-sharing scheme based on G' with respect to the second approach there is at least one set of $d^\perp$ shares that determines the secret.*

Proof: By assumption the dual code has at least one codeword c with Hamming weight $d^\perp$. If the first component of c is not zero, by shifting c a number of times we get a codeword of the dual code

$$c' = (c_0, 0, \cdots, 0, c_{i_1}, 0, \cdots, 0, c_{i_{d^\perp-1}}, 0, \cdots, 0),$$

where $c_0, c_{i_1}, \cdots, c_{i_{d^\perp}}$ are nonzero. By Lemma 2 the set of shares $\{t_{i_1}, \cdots, t_{i_{d^\perp}}\}$ determines the secret. □

In what follows in this section we consider now secret-sharing schemes based on codes with respect to the second approach. By deleting the first element of the support of each codeword

$$c = (b_0, b_1, \cdots, b_{n-1}) \in C^\perp, \quad b_0 \neq 0$$

we get a group of parties that can determine the secret, i.e., an element γ of the access structure Γ. Such a γ is said to be *codeword-derived*. It is clear that a number of codewords may give the same element of Γ. In the sequel we use $\Gamma(C)$ to denote all the codeword-derived elements of Γ. Note that every element in $\Gamma \setminus \Gamma(C)$ must contain an element of $\Gamma(C)$.

Theorem 11. *Let C be an $[n, k, d]$ code over $GF(q)$ with $d \geq 3$. Then in the secret-sharing scheme based on C with respect to the second approach $|\Gamma(C)| \leq q^{n-k-2}$ and every party belongs to at least one $\gamma \in \Gamma(C)$.*

Proof: Let $C^\perp$ be the $[n, n-k, d^\perp]$ dual code of C. Let V_0 be the set of all the codewords of $C^\perp$ with the first coordinate being zero. Then V_0 is a subspace of $C^\perp$. Since $C^\perp$ has dimension $n-k$, V_0 has dimension $n-k-1$. By the Singleton bound in coding theory we have $n-k-2 \geq d-3 \geq 0$.

It is not hard to prove there is a codeword $\mathbf{a} = (1, a_1, \cdots, a_{n-1}) \in C^\perp$ such that the coset decomposition of $C^\perp$ with respect to V_0 is

$$C^\perp = \cup_{x \in GF(q)}(V_0 + \mathbf{a}x).$$

Since V_0 is a subspace, all nonzero multiples of c belong to V_0 if $c \in V_0$. Note that all the nonzero multiples of a codeword have the same support, it then follows from Lemma 2 that $|\Gamma(C)| \leq q^{n-k-2}$.

By the assumption $d \geq 3$, there is no other zero coordinate in the vectors of V_0. It follows that each party belongs to at least one group of parties that can determine the secret. $\qquad \Box$

Secret-sharing schemes based on binary codes can only be used for sharing a one-bit secret, but since many secrets can be expressed as a finite binary sequence, it is easy to extend any secret-sharing scheme for the share of a one-bit secret into one for sharing secrets of arbitrary size. Thus, the importance of one-bit secret-sharing schemes follows. In what follows we prove some results about one-bit secret-sharing schemes based on some binary codes.

In the case of $q = 2$ Theorem 11 can be strengthened as follows.

Theorem 12. *Let C be a binary $[n, k, d]$ code with $d \geq 3$, and $\Gamma(C)$ be the same as before. Then $|\Gamma(C)| = 2^{n-k-1}$, and every party appears in exactly 2^{n-k-2} subgroups of parties $\gamma \in \Gamma(C)$.*

Proof: By the proof of Theorem 11 we have

$$C^\perp = V_0 \cup (V_0 + \mathbf{a}).$$

Note that for binary codes two codewords have the same support if and only if they are identical. It follows that $|\Gamma(C)| = 2^{n-k-1}$.

When putting all the vectors of V_0 into an $2^{n-k-1} \times n$ matrix, each column other than the first of the matrix has the equal number of ones and zeros since $d \geq 3$ and V_0 is a linear space. Thus, each party appears in exactly 2^{n-k-2} subgroups of parties $\gamma \in \Gamma(C)$. $\qquad \Box$

Theorem 13. *Let $C^\perp$ be the $[n, n-k, d^\perp]$ dual code of a binary $[n, k, d]$ code, and let $\Gamma(C)$ be the same as before. If $d^\perp \geq n/2$ then*

1. $|\Gamma(C)| = 2^{n-k-1}$;
2. no element $\gamma_1 \in \Gamma(C)$ with $|\gamma| < n$ can contain another one of $\Gamma(C)$;
3. for any two elements γ_1 and γ_2 of $\Gamma(C)$

$$|\gamma_1 \cap \gamma_2| \geq d^\perp - w_{max}/2 - 1,$$

where w_{max} is the maximum Hamming weight of the codewords of $C^\perp$.

Proof: Note first that two binary codewords have the same support if and only if they are identical. Thus, $|\Gamma(\mathcal{C})| = 2^{n-k-1}$. This proves part one.

A vector $\mathbf{a} = (a_0, \cdots, a_{n-1})$ is said to cover $\mathbf{c} = (c_0, \cdots, c_{n-1})$ if $c_i \neq 0$ whenever $a_i \neq 0$. Let $\mathbf{x}, \mathbf{y} \in \mathcal{C}^{\perp}$. Suppose that $\mathbf{x}$ covers $\mathbf{y}$, then

$$wt(\mathbf{x} - \mathbf{y}) = wt(\mathbf{x}) - wt(\mathbf{y}) \geq d^{\perp},$$

where $wt(\mathbf{x})$ denotes the Hamming weight of $\mathbf{x}$. It follows that $wt(\mathbf{x}) \geq d + wt(\mathbf{y}) \geq 2d^{\perp}$. Therefore if $d^{\perp} > n/2$ then $wt(\mathbf{x}) > n$, which is impossible. If $d^{\perp} = n/2$, then $\mathbf{x}$ must be the all-one codeword of $\mathcal{C}^{\perp}$. This proves part two.

Let $\gamma_i \in \Gamma(\mathcal{C})$ be derived from codeword $\mathbf{c}_i \in \mathcal{C}^{\perp}$ for $i = 1$ and 2. It is not hard to see that

$$2|\gamma_1 \cap \gamma_2| \geq wt(\mathbf{c}_1) + wt(\mathbf{c}_2) - wt(\mathbf{c}_1 + \mathbf{c}_2) - 2$$
$$\geq 2d^{\perp} - w_{max} - 2$$

which proves the last part. □

4   Concluding remarks

For an $[n, k]$ linear code let A_i be the number of codewords of Hamming weight i, the numbers $A_0, A_1, \cdots, A_n$ are called the weight distribution of the code. The determination of the access structure of secret-sharing schemes based on linear codes is more complicated than that of the weight distribution of linear codes. When determining the weight distribution of codes we are only concerned with the weight of codewords, while we are concerned with not only the weight of codewords, but also the pattern of codewords when determining the access structure of secret-sharing schemes based on linear codes. In general the determination of the access structure of secret-sharing schemes based on codes with respect to the two approaches is a very hard problem.

Coding theory tells us that the determination of the weight distribution of codes is generally quite hard, let alone the determination of the access structure of secret-sharing schemes based on codes. The weight of MDS codes and the access structure of secret-sharing schemes based on MDS codes are clear due to the speciality of codes. We have been able to determine the access structure of some secret-sharing schemes based on some codes since these codes are closely related to MDS codes.

However, in many applications the number n of parties involved in a secret sharing is quite small. In those cases we need only short linear codes, so it is possible to determine the access structure of those secret-sharing schemes based on short codes by computer.

References

1. G. R. Blakley, *Safeguarding cryptographic keys*, Proc. NCC AFIPS 1979, 313-317.
2. E. F. Brickell and D. M. Davenport, *On the classification of ideal secret sharing schemes*, J. Cryptology **4** (1991), 105-113. [Preliminary version in "Advances in Cryptology—Crypto'89, LNCS **435** (1990), 278-285.]
3. E. F. Brickell, *Some ideal secret sharing schemes*, in "Advances in Cryptology — Eurocrypt'89", LNCS **434** (1990), 468-475.
4. C. Charnes, J. Pieprzyk and R. Safavi-Naini *Conditionally secure secret sharing scheme with disenrollment capability*, in "2nd ACM Conference on Computer and Communications Security," ACM Press, 1994, 89-95.
5. E. Dawson, E. S. Mahmoodian and A. Rahilly, *Orthogonal arrays and ordered threshold schemes*, Australian Journal of Combinatorics **8** (1993), 27-44.
6. M. Van Dijk, *A linear construction of Perfect secret sharing schemes*, in "Advances in Cryptology—Eurocrypt'94", A. De Santis, ed., LNCS **950** (1995), 23-34.
7. E. D. Karnin, J. W. Green and M. Hellman, *On secret sharing systems*, IEEE Trans Inform. Theory, Vol. IT-29 (1983), 644-654.
8. F. J. MacWilliams and N. J. A. Sloane, *The Theory of Error-Correcting Codes*, North-Holland, 1978.
9. K. M. Martin, *New secret sharing schemes from old*, Journal of Combinatorial Mathematics and Combinatorial Computing **14** (1993), 65-77.
10. J. L. Massey, *Minimal codewords and secret sharing*, Proc. 6th Joint Swedish-Russian Workshop on Inform. Theory, Mölle, Sweden, August 22-27, 1993, 276-279.
11. J. L. Massey, *Some applications of coding theory in cryptography*, in "Codes and Cyphers: Cryptography and Coding IV" (Ed. P. G. Farrell). Esses, England: Formara Ltd., 1995, 33-47.
12. R. J. McEliece and D. V. Sarwate, *On sharing secrets and Reed-Solomon codes*, Comm. ACM **24** (1981), 583-584.
13. A. Shamir, *How to share a secret*, Comm. ACM **22**, 1979, 612-613.
14. G. J. Simmons, *How to (really) share a secret*, in "Advances in Cryptology — Crypto'88", Goldwasser, ed., LNCS **403** (1989), 390-448.
15. G. J. Simmons, *Geometric shared secret and/or shared control schemes*, in "Advances in Cryptology — Crypt'90", LNCS **537** (1991), 216-241.
16. Y. Zheng, T. Hardjono and J. Seberry, *Reusing shares in secret sharing schemes*, The Computer Journal **37** (1994), 199-205.

On Construction of Resilient Functions

Chuan-Kun Wu and Ed Dawson

Information Security Research Centre
Queensland University of Technology
GPO Box 2434, Brisbane QLD 4001
AUSTRALIA
E-mail: {wu, dawson}@fit.qut.edu.au

Abstract. An (n, m, t) resilient function is a function $f : GF(2)^n \longrightarrow GF(2)^m$ such that every possible output m-tuple is equally likely to occur when the values of t arbitrary inputs are fixed by an opponent and the remaining $n - t$ input bits are chosen independently at random. The existence of resilient functions has been largely studied in terms of lower and upper bounds. The construction of such functions which have strong cryptographic significance, however, needs to be studied further. This paper aims at presenting an efficient method for constructing resilient functions from odd ones based on the theory of error-correcting codes, which has further expanded the construction proposed by X.M.Zhang and Y.Zheng. Infinite classes of resilient functions having variant parameters can be constructed given an old one and a linear error-correcting code. The method applies to both linear and nonlinear resilient functions.

1 Introduction

Let $F: GF(2)^n \longrightarrow GF(2)^m$ be an (n, m) Boolean function. F is called an (n, m, t) resilient function if every possible m-tuple in $GF(2)^m$ is equally likely to occur in its output when an opponent chooses t arbitrary input bits and fixes their values beforehand, and the remaining $n - t$ input bits are chosen independently at random. This concept was introduced by Chor *et al.* in [3] and Bennett *et al* in [1] independently. This is actually a natural generalization of the concept of correlation immune function which was introduced by Siegenthaler in [7]. The applications of resilient functions include stream ciphers, key distribution especially for quantum cryptography, secret sharing, fault-tolerance, and equivalence to the combinatorial structure of orthogonal arrays (see [3, 4]). Further application of resilient functions could be developed from the application of correlation immune functions since a resilient function is actually a collection of some balanced correlation immune functions with certain interrelations.

The problem of determining the largest value of t such that an (n, m, t) resilient function exists given n and m has been largely studied (see for example [1, 2, 3, 5, 8, 9]). Considering the practical application of resilient functions, efficient methods are required to generate a large variety of such functions which satisfy certain properties from which a particular one can easily be chosen. The problem on how to construct linear resilient functions was completely solved in [3] with an explicit expression. Normally linear functions are not strongly required in cryptographic applications because of their weakness under cryptanalysis. The construction of nonlinear resilient functions, however, is far from being solved. There have been only a few papers where the construction of nonlinear resilient functions was studied (see for example [4], [5], [9] and [11]). In [9] the construction of nonlinear resilient functions was discussed such that linear ones with the same parameters do not exist, while [11] was devoted to the construction of nonlinear resilient functions of which other cryptographic properties still remain. Resilient functions which were studied in [5] are actually balanced correlation immune functions since m=1 there. This paper develops further the method in [11]. It will be shown how to generate a large variety of resilient functions with good parameters of which other cryptographic properties can easily be controlled.

2 Preliminaries

A function $f: GF(2)^n \longrightarrow GF(2)$ is called a *Boolean function* of n variables. If there exist $a_0, a_1, \ldots, a_n \in GF(2)$ such that $f(x) = a_0 + a_1 x_1 + \cdots + a_n x_n$, where $x = (x_1, \ldots, x_n) \in GF(2)^n$, then $f(x)$ is called an *affine* function. In particular, $f(x)$ is also called a *linear* function if $a_0 = 0$. The input $x = (x_1, \ldots, x_n)$ will be treated as *unbiased*, i.e., for every $(a_1, \ldots, a_n) \in GF(2)^n$, we have $Prob((x_1, \ldots, x_n) = (a_1, \ldots, a_n)) = \frac{1}{2^n}$. The *truth table* of $f(x)$ is a binary vector of length 2^n generated by $f(x)$. The *Hamming weight* of a binary vector α, denoted by $W_H(\alpha)$, is the number of ones in α, and similarly the Hamming weight of a Boolean function $f(x)$, denoted by $W_H(f)$, is then the number of ones in its truth-table. If $W_H(f) = 2^{n-1}$, then $f(x)$ is called *balanced* . The *degree* of $f(x)$, denoted by $\deg(f)$, is the largest number of variables appearing in one product term of its polynomial expression (in some papers it is also called algebraic normal form). The function $f(x)$ is called *statistically independent* of $(x_{i_1}, \ldots, x_{i_t})$, where $\{i_1, \ldots, i_t\} \subseteq \{1, 2, \ldots, n\}$, if for every $(a_1, \ldots, a_t) \in GF(2)^t$ we have

$$Prob(f(x) = 1 | (x_{i_1}, \ldots, x_{i_t}) = (a_1, \ldots, a_t)) = Prob(f(x) = 1).$$

The function $f(x)$ is called *correlation immune* (CI) of order t (see [7]) if for every subset $\{i_1, \ldots, i_t\}$ of $\{1, 2, \ldots, n\}$, $f(x)$ is statistically independent of $(x_{i_1}, \ldots, x_{i_t})$. We will denote by $\mathcal{F}_n$ the set of all Boolean functions of n variables and by $\mathcal{L}_n$ the set of affine ones.

A function $F : GF(2)^n \longrightarrow GF(2)^m$ is called an (n, m) Boolean function. Functions like this will be considered under the case when $m \leq n$. It is well known that an (n, m) Boolean function F can always be written as $F = [f_1, \ldots, f_m]$, where $f_i \in \mathcal{F}_n$. F is called *unbiased* with respect to $\{i_1, \ldots, i_t\} \subseteq \{1, 2, \ldots, n\}$ if for every $(a_1, \ldots, a_t) \in GF(2)^t$ and for every $(y_1, \ldots, y_m) \in GF(2)^m$, we have

$$Prob(F(x) = (y_1, \ldots, y_m) | (x_{i_1}, \ldots, x_{i_t}) = (a_1, \ldots, a_t)) = \frac{1}{2^m}.$$

The function $F = [f_1, ..., f_m]$ is called an (n, m, t) *resilient function* if and only if F is unbiased with respect to any subset T of $\{1,2,...,n\}$ with $|T| = t$, where $|.|$ means the cardinal number of a set. Obviously an (n, m, t) resilient function exists only if $t \leq n - m$. It is easy to check that each component of an (n, m, t) resilient function is a Boolean function of n variables with CI order t. The following lemma of which the proof can be found in [11] describes more precisely the relationship between resilient functions and CI functions:

Lemma 1. *Let* $F = [f_1, ..., f_m]$, $f_i \in \mathcal{F}_n$. *Then* $F(x)$ *is an* (n, m, t) *resilient function if and only if every nonzero linear combination of* $f_1, ..., f_m$ *is a balanced CI function of order* t, *or equivalently an* $(n, 1, t)$ *resilient function.*

It should be indicated that, although the definition of both correlation immune function and resilient function have no restriction on the value of t, usually the largest value of t is pursued.

What follows in this section is a brief introduction of a Walsh transformation which can help the forthcoming discussions. Let $f(x) \in \mathcal{F}_n$, then

$$S_f(\omega) = \sum_x f(x)(-1)^{\omega \cdot x}$$

is called a *Walsh transformation* of $f(x)$ which is a real valued function, where $\omega \cdot x = \omega_1 x_1 \oplus \cdots \oplus \omega_n x_n$ is the *inner product* of ω and x. Accordingly, the inverse transformation can be expressed as

$$f(x) = 2^{-n} \sum_\omega S_f(\omega)(-1)^{\omega \cdot x}.$$

When the nonzero points of $f(x)$ form a subspace of $GF(2)^n$, we have the following result:

Lemma 2. *Let* V *be a subspace of* $GF(2)^n$ *with dimension* ≥ 1. *Then*

$$\sum_{x \in V}(-1)^{\omega \cdot x} = \begin{cases} |V| & \text{if } \omega = 0, \\ 0 & \text{otherwise.} \end{cases}$$

3 Known constructions

This section mainly summarizes some known results related to the construction of resilient functions on which the forthcoming discussion will be based.

Denote by $E(i_1, ..., i_t) = \{x \in GF(2)^n : x_i = 0 \text{ if } i \notin \{i_1, ..., i_t\}\}$. The proof of the following lemma can be found in [10]. For completeness we have included this as well here.

Lemma 3. *Let* $f(x) \in \mathcal{F}_n$. *Then* $f(x)$ *is statistically independent of* $(x_{i_1}, ..., x_{i_t})$ *if and only if for every nonzero vector* $\omega \in E(i_1, ..., i_t)$, *we have* $S_f(\omega) = 0$.

Proof. The following conclusion is proved in [10]: Let A be a set of k-dimensional binary vectors. Then $\{(y_{i_1}, ..., y_{i_j}) : y \in A\}$ contains an equal number of even weight vectors and odd weight vectors for every $1 \leq j \leq k$ and $1 \leq i_1 < i_2 < \cdots < i_j \leq k$, if and only if A is $GF(2)^k$ or contains multiple copies of $GF(2)^k$.

By the conclusion above we see that, for every nonzero vector $\omega \in E(i_1, ..., i_t)$, we have

$$S_f(\omega) = \sum_{f(x)=1}(-1)^{\omega \cdot x} = 0$$
$$\Longleftrightarrow \{(x_{i_1}, ..., x_{i_t}) : f(x) = 1\} \text{ has the property as that of the set } A$$
$$\Longleftrightarrow \text{ for every } (a_1, ..., a_t) \in GF(2)^t, \ Prob((x_{i_1}, ..., x_{i_t}) = (a_1, ..., a_t)| f(x) = 1)$$
$$\qquad = 2^{-t} = Prob((x_{i_1}, ..., x_{i_t}) = (a_1, ..., a_t))$$
$$\Longleftrightarrow Prob(f(x) = 1| (x_{i_1}, ..., x_{i_t}) = (a_1, ..., a_t)) = Prob(f(x) = 1)$$
$$\Longleftrightarrow f(x) \text{ is statistically independent of } (x_{i_1}, ..., x_{i_t}).$$

$\square$

By lemma 2, for a nonzero ω we have

$$\sum_x (-1)^{f(x) \oplus \omega \cdot x} = \sum_x [1 - 2f(x)](-1)^{\omega \cdot x} = -2S_f(\omega).$$

On the other hand,

$$\sum_x (-1)^{f(x) \oplus \omega \cdot x} = \sum_x [1 - 2(f(x) \oplus \omega \cdot x)] = 2^n - 2W_H(f(x) \oplus \omega \cdot x).$$

So we have

$$S_f(\omega) + W_H(f(x) \oplus \omega \cdot x) = 2^{n-1}.$$

Note that for each nonzero $\omega \in E(i_1, ..., i_t)$, $\omega \cdot x$ is just a nonlinear combination of $(x_{i_1}, ..., x_{i_t})$. By lemma 3 we have

Lemma 4. *Let $f(x) \in \mathcal{F}_n$. Then $f(x)$ is statistically independent of $(x_{i_1}, ..., x_{i_t})$ if and only if for every nonzero vector $(c_1, ..., c_t) \in GF(2)^t$, $f(x) \oplus c_1 x_{i_1} \oplus \cdots \oplus c_t x_{i_t}$ is balanced.*

The conclusion of lemma 4 is much like that of lemma 5 in [11], but it has been slightly generalized and the proof is simpler.

Corollary 5. *Let $f(x) \in \mathcal{F}_n$. Then $f(x)$ is CI of order t if and only if for every $\alpha = (a_1, ..., a_n) \in GF(2)^n$ with $1 \leq W_H(\alpha) \leq t$, $f(x) \oplus a_1 x_1 \oplus \cdots \oplus a_n x_n$ is balanced.*

The existence and construction of linear resilient functions can be described by the following lemma of which the proof can be found in [3]:

Lemma 6. *Let G be a generating matrix of an $[n, m, d]$ linear code. Then $f(x) = xG^T$ is an (n, m, t) resilient function if and only if $d = t + 1$, where G^T is the transposed matrix of G.*

In lemma 6 the value of t should be the largest possible value although it is not indicated in particular. Lemma 6 implies that if there is an $(n, m, n-m+1)$ maximum distance separable (MDS) code, then an $(n, m, n-m)$ resilient function exists. But MDS binary codes exist only if $m = 1$ or $m = n - 1$. This is in agreement with the results in [1] where it was shown that $(n, m, n-m)$ resilient functions do not exist for $1 < m < n - 1$.

The following are some known results relating to the construction and/or existence of nonlinear resilient functions.

Lemma 7 [11]. *Given an (n, m, t) resilient function, there is an iterative method to construct a $((h+1)^k n, h^k m, 2^k(1+t) - 1)$ resilient function for all $h = 2, 3, \cdots$ and $k = 1, 2, \ldots$.*

The conclusion of lemma 7 is actually a consequent result by repeating the following base construction: Let $F(x) = [f_1(x), ..., f_m(x)]$ be an (n, m, t) resilient function. Then $G(x, y, z, u) = [F(x) \oplus F(y), F(y) \oplus F(z), F(z) \oplus F(u)]$ is a $(4n, 3m, 2t + 1)$ resilient function.

Lemma 8 [11]. *Let F be an (n, m, t) resilient function and G be a permutation of $GF(2)^m$. Then $P(x) = G(F(x))$ is also an (n, m, t) resilient function. Moreover, different permutations lead to different resilient functions.*

From lemma 8 it can be seen that given an (n, m, t) resilient function, there must exist at least 2^m different resilient functions with the same parameters. We would like to include here another result presented by D.R.Stinson and J.L.Massey in [9] of which good parameters of constructed resilient functions rather than their variety are pursued.

Lemma 9. *If a systematic (n, K, d) code having dual distance d' exists, then there is an $(n, k, d' - 1)$ resilient function, where $K = 2^k$.*

Lemma 10. *For odd integer $r \geq 3$, there is a nonlinear $(2^{r+1}, 2^{r+1} - 2r - 2, 5)$ resilient function and a nonlinear $(2^{r+1}, 2r+2, 2^r - 2^{(r-1)/2} - 1)$ resilient function.*

4   New constructions

In this section a method for efficiently constructing resilient functions will be described which can be obtained by performing a linear transformation on the repeated concatenation of a resilient function by feeding different inputs. This will be described in theorem 12. First we include a result from [11].

Lemma 11 [11]. *Let $F = [f_1, ..., f_m]$ be an (n_1, m, t_1) resilient function and $G = [g_1, ..., g_m]$ be an (n_2, m, t_2) resilient function. Then*

$$P(x, y) = F(x) \oplus G(y) = [f_1(x) \oplus g_1(y), ..., f_m(x) \oplus g_m(y)]$$

is an $(n_1 + n_2, m, t_1 + t_2 + 1)$ resilient function, where $x \in GF(2)^{n_1}$ and $y \in GF(2)^{n_2}$.

By using lemma 11 we can prove the following theorem:

Theorem 12. *Let $F = [f_1, ..., f_m]$ be an (n, m, t) resilient function, G be a generating matrix of an $[N, K, d]$ linear code. Then*

$$[F(\underline{x}_1), F(\underline{x}_2), ..., F(\underline{x}_N)]G^T$$

is an $(nN, mK, d(t+1)-1)$ resilient function, where $\underline{x}_i \in GF(2)^n$, $i = 1, 2, ..., N$.

Proof. Let

$$G = \begin{bmatrix} c_{11} & c_{12} & \cdots & c_{1N} \\ c_{21} & c_{22} & \cdots & c_{2N} \\ \cdots & \cdots & \cdots & \cdots \\ c_{K1} & c_{K2} & \cdots & c_{KN} \end{bmatrix}.$$

Then we have

$$[F(\underline{x}_1), F(\underline{x}_2), ..., F(\underline{x}_N)]G^T$$

$$= \left[(f_1(\underline{x}_1), ..., f_1(\underline{x}_N)) \begin{pmatrix} c_{11} \\ \vdots \\ c_{1N} \end{pmatrix}, ..., (f_m(\underline{x}_1), ..., f_m(\underline{x}_N)) \begin{pmatrix} c_{11} \\ \vdots \\ c_{1N} \end{pmatrix}, ..., \right.$$

$$\left. (f_1(\underline{x}_1), ..., f_1(\underline{x}_N)) \begin{pmatrix} c_{K1} \\ \vdots \\ c_{KN} \end{pmatrix}, ..., (f_m(\underline{x}_1), ..., f_m(\underline{x}_N)) \begin{pmatrix} c_{K1} \\ \vdots \\ c_{KN} \end{pmatrix} \right].$$

For a nonzero vector $(a_{11}, ..., a_{1m}, a_{21}, ..., a_{2m}, ..., a_{K1}, ..., a_{Km}) \in GF(2)^{mK}$, we have the following corresponding linear combination:

$$a_{11}(f_1(\underline{x}_1), ..., f_1(\underline{x}_N)) \begin{pmatrix} c_{11} \\ \vdots \\ c_{1N} \end{pmatrix} \oplus \cdots \oplus a_{1m}(f_m(\underline{x}_1), ..., f_m(\underline{x}_N)) \begin{pmatrix} c_{11} \\ \vdots \\ c_{1N} \end{pmatrix} \oplus \cdots$$

$$\oplus a_{K1}(f_1(\underline{x}_1), ..., f_1(\underline{x}_N)) \begin{pmatrix} c_{K1} \\ \vdots \\ c_{KN} \end{pmatrix} \oplus \cdots \oplus a_{Km}(f_m(\underline{x}_1), ..., f_1(\underline{x}_N)) \begin{pmatrix} c_{K1} \\ \vdots \\ c_{KN} \end{pmatrix}$$

$$= [f_1(\underline{x}_1), ..., f_m(\underline{x}_1)] \begin{bmatrix} a_{11} & a_{21} & \cdots & a_{K1} \\ a_{12} & a_{22} & \cdots & a_{K2} \\ \cdots & \cdots & \cdots & \cdots \\ a_{1m} & a_{2m} & \cdots & a_{Km} \end{bmatrix} \begin{pmatrix} c_{11} \\ c_{21} \\ \vdots \\ c_{K1} \end{pmatrix}$$

$$\oplus [f_1(\underline{x}_2), ..., f_m(\underline{x}_2)] \begin{bmatrix} a_{11} & a_{21} & \cdots & a_{K1} \\ a_{12} & a_{22} & \cdots & a_{K2} \\ \cdots & \cdots & \cdots & \cdots \\ a_{1m} & a_{2m} & \cdots & a_{Km} \end{bmatrix} \begin{pmatrix} c_{12} \\ c_{22} \\ \vdots \\ c_{K2} \end{pmatrix}$$

$$\oplus \cdots \cdots$$

$$\oplus [f_1(\underline{x}_N), ..., f_m(\underline{x}_N)] \begin{bmatrix} a_{11} & a_{21} & \cdots & a_{K1} \\ a_{12} & a_{22} & \cdots & a_{K2} \\ \cdots & \cdots & \cdots & \cdots \\ a_{1m} & a_{2m} & \cdots & a_{Km} \end{bmatrix} \begin{pmatrix} c_{1N} \\ c_{2N} \\ \vdots \\ c_{KN} \end{pmatrix}.$$

$$\text{Denote } A = \begin{bmatrix} a_{11} & a_{21} & \cdots & a_{K1} \\ a_{12} & a_{22} & \cdots & a_{K2} \\ \cdots & \cdots & \cdots & \cdots \\ a_{1m} & a_{2m} & \cdots & a_{Km} \end{bmatrix}. \text{ By assumption, } A \text{ must be a nonzero}$$

$m \times K$ binary matrix and AG has at least one nonzero row vector. Recall that the minimum distance of the code which is generated by G is d. The nonzero row vector of AG then has Hamming weight at least d, and hence AG has at least d nonzero column vectors. Notice that each nonzero column of AG corresponds to a nonzero linear combination of $f_1(\underline{x}_i), \ldots, f_m(\underline{x}_i)$. By lemma 1, this leads to a balanced CI function of order t, or equivalently an $(n, 1, t)$ resilient function, and by repeated use of lemma 11 we know that the linear combination above leads to an $(nN, 1, d(t+1)-1)$ resilient function. Again by lemma 1 we know that $[F(\underline{x}_1), F(\underline{x}_2), \ldots, F(\underline{x}_N)]G^T$ is indeed an $(nN, mK, d(t+1)-1)$ resilient function. □

Corollary 13. *If there exist an (n, m, t) resilient function and an $[N, K, d]$ linear code, then an $(nN, mK, d(t+1)-1)$ resilient function must exist.*

It should be noted that by setting $G = \begin{bmatrix} 1100 \\ 0110 \\ 0011 \end{bmatrix}$, it yields from theorem 12 the base construction in [11]. Furthermore, given a generating matrix G of an $[n, k, d]$ linear code, by replacing every 1 in G by G itself and replacing every 0 in G by an all-zero $k \times n$ matrix we got a new matrix which generates an $[n^2, k^2, d^2]$ linear code. Repeat the procedure we know that an $[n^j, k^j, d^j]$ code for each $j = 1, 2, \ldots$ can be obtained. When $d=2$, an even weight $[h+1, h, 2]$ linear code exists for every $h = 2, 3, \ldots$, and consequently so does an $[(h+1)^j, h^j, 2^j]$ linear code. By corollary 13 we have the conclusion of lemma 7, which is an important result in [11]. Obviously the result can be largely improved by adopting a linear code with better parameters. It is suggested to refer to the existence of resilient functions by combining corollary 13 and lemma 9.

The conclusion of theorem 12 can be slightly generalized with a similar proof. We write the result out as follows, which seems to be more flexible.

Theorem 14. *Let $F_i(x)$ be an n_i, m, t_i resilient function, $i = 1, 2, \ldots, N$, G be a generating matrix of an $[N, K, d]$ linear code. Then*

$$[F_1(\underline{x}_1), F_2(\underline{x}_2), \ldots, F_N(\underline{x}_N)]G^T$$

is an (n', m', t') resilient function, where $x_i \in GF(2)^{n_i}$, $i = 1, 2, \ldots, N$, $n' = \sum_{i=1}^N n_i$, $m' = mK$, $t' = min\{t_{i_1} + t_{i_2} + \cdots + t_{i_d} + d - 1 : 1 \leq i_1 < i_2 < \cdots < i_d \leq N\}$.

Proof. Denote by $F_i(x) = [f_{i1}(\underline{x}_i), f_{i2}(\underline{x}_i), \ldots, f_{im}(\underline{x}_i)]$, $i = 1, 2, \ldots, N$. Then the proof of theorem 14 can be formulated from that of theorem 12 by simply replacing $f_j(\underline{x}_i)$ with $f_{ij}(\underline{x}_i)$ for all $i \in \{1, 2, \ldots, N\}$ and $j \in \{1, 2, \ldots, m\}$. □

As well, lemma 11 is a special case of theorem 14 when $G = [1, 1]$ is a generating matrix of a $[2, 1, 2]$ linear code.

5 Conclusion

We have developed a method to construct resilient functions. The main feature of this method is that it can generate a large variety of resilient functions from which one can choose for a practical use. The method in this paper is a further expansion of part of the results in [11] and is more powerful in terms of parameters. However, parameters of resilient functions which are constructed by the method of this paper are still not as good as that in [9]. It should be noted that in this paper as well as in [11] methods for constructing new resilient functions from odd ones are stressed, while in [9] construction of nonlinear resilient functions with good parameters (which exceed those of linear such functions) is devised. Future research needs to be conducted into combining methods in this paper with those of [11] and [9] which may construct other infinite classes of resilient functions with better parameters by exploiting nonlinear codes and good known resilient functions.

References

1. C.H.Bennett, G.Brassard and J.M.Robert, "Privacy amplification by public discussion", *SIAM J. Comput.* **17** (1988), 210-229.
2. J.Bierbrauer, K.Gopalakrishnan, and D.R.Stinson, "Bounds on resilient functions and orthogonal arrays", *Advances in Cryptology - CRYPTO'94*, Springer-Verlag, (1994) 247-256.
3. B.Chor, O.Goldreich, J.Hastad, J.Friedman, S.Rudich and R.Smolensky, "The bit extraction problem or t- resilient functions", *Proc. of 26th IEEE Symp. on Foundations of Computer Science*, (1985) 396-407.
4. P.Camion and A.Canteaut, "Construction of *t*- resilient functions over a finite alphabet", *Proc. Eurocrypt'96*, Verlin: Springer-Verlag, 1996.
5. K.Gopalakrishnan, D.G.Hoffman, and D.R.Stinson, "A note on a conjecture concerning symmetric resilient functions", *Information Processing Letters* **47** (1993), 139-143.
6. J.H. van Lint, *Introduction to Coding Theory*, Springer-Verlag, 1992.
7. T.Siegenthaler, "Correlation-immunity of nonlinear combining functions for cryptographic applications", *IEEE Trans. on Infor. Theory*, Vol. IT-**30** (1984) 5, 776-780.
8. D.R.Stinson, "Resilient functions and large sets of orthogonal arrays", *Congressus Numerantium* **92** (1993), 105-110.
9. D.R.Stinson and J.L.Massey, "An infinite class of counterexamples to a conjecture concerning nonlinear resilient functions", *J. of Cryptology*, **8** (1995), 167-173.
10. C.K.Wu, "On the independency of Boolean functions of their variables", *J. of Xidian University*, (1988) 73-81. (in Chinese)
11. X.M.Zhang and Y.Zheng, "On nonlinear resilient functions (extended abstract)", *Proc. of Eurocrypt'95*, Springer-Verlag (1995) 274-288; See also "Cryptographically resilient functions", *IEEE Trans. on Inform. Theory*, (1996/97) to appear.

Another Approach to Software Key Escrow Encryption

Ed Dawson and Jingmin He

Information Security Research Centre
Queensland University of Technology
Gardens Point, GPO Box 2434
Brisbane QLD 4001, Australia

Abstract. Key escrow cryptography has gained much attention in the last two years. A key escrow system can provide cryptographic protection to unclassified, sensitive data, while at the same time, allows for the decryption of encrypted messages when lawfully authorized.

In this paper we propose a protocol for supporting the implementation of key escrow systems in software. Our protocol is an improvement on two previous protocols proposed in [1].

Key words: Key escrow, law enforcement, wiretapping, public key.

1 Introduction

Key escrow cryptography has been becoming popular recently. The concept was initiated by an US government announcement in April 1993 (the Escrowed Encryption Standard or EES, also commonly known as the "Clipper proposal"), in which a new federal standard encryption system with key escrow capability was proposed [8]. But the original idea came earlier, at least as early as in 1992 when Micali introduced the concept of *fair public-key cryptosystems* [6].

In a key escrow system, every session key used in every communication between users is escrowed and thus is recoverable by a law enforcement agent under necessary legal conditions. This will allow the government to enforce laws in special circumstances under appropriate legal safeguards.

In August 1993, NIST (National Institute for Standards and Technology) of US announced a cooperative program with industry to explore the possibilities of performing key escrow cryptography using software-only techniques. Balenson et al [1] proposed two protocols to support the implementation of key escrow cryptography in software. As already pointed out by the authors themselves, their protocols suffer from some drawbacks. Furthermore, we will show that their protocols suffer from some additional drawbacks as well.

The most controversial issue about software key escrow is that it is difficult to ensure that the key escrow software will function correctly and not be modified by the user to bypass or corrupt the escrow process. In other words, to ensure the proper functioning of the escrow process, there must be some codes in the software which cannot be bypassed easily. We call this the *software protection assumption* (SPA). Our protocol has used this assumption, as have the original protocols proposed in [1]. In [3], we described a hardware based key escrow system which provides similar attributes as the software system described in this paper, and does not rely on SPA.

In this paper we propose another protocol for software key escrow encryption. Our protocol is an improvement over the two protocols of [1]. Particularly, we have tried to eliminate some weaknesses of those protocols and add some new features.

The paper is organized as follows. Section 2 reviews and analyses the two protocols of [1]. Our protocol is presented in Section 3. Section 4 contains some concluding remarks.

2 Two Known Protocols for Software Key Escrow Encryption

Balenson, Ellison, Lipner and Walker [1] proposed two protocols to implement a software key escrow system. We call them protocols BELW1 and BELW2. In this section we review and analyse these two protocols and point out their weaknesses.

Throughout the rest of the paper, we assume the use of a public key cryptosystem with an encryption/decryption pair (E, D). For any user A, there is a public key KP_A and a corresponding secret key KS_A. The encryption of a message m under KP_A is denoted by $E_{KP_A}(m)$ which gives a ciphertext c which in turn can be decrypted as $D_{KS_A}(c) = m$. The secret key KS_A will also be used by A to sign digital signatures which are easy to verify. For simplicity, we denote A's signature on a message M by $S_A(M)$.

We will use the notation e and d to denote the encryption/decryption function of a standard symmetric encryption algorithm (e.g., DES or Skipjack). The encryption of a message m under this algorithm and a key k is denoted by $e_k(m)$ and the decryption of a ciphertext c is denoted by $d_k(c)$. Finally, we will use h to denote a secure one-way hash function.

2.1 Protocol BELW1

Protocol BELW1 is similar to the original EES proposal, except that a public key cryptosystem, instead of Skipjack, is used. It is actually a simulation of the EES proposal. It can also be considered as a special implementation of a fair public key system as suggested by Micali [6].

A Key Escrow Programming Facility (KEPF) holds a pair of master public/private keys (KEPFP, KEPFS) and a public family key KFP. For each program instance (a software "chip"), a unique program identifier UIP and a pair of public/private keys (KUP,KUS) are generated, where KUS=KUS1$\oplus$ KUS2 is the sum of two random numbers KUS1 and KUS2. The following parameters are installed in the program instance:

KEPFP, KFP, UIP, KUP, S_{KEPF}(UIP,KUP).

At the same time, UIP, KUS1 are sent to the first key escrow agent KEA1 and UIP, KUS2 are sent to the second key escrow agent KEA2. The private family key KFS is held by the law enforcement agency.

When a user A wants to send a message M to another user B, A and B first negotiate a common session key KS. Then A sends a ciphertext C, a LEAF (Law Enforcement Access Field), and an extra value Escrow Verification String (EVS) to B:

$$C = e_{KS}(M), LEAF = E_{KFP}(E_{KUP}(KS)|UIP),$$

$$EVS = e_{KS}(UIP, KUP, S_{KEPF}(UIP|KUP)),$$

where | denotes concatenation operation.

Unlike the EES the LEAF does not contain an escrow authenticator. This is because the receiver can validate the LEAF by recomputing it. This is accomplished by the sender transmitting the EVS. The EVS contains the encrypted versions of UIP, KUP and the certificate for UIP, KUP signed using the private key component of the key escrow program facility.

B will decrypt EVS with KS, compute $E_{KFP}(E_{KUP}(KS)|UIP)$ to verify the LEAF, and then finally decrypt C.

To wiretap a communication transmitted by A, the investigator LED (Law Enforcement Decryptor) will decrypt LEAF with the private family key KFS. obtain a court order, and submit the court order and the UIP to the two key escrow agents. The two agents will give him the two private unit subkeys KUS1 and KUS2. The investigator can then obtain the secret unit key KUS=KUS1$\oplus$ KUS2, use it to decrypt $E_{KUP}(KS)$ to obtain KS, and finally, use KS to decrypt C to obtain the message M.

2.2 Protocol BELW2

We still have two key escrow agents KEA1 and KEA2. KEA1 has a pair of public/secret keys (KEA1P, KEA1S). Similarly, KEA2 has a pair of public/secret keys (KEA2P, KEA2S). The two public keys KEA1P and KEA2P are given to the software vendor.

When a user, say A, wants to purchase a software key escrow product from the vendor, all the vendor does is to integrate the two public keys KEA1P, KEA2P with the encryption/decryption algorithms, and gives the whole package to the user. We denote such a package by SP_A (A's software program).

Suppose user A wants to send a message M to user B. First, A and B negotiate a common session key KS. Then A feeds the two inputs M, KS to her key escrow program SP_A which will output the following information:

1. The encrypted message $C = e_{KS}(M)$.
2. LEAF=$E_{KEA1P}(KS1)$, $E_{KEA2P}(KS2)$ where KS=KS1$\oplus$ KS2.
3. EVS=$e_{KS}(KS1, KS2)$.

Upon receiving the above messages from user A, user B will feed them and the session key KS to his key escrow program SP_B which will work as follows:

1. Use KS to decrypt EVS to obtain two components a and b.

2. Check if $a \oplus b = \text{KS}$. If not, stop and reject A's transmission.

3. Use the two public keys KEA1P, KEA2P to encrypt a and b respectively,

$$c = E_{\text{KEA1P}}(a), d = E_{\text{KEA2P}}(b),$$

and check if c and d are equal to the two components of LEAF respectively. If not, stop and reject A's transmission.

4. Finally, use KS to decrypt C to obtain the message M.

To wiretap a communication, the LED will present a court order and the LEAF information to the two KEAs. They will decrypt the corresponding components of LEAF with their secret keys, and give KS1 and KS2 back to the LED. The LED uses KS1$\oplus$KS2 to decrypt the ciphertext C.

2.3 Analysis

As pointed out by the authors themselves, one drawback of the first protocol BELW1 is that it exposes a communication user's escrow information (UIP and KUP) to programs controlled by other users. For example, Bob can modify his copy of the product to memorize or "harvest" Alice's UIP and KUP (and possibly also those of other parties with whom he communicates). He can then communicate with his modified program using LEAFs formed with Alice's (or other users') program identifier and escrow public keys. Consequently, when a wiretapping is initiated against Bob, it is Alice's secret key that will be withdrawn by the law enforcement agent. This possibility is unacceptable by any honest users.

Another drawback of BELW1 is that when the LED obtains the necessary information (the two parts of the secret key) to wiretap someone, he can decrypt everything transmitted by that person at anytime, whether that transmission was conducted a long time ago or will be conducted in the future. This gives the LED too much power.

The second protocol BELW2 does not have the above drawbacks, but it requires the two escrow agents to be on-line and involved with every decryption of a new session key used by the one who is under investigation. This is a burden to escrow agents because we may have only a few (possibly two) key escrow agents, each of them is involved in every transaction and there are of course a huge number of transactions everyday. Besides, in many cases the law enforcement agent must act in an off-line way so that he can wiretap a suspect's communication and take necessary actions immediately.

BELW2 suffers from other drawbacks as well. Note that the software products sold out by the vendor are all the exact same (all with the same encryption/decryption algorithms and the same pair of public keys KEA1P and KEA2P). Thus, one person can use another person's SP without any modification, and consequently, a user can sell copies of his SP to other people without spending any extra efforts of his own.

More specifically, BELW2 has the following problems:

1. No identity of any party appears in the transmission from A to B. This of course will cause much inconvienience for a law enforcement agent. Note that

one of the main objectives of key escrow cryptography is to help criminal investigation. The identity of a message sender must be identifiable, and preferably the receiver's identity as well.

2. The setting as suggested in the above protocol does not help software protection. A software key escrow system benefits from it's flexibility and market competitive power (cheap cost). But this also enables easy software reproduction. As we can see from the description in the last subsection, an exact same version of software product is sold to everyone, and one person can use another person's merchandise to do his own job. This is too bad for the vendor.

3. The LED may abuse the system. When the LED tries to obtain a court order from a judge, he may claim that he is investigating a dangrous serial killer, although he actually is wiretapping someone else. The judge cannot tell any difference because nothing about the identities of a message sender/receiver appears in an intercepted transmission. The two KEAs are in a similar position as the judge.

3 Our Approach

In this section we propose another protocol to support software implementation of key escrow systems. Our approach combines the advantages of protocols BELW1 and BELW2. Besides, we add some new features.

1. We argue that the negotiation of a common session key before a communication is not necessary. Since a software key escrow system uses public key cryptosystems, it would be strange and wasteful not to use it to do it's best job, i.e., reaching key agreement.

2. Our use of public key crytosystems naturally helps the law enforcement agent to identify both the sender and receiver of a communication, which is very important in some cases. Note that identifying a receiver does no harm to him because his secret will not be exposed.

3. We bind the identity of a user to its SP so that only he himself can use his SP.

4. The LED, once given a permission from the judge, can only wiretap what he has requested. In other words, he cannot misuse his power.

3.1 The Protocol

In all known key escrow systems, when the LED wants to conduct a wiretapping, he must obtain a court order from a judge first. The format and content of such a court order has never been specified very clearly. In our case, we assume the following. Suppose the LED has presented his request in a document M. The judge will first verify the validity of M. Then he will sign the hash value of M (and other possible complementary materials) and give the signed document back to the LED.

The Registration of a User

As before, we have two key escrow agents, KEA1 and KEA2. Each of them has a pair of public/secret keys. Their secret keys can be used to sign signatures which are easy to verify.

Suppose a user A wants to join the system. A must first go to each of the two agents to register himself. As a result, A is given a certified unique identification number UID_A. A is also given two pairs of public encryption/decryption keys (KAP1, KAS1) and (KAP2, KAS2). KAP1 (KAP2) is certified by KEA1 (KEA2). KEA1 (KEA2) keeps a copy of KAS1 (KAS2).

Note that it is commonly believed that a proper and safe use of public key systems should involve some certification process. For example, a user can only use a certified public key to do a transmission. This is about the management of public keys and we will not pursue this topic any further here.

Purchase a Software Product

User A then goes to purchase a SP from the vendor. A must show his unique UID_A. He must also help the vendor to sign his identity. The vendor generates a parameter (password) r_A which is related to user A in a unique way. For example, $r_A = (r1_A, r2_A)$ where $r1_A$ ($r2_A$) is the signature of UID_A under A's first (second) secret key. The hash value, $h(r_A)$, will be embedded into the product. The vendor integrates UID_A, KAP1, KAP2 and the necessary cryptographic algorithms into a software product SP_A and gives it to A.

Encryption/Decryption

Suppose user A wants to send a message M to user B. A will feed the following inputs to SP_A:

1. The identities of the sender and the receiver, UID_A, UID_B.

2. His secret keys and password r_A.

3. The certified public keys KBP1, KBP2 of B.

4. A random number t which serves as a logic timestamp for this particular transmission. A should use different timestamps for different transmissions for his own safety.

5. Three random numbers K, K1, K2 such that K=K1$\oplus$ K2. K is going to be used as the encryption key for this transmission.

6. The message M.

The encryption is done by the SP_A as follows:

1. Checks the identity to see if it is equal to the embedded identification number within the SP_A. If not, stop.

2. Checks if the password is correct (in the standard way). This step includes signing UID_A and comparing it with r_A.

3. Computes and outputs the following quantities:

$$C = e_K(M)$$
$$a = E_{KAP1}(K1), E_{KAP2}(K2)$$

$$b = E_{KBP1}(K1), E_{KBP2}(K2)$$
$$H = h(UID_A, UID_B, t, K1, K2)$$
$$S = S_A(H)$$
$$LEAF = UID_A, UID_B, KAP1, KAP2, a, b, t, S$$
$$EVS = e_K(UID_A, UID_B, t, H, S).$$

A will send C, LEAF and EVS to B.

At the receiving side, the sender A's public keys, B's identity UID_B, B's secret keys KBS1,KBS2, his password r_B, his signing key, C, LEAF and EVS are fed into SP_B which works as follows:

1. Checks the correctness of the password and UID_B. This includes signing UID_B and comparing it with r_B.

2. Decrypts b to obtain $K1'$ and $K2'$.

3. Uses A's public keys (contained in LEAF) to compute $E_{KAP1}(K1')$ and $E_{KAP2}(K2')$ and compares them with a. If no match, stop.

4. Computes K=K1$\oplus$ K2.

5. Uses K to decrypt EVS.

6. Compares the corresponding quantities contained in both the decrypted version of EVS and LEAF (i.e., UID_A, UID_B, t, and S). If a mismatch is found, stop.

7. Computes H and verifies that S is the correct signature of H by A.

8. Finally, uses K to decrypt C to obtain M.

Wiretapping

To wiretap the communication from A (to B), as usual, the LED must first obtain a court order. In our case, he must choose between two alternatives: either to wiretap only one session on-line, or wiretap possibly many sessions off-line. To obtain the permission for an off-line wiretap, the LED must provide strong reasons for doing so.

If an on-line wiretapping is requested (here on-line means that the two escrow agents are on-line), the procedure is similar to the protocol BELW2. More specifically, the following steps are followed:

1. The LED obtains a court order.

2. The LED presents the court order and LEAF to the two KEAs.

3. The first key escrow agent KEA1 decrypts the first component of a (using KAS1) and sends K1 back to the LED. Similarly, the second key escrow agent KEA2 decrypts the second component of a (using KAS2) and sends K2 back to the LED.

4. The LED computes $K = K1 \oplus K2$ and uses it to decrypt the ciphertext C to obtain M.

If an off-line wiretapping is requested (here off-line means that the two escrow agents are off-line during the actual wiretapping), the following steps are followed:

1. The LED obtains a court order which should include an expiration date T.

2. The LED presents the court order and LEAF to the two KEAs.

3. The first (second) key escrow agent KEA1 (KEA2) gives KAS1 (KSA2) back to the LED.

4. The LED uses KAS1 and KAS2 to do necessary decryption, as in BELW2.

5. At time T, the judge informs user A that his current public/secret keys have expired.

6. At time T, each of the two escrow agents generates a public/secret key pairs, registers them for A, and gives them back to A to replace A's old keys.

3.2 Analysis

Our protocol described in the last subsection has some desirable attributes.

1. The identities of both the sender and receiver are binded to the content of every transaction. Thus, the LED cannot misuse the system by requesting one wiretapping but actually conducting another wiretapping.

2. Suppose a user, say B, has got a copy of SP_A of another user A. If B stole SP_A from A, he cannot use it because he does not have the correct password. On the other hand, if SP_A, together with r_A, were given to B by A herself, she must also give all her secret keys to B. This will allow B to communicate with other people in the name of A, which will certainly draw A in big trouble. In other words, if A tries to sell her SP to other people, A will have to sell everything to that person, and it is her, not the buyer, that will be responsible for tne future use of SP_A.

3. As we have argued before, the use of public key systems should be supplemented with a certification system. For example, if a user A's public key is stored in a public directory, that key must be certified so that when other people want to use it, they know they are using the correct key. Our protocol naturally fits into and makes good use of this requirement.

4. The two alternatives of on-line or off-line wiretapping provided by our protocol, although a simple combination of BELW1 and BELW2, are indeed useful. For the off-line wiretapping, a valid period of time is attached so that the LED cannot conduct a wiretapping against someone forever. Note again, that the key escrow agents will stop the LED by giving the wiretapped user a new pair of keys. This is consistent with the requirement of the registration of public/secret keys. One possible drawback is that in this case, a user will know (afterwards) that he has been wiretapped.

5. A session key is established during a communication, instead of a a priori negotiation. For simplicity, we have allowed a message sender to choose the session key. By a simple modification, a mutual agreement protocol could be used. In some applications, a mutual agreement is not necessary and is actually inconvenient, because that requires an on-line receiver.

4 Conclusion

The concept of key escrow cryptography is very new and very controversial at this stage. We believe it could be acceptable to our society, although much more work has to be done in this general area. In this paper we have proposed another approach to software key escrow encryption. Our protocol is an improvement over two known protocols.

Both the software key escrow systems in this paper and in [1] depend on the software protection assumption (SPA) that there are some codes in the software which cannot be bypassed easily. Whether or not the SPA is true, or how it could become true, is another "big" question.

In fact a recent paper [3] has seriously questioned the truth of the SPA in relation to software based key escrow cryptosystems. In this paper it was demonstrated how to bypass the protocol and mislead the authorities in a software key escrow system proposed by Desmedt [4]. In order to overcome such attacks we believe some combination of hardware and software is a possible solution. We are investigating further into this possibility.

References

1. D.M.Balenson, C.M.Ellison, S.B.Lipner, and S.T.Walker, A new approach to software key escrow encryption, manuscript, 1994.
2. M.Blaze, Protocol failure in the escrowed encryption standard, Proceedings of the 2nd ACM Conference on Computer and Communications Security, 1994, pp.59-67.
3. J.He and E.Dawson, A new key escrow cryptosystem, in Cryptography: Policy and Algorithms, Springer-Verlag 1995, LNCS Vol.1029,pp.105-114.
4. Y.Desmedt, Securing traceability of ciphertexts - towards a secure software key escrow system, Advances in Cryptology, L.C.Guillon and V.Quisquater (editors), LNCS 921, Springer-Verlag 1995, pp.147-157.
5. L.Knudsen and T.Pedersen, On the difficulty of software key escrow, Advances in Cryptology: EUROCRYPT'96, Springer-Verlag 1996, LNCS 1070, pp.237-244.
6. S.Micali, Fair public-key cryptosystems, Advances in Cryptology: CRYPTO'92, Springer-Verlag 1993, LNCS 740, pp.113-138.
7. National Institute for Standards and Technology, The digital signature standard, Comm. ACM, Vol.35,No.7, 1992, pp.36-40.
8. National Institute for Standards and Technology, Escrowed Encryption Standard, Federal Information Processing Standards Publication 185, U.S. Dept. of Commerce, 1994.

Cryptography Based on Transcendental Numbers *

Josef Pieprzyk, Hossein Ghodosi, Chris Charnes and Rei Safavi-Naini

Department of Computer Science
University of Wollongong
PO Box 1144
Wollongong, NSW 2500, AUSTRALIA
e-mail: *josef/hossein/charnes/rei@cs.uow.edu.au*

Abstract. We investigate irrational numbers as a source of pseudorandom bits. We suggest two secure pseudorandom bit generators based on transcendental numbers. These two classes of transcendentals are applied to construct novel encryption algorithms. Properties of the encryption algorithms are studied and preliminary cryptanalysis is given.

1 Introduction and Motivation

Assume that you are living in a country where all crypto algorithms such as DES, FEAL, LOKI, RSA have been banned by the presidential decree. The use and possession of cryptographic software is illegal. Those who have broken the decree (the law) are subject to the prosecution and severe punishment. Even worse, the president formed a special Crypto Commission to investigate publicly accessible software to check its applicability for encryption and decryption. Based on the Commission recommendations, all hashing and signature schemes have been declared illegal. However, if citizens want to use the banned cryptographic algorithms, they can apply to the government (via its Crypto Agency) for licences. Suppose now that a citizen Bob has lodged his application to the Crypto Agency. The agency checks the Bob's credentials and issues a licence. The licence specifies that Bob can use cryptographic algorithms with secret key of length up to 25 bits. After some time, Bob realizes that the key he is allowed to use is too short and that everybody can read his cryptograms. As Bob is a law-obeying citizen, he wonders how he can use still publicly accessible software for encryption without infringing existing law.

It could be argued that the scenario depicted above is unrealistic. Unfortunately, there are countries that already consider cryptographic software, hardware and the law governing their usage as part of their national defence. Besides, it is interesting to investigate other methods and techniques which can be used to design new cryptographic algorithms. Note that without exceptions, all cryptographic algorithms are based on computations in finite algebraic structures

---

* Support for this project was provided in part by the Australian Research Council under the reference number A49530480 and the ATERB grant (groups, rings, fields, and vector spaces). The results of these computations are always exact – no rounding or approximation is involved. On the contrary, a single-bit error in the input of any cryptographic algorithm, generates an unintentional result. Modern cryptography deals with integers only; this principle is widely accepted in the crypto community. A new and efficient algorithm for multiplication in Galois fields is seen as a significant contribution to Cryptology, but a paper which describes a new and efficient algorithm for multiplication of 'floating-point' numbers, is considered to be irrelevant to Cryptology.

We will show how to design the necessary cryptographic algorithms using floating-point arithmetic with an arbitrary precision. Note that floating-point arithmetic is never used in cryptographic algorithms.

We define a new paradigm in the design of cryptographic algorithms. The paradigm is based on irrational numbers and floating-point arithmetic. We firstly investigate cryptographic properties of irrational numbers and methods for fast generation of irrationals. We propose a number of encryption/decryption schemes based on this paradigm and make a preliminary cryptanalysis of these schemes.

2 Irrational Numbers

The decimal representation of a rational number is either finite or periodic. By contrast the decimal representation of every irrational number is infinite and non-periodic. Real numbers satisfying an equation of the form

$$a(x) = a_0 + a_1 x + \ldots + a_d x^d = 0$$

with integral coefficients, are either integers or irrationals. Such numbers are called *algebraic*. Irrational numbers which satisfy no such equation are called *transcendental*.

An attractive feature of the reals is that they can represent any (infinite or finite) sequence of integers. Consider an experiment in which an unbiased coin is flipped an 'infinite' number of times. It is clear that the resulting random sequence is equivalent to some real number. Obviously, this sequence (the real) must not be either a rational or algebraic number (see Section 3.1), as in both cases a finite subsequence uniquely determines the rest of the (infinite) sequence. All infinite sequences of truly random integers fall into the broad class of transcendentals. Algebraic irrationals may look 'random' but their 'randomness' is limited to a finite subsequence.

Borel [3] introduced the notion of *normal* reals. A real is called to be *normal* with respect to the base p if for any natural number k all p^k possible strings of length k occur with equal probability.

3 Binary Sequences from Expansions of Irrational Numbers

We now discuss whether irrational numbers can be used to generate (at least in principle) random numbers (or bits). The importance of random and pseudorandom number generation in Cryptology as well as their relation to the complexity theory is discussed in [12] and [14].

We say that a random bit generator passes *the strong next bit test* if the k first consecutive bits do not provide any information about the $(k + 1)$-th bit (where k is an arbitrary natural number).

Assume that we have a number (bit) generator G which allows the user to select an irrational number randomly with a uniform probability distribution from the interval $(0, 1)$. The irrational number is kept secret. An attacker who does not know the irrational but can observe the output of the generator G, wants to pinpoint it from the observed expansion.

Lemma 1. *There is at least one irrational number in whose binary expansion the finite binary sequence $S = 0.s_1 s_2 \cdots s_k$ occurs.*

Proof. The sequence $S010010001\ldots$, is an infinite non-periodic sequence which contains the sequence S.

Theorem 2. *There are infinitely many irrational numbers in whose binary expansions the finite binary sequence $S = 0.s_1 s_2 \cdots s_k$ occurs.*

Proof. Assume that there are a finite number M of irrational numbers whose binary expansions contain the sequence S. Extending S by just one bit gives two different sequences $S \parallel 0$ and $S \parallel 1$. Thus extending S by i bits gives 2^i different finite length sequences (each of length $k + i$). We choose an i such that $2^i > M$. Hence there is a finite length sequence which can not appear in the binary expansion of any irrational number, a contradiction to Lemma 1

We show that the binary expansion of irrational numbers passes the strong next bit test.

Theorem 3. *Binary expansion of irrational numbers passes the strong next bit test.*

Proof. Let $\alpha_1, \alpha_2, \cdots, \alpha_k$ be the first k bits of the binary expansion of an irrational number and let the rational number $b = 0.\alpha_1 \alpha_2 \cdots \alpha_k$ be an approximation of all irrational numbers whose binary expansions begin with sequence $\alpha_1, \alpha_2, \cdots, \alpha_k$ (there are infinitely many of such irrational numbers). It is clear that those irrational numbers, which produce zero as the next bit belong to the interval $D_0 = (b, b + 2^{-(k+1)})$. Let $\gamma = \alpha_1, \alpha_2, \cdots, \alpha_k, 0, \cdots$ be one of such irrational numbers. No matter what the bits following 0 are, the value of this irrational number is greater than b but less than $b + 2^{-(k+1)}$ and hence $\gamma \in D_0$.

Irrational numbers of the interval D_0 will generate the sequence $\alpha_1, \alpha_2, \cdots, \alpha_k, 0$. Because for any irrational number $\beta \in D_0$ the irrational number $(\beta - b) < 2^{-(k+1)}$ has the expansion $0.\overbrace{00 \cdots 0}^{k+1} \cdots$ (but not all zeros because it is an irrational number). Hence the binary expansion of β is of the form:

$$\alpha_1, \alpha_2, \cdots .\alpha_k, 0, \cdots + 0.\overbrace{00 \cdots 0}^{k+1} \cdots = \alpha_1, \alpha_2, \cdots, \alpha_k, 0, \cdots.$$

This proves that D_0 is the set of irrational numbers of the form $\alpha_1, \cdots, \alpha_k, 0, \cdots$ Similarly, it can be shown that all irrational numbers which generate one as the next bit are belong to the interval $D_1 = (b + 2^{-(k+1)}, b + 2^{-k})$, and conversely.

We assume that the set of real numbers with initial sequence $\alpha_1, \alpha_2, \cdots, \alpha_k$ is uniformly distributed. Then the probability that α_{k+1} be zero, given the knowledge of $\alpha_1, \alpha_2, \cdots, \alpha_k$, is the measure of the set D_0, which is $\frac{1}{2^{k+1}}$. Likewise, the probability that α_{k+1} be one is $\frac{1}{2^{k+1}}$. Hence the probability of either bit occurring is the same.

All truly random sequences have to be among the irrationals. This phenomenon is related to well-known properties of reals. The set of all reals can be split into the subset of all rationals $\mathcal{R}$ and the subset of all irrationals $\mathcal{I}$. The cardinality of the rationals is $|\mathcal{R}| = \aleph_0$ and rational numbers are equinumerous with all natural numbers. The cardinality of all the irrationals is $|\mathcal{I}| = 2^{\aleph_0}$. The set $\mathcal{I}$ is equinumerous with the interval $(0,1)$ and as a matter of fact with any interval $(a, b); a, b \in \mathcal{R}, a < b$. Cantor's theorem gives that $\aleph_0 < 2^{\aleph_0}$.

If there is a measure μ which assigns a positive real to a measurable subset from the interval $\mathcal{S} = (0, 1)$ with the condition that $\mu(\mathcal{S}) = 1$, then the measure of any subset of rationals in $\mathcal{S}$ equals zero.

There arises an important practical question: how to implement a random selection of irrationals? A possible solution is to flip an unbiased coin an infinite number of times – this solution is not practical. On the other hand, if the selection of irrationals is to be done within a limited period of time, we need to use a finite number of indices to mark all the possible outcomes (irrationals). So we can randomly choose an index and use the corresponding irrational. This gives a practical irrational number generator, whose security is determined by the size of the index applied and the difficulty of finding the selected irrational from its approximation (an attacker can see polynomially many output bits). This brings us to the well-known problem of designing pseudorandom bit generators (PBG).

Pseudorandom sequences have to satisfy the following conditions:

1. they have to be generated 'easily' (the corresponding generation problem should belong to the class P - see Garey and Johnson [6]),
2. they have to be indexed by a finite seed which is selected randomly and uniformly from all possible values of the seed (or index),
3. they have to pass *the next bit test* (i.e. it is computationally infeasible to determine $(k + 1)$-bit from the sequence of k polynomially many bits – see [2]).

3.1 Weak Irrational Numbers

Algebraic numbers form a subset of irrationals that can be efficiently indexed using coefficients of their minimal polynomials. However, the resulting sequences are not secure.

An algebraic irrational can be generated by randomly selecting a minimal polynomial $a(x) = a_0 + a_1 x + a_2 x^2 + \cdots + a_d x^d$ and one of the d roots. The polynomial and the root uniquely define the irrational α. The security parameters of a such generator is the pair (d, H) where d is the degree of $a(x)$ and H is the height of $a(x)$, i.e. all coefficients a_i $(i = 1, \ldots, d)$ are selected from the set of integers from the interval $[-H, H]$. Kannan, Lenstra and Lovász have shown in [7] that given the knowledge of the first polynomialy many bits of an algebraic number, that number can be determined, i.e. its defining polynomial can be identified.

They use the LLL-algorithm [9] to efficiently determine the minimal polynomial $a(x)$ of an algebraic irrational α from its approximation $\tilde{\alpha}$. More precisely, their method (the KLL attack) identifies (in polynomial time) the polynomial $a(x)$ given the first $O(d^2 + d \log H)$ bits of the binary expansion of the algebraic number. As we have to assume that both parameters d and H are known to potential cryptanalysts, the binary expansions of algebraic irrationals are not secure pseudorandom sequences.

Observe that if the minimal polynomial is restricted to binomials $a(x) = a_0 + x^d$, the KLL attack can identify the irrational from a sufficiently long sequence of consecutive bits (which do not necessarily start at the beginning). Since a sufficiently long sequence of consecutive bits pinpoints the unique irrational with its minimal polynomial $b(x) = b_0 + b_1 x + b_2 x^2 + \cdots + b_d x^d$. The polynomial $b(x)$ can then be reduced to the corresponding binomial $a(y)$ by applying the transformation $x = y + N$, where N is an integer. Note also that if the analysis of bits in the KLL algorithms does not start from the beginning of a number, the parameter H increases. But this has a negligible impact on efficiency of the algorithm for a 'reasonable' values of N.

The approximation described in [7] extends to certain classes of transcendental numbers. Examples of such numbers are: $cos^{-1}(\alpha)$, $sin^{-1}(\alpha)$ and $log(\alpha)$ where α is an algebraic irrational. This follows from Theorem 2.3 of [7]. We examine closely the hypothesis of this theorem.

The assumption that there is a complex valued *approximable* function f which satisfies the *uniform Lipschitz condition*, is central to the hypothesis of Theorem 2.3 of [7]. (For the precise definitions see [7].) Examples of such functions are: $sin(z)$, $cos(z)$. Kannan et al. have extended their algorithm to this setting. Theorem 2.3 of [7] states that there is an (efficient) algorithm which takes as input such a function f, parameters (d, H), a complex number $\tilde{\beta}$, and outputs a complex number β which is approximated by $\tilde{\beta}$, and an algebraic number $f(\beta)$ of degree at most d and height at most H. The assertion about the transcendental numbers $cos^{-1}(\alpha)$, etc., follows from this theorem.

3.2 Transcendentals Immune to the KLL attack

It is obvious that algebraic irrationals should be avoided as they can be 'easily' identified. The class of transcendentals looks like the only choice. To avoid the KLL attack we should choose transcendental numbers not of the type $f^{-1}(\beta)$, where f and β satisfy the hypothesis of Theorem 2.3 of [7]. We will consider two classes of transcendental numbers which seem suitable for cryptographic applications.

Class 1 - Simple exponentiation. Consider irrationals of the following form α^β, where α is a positive integer $\neq 0, 1$ and β is a real quadratic irrational. They are known to be transcendental, see [1]. In particular $2^{\sqrt{2}}$, $2^{\sqrt{5}}$ are transcendental. Moreover transcendental numbers of the type α^β cannot be determined with the help of Theorem 2.3 of [7] as long as α and β are not revealed. Hence we propose to use the binary expansions of numbers of the form α^β, as (secure) pseudorandom sequences. Both α and β are kept secret and create a seed (index) of the pseudorandom number generator.

Class 2 - Composite exponentiation. For our second class we require a theorem of Baker [1]. It is proved in [1], that numbers of the form $\alpha_1^{\beta_1} \ldots \alpha_n^{\beta_n}$, such that: the $\alpha_i \neq 0, 1$ and β_i are algebraic, and $1, \beta_1, \ldots, \beta_n$ are linearly independent over the rationals, are transcendental. The simplest numbers of this type have the form $\alpha_1^{\beta_1} \alpha_2^{\beta_2}$, where $\beta_1 = \sqrt{m}$, $\beta_2 = \sqrt{n}$ and m, n are natural non-square numbers such that mn is a non-square. We take these numbers as our second source. It is easily shown that if mn is a non-square, then $1, \beta_1, \beta_2$ are linearly independent over the rationals. Hence for such β_i, $\alpha_1^{\beta_1} \alpha_2^{\beta_2}$ is transcendental. An example of these numbers is $2^{\sqrt{2}} 3^{\sqrt{3}}$. Clearly, the numbers $\alpha_1, \ldots, \alpha_n$, $\beta_1, \ldots, \beta_n$ need to be secret and constitute a seed (index) of the pseudorandom number generator.

3.3 Security Considerations

Assume that $\mathcal{A}$ is a subset of positive integers and $\mathcal{B}$ is a subset of algebraic numbers. Class 1 transcendentals are indexed by two functions. The first, $I_\alpha : \{0, 1\}^{n_\alpha} \to \mathcal{A}$ assigns a positive integer $\alpha \in \mathcal{A}$ to a binary string of length n_α. The second, $I_\beta : \{0, 1\}^{n_\beta} \to \mathcal{B}$ assigns an algebraic number $\beta \in \mathcal{B}$ to a binary string of length n_β (or equivalently its minimal polynomial). We can apply the KLL algorithm to identify the transcendental $T = \alpha^\beta$. Note that T is identified if both α and β are found from a sufficiently long sequence of bits of T. Our identification algorithm is the following. First randomly select $\alpha' \in \mathcal{A}$ and calculate

$$\beta' = \log_{\alpha'} T .$$

The KLL algorithm is used to determine β'. This attack succeeds if $\alpha' = \alpha$, for then the KLL algorithm will identify the irrational algebraic number from the approximation of β. The complexity of this algorithm is $O(2^{n_\alpha} p_{KLL}(n_\beta))$, where $p_{KLL}(n_\beta)$ is the time complexity function of the KLL algorithm required to identify the algebraic number β.

Proposition 4. *The parameters α and β of Class 1 transcendentals can be identified by an algorithm whose time complexity function is*

$$p_{class1}(n) \leq O(2^{n_\alpha - 1} p_{KLL}(n_\beta)) .$$

For certain indices $E(\beta) = \alpha^\beta$ is *multiplicative*, i.e., satisfies the equation $E(\beta_1 \cdot \beta_2) = E(\beta_1) \cdot E(\beta_2)$. If β_1 and β_2 are such that $E(\beta_1 \cdot \beta_2) = E(\beta_1) \cdot E(\beta_2)$ (which holds whenever $\beta_1 \cdot \beta_2 = \beta_1 + \beta_2$), then an adversary who could determine a polynomial fraction of the key pairs (α, β) of Class 1 numbers from their approximations in polynomial time, could also using the multiplicative property, determine all such key pairs in random polynomial time. (Rivest makes this remark about RSA in [10].) The condition under which $E(\beta)$ is multiplicative is the following.

Proposition 5. *Let $\beta_1 = \sqrt{m}$ and $\beta_2 = \sqrt{n}$ be two real irrationals, then $E(\beta_1 \cdot \beta_2) = E(\beta_1) \cdot E(\beta_2)$ is equivalent to the solution of the Diophantine equation*

$$(mn)^2 + m^2 + n^2 = 2mn(1 + m + n).$$

An easy argument using Proposition (5) gives that the necessary condition that $E(\beta)$ to be multiplicative is that m and n are even positive integers. Hence we should avoid these indices when using Class 1 transcendentals.

3.4 Efficiency

First let us consider the problem of generating of the expansions of algebraic irrationals. This problem can be restated as the well-known problem of root-finding of a minimal polynomial $a(x)$ (the algebraic irrational is a root of some $a(x)$). Newton's method (and its modifications such as the secant method) gives a very efficient algorithm whose time complexity function is $O(n)$ where n is the degree of $a(x)$ and the underlying computer has fixed precision floating-point arithmetic (see any standard book on numerical analysis such as [13],[4],[8]).

We, however, would like to select the precision parameter as an argument of the algorithm. In this setting, the best known algorithm called the *splitting circle method* runs in time bounded by $O(n^3 \log n + sn^2) \log(ns) \log \log(ns)$, where $n = deg(a(x))$ and s represents the required precision of computations (the expected error has to be smaller than 2^{-s}). Readers in.erested in details of the splitting circle method are referred to Schönhage [11].

To generate a pseudorandom sequence (of Class 1 or 2), one has to use floating-point arithmetic with arbitrary precision. This facility is available with mathematical packages such as Maple or Mathematica. The requested precision automatically determines the length of the generated pseudorandom sequence. We experimented with Maple and generated transcendentals of both classes. On the average, it took 10 minutes to produce a pseudorandom string of length of 30 kbits. Obviously dedicated algorithms will generate pseudorandom sequences significantly quicker.

4 Encryption Primitives Based on Transcendentals

We can use transcendentals of the form α^β for encryption in at least two ways. In *stream cipher mode*, the binary sequence generated by the expansion of the transcendental is used as a pseudorandom sequence and is XORed bitwise with the binary representation of the message. That is, an n-bit message $M = (m_1, \ldots, m_n)$ is encrypted to $C = (c_1, \ldots, c_n)$ where $c_i = m_i \oplus r_i$ and r_i is the ith bit of the string $R = (r_1, \ldots, r_n)$, generated from the expansion of a transcendentals of either Class 1 or 2. To decrypt a message the receiver needs to know the seed of the PBG, which is used recreate the pseudorandom string R. For a known-plaintext attack, encryption is as secure as R. (If the system is used once only.) We note that unpredictability of R (passing the next-bit test) is not proven but we do not know of any algorithm which can predict the next bit of the sequence either.

We consider now another approach to encryption. Let $\mathcal{Z}_N = \{a \mid a \in \mathcal{Z}; 1 < a < N\}$, i.e. $\mathcal{Z}_N$ is the set of all positive integers smaller than N and bigger than 1. The message space is $\mathcal{M} = \mathcal{Z}_N$. The following notation is used. If there are two irrationals γ and δ, then an equation $\gamma = \delta$ means exact equality. We write $\gamma \stackrel{s}{=} \delta$ if $\mid \gamma - \delta \mid < 10^{-s}$.

The basic encryption/decryption algorithm is defined as follows. Let the message $M \in \mathcal{M}$. The cryptographic key is $K = (\alpha, \beta)$ where $\alpha \in \mathcal{Z}$ and β is a quadratic irrational from the set $\Gamma = \{\sqrt{u} \mid u \in \mathcal{Z}_N; u \text{ is square free }\}$. The cryptogram is generated as

$$C \stackrel{s}{=} (M + \alpha)^\beta. \tag{1}$$

Cryptograms are positive rationals which are approximations of transcendentals. Both α and β are chosen according to the requirements for Class 1 transcendentals considered in Section (3.2). The security parameter of this scheme is the number of different keys which can be used (or the size) of the key space $\mathcal{K}$. The size of the key space relates to the sizes of the spaces from which the pair (α, β) is chosen.

In decryption the reverse operations are used to give

$$M \stackrel{s}{=} C^{\frac{1}{\beta}} - \alpha.$$

The message M can be recovered from the cryptogram since it is an integer, which can be determined from the s-place approximation of C.

The remarks made in Section 3.3 are also relevant to this encryption scheme.

Proposition 6. *The encryption scheme defined by (1) has the multiplicative property for a given pair (α, β) whenever the the following Diophantine equation has a solution*

$$(M_1 + M_2) + \alpha = 1.$$

Since the messages and the key α are always positive integers, it follows from Proposition 6 that the encryption scheme defined by (1) is never multiplicative.

4.1 Attacks on Class 1 sources

In a *ciphertext only attack*, the attacker knows only the ciphertext blocks and is required to find the plaintext (or the key). Using the equations which describe the system, the cryptanalyst must derive α and β from the following set of equations in which the M_i's are unknown.

$$C_1 \overset{\text{?}}{=} (M_1 + \alpha)^\beta$$
$$C_2 \overset{\text{?}}{=} (M_2 + \alpha)^\beta$$
$$\cdots\cdots\;\cdots\;\cdots\cdots$$
$$C_t \overset{\text{?}}{=} (M_t + \alpha)^\beta$$

This might seem impossible as the number of equations is always smaller than the number of unknowns. However properties of α^β might help a cryptanalyst to draw some conclusions about the value of the key. In particular for a sufficiently long sequence of cipher blocks the enemy can use the monotonicity of x^β to order the cipher blocks and use three consecutive values to solve for α, β and M_1. That is solve

$$C_1 \overset{\text{?}}{=} (M_1 + \alpha)^\beta,$$
$$C_2 \overset{\text{?}}{=} (M_1 + 1 + \alpha)^\beta,$$
$$C_3 \overset{\text{?}}{=} (M_1 + 2 + \alpha)^\beta.$$

This can be repeated for a number of triplets C_i, C_j, C_k that are likely to correspond to consecutive message blocks.

Next, we outline a *plaintext attack* on the system defined by (1). In this situation the attacker knows both the M_i's and C_i's in the above set of equations.

Let $(C_1, M_1), (C_2, M_2)$ be a pair of known plaintexts/ciphertexts. Then we have

$$\ln C_1 = \beta(\ln(1 + \alpha/M_1) + \ln M_1),$$
$$\ln C_2 = \beta(\ln(1 + \alpha/M_2) + \ln M_2).$$

Now we will assume that $|\alpha/M| < 1$ for any key/message pair (α, M).

We can approximate $\ln(1 + \alpha/M_1)$ and $\ln(1 + \alpha/M_2)$ as polynomials $P(\alpha/M_1)$ and $Q(\alpha/M_2)$, by taking sufficiently many terms in the power series expansion of $\ln(1+x)$. (With remainder a known function of x.) We obtain the approximations

$$\ln C_1 \approx \beta(P(\alpha/M_1) + \ln M_1),$$
$$\ln C_2 \approx \beta(Q(\alpha/M_2) + \ln M_2).$$

By dividing these two expressions we can eliminate β. Now one obtains a polynomial in α with known (approximate) coefficients. This can be solved for α and determined exactly, since α is an integer. Having the α, one can check if the approximation to $\ln(1+\alpha/M_1)$ and $\ln(1+\alpha/M_2)$ by polynomials was sufficiently accurate. If not, more terms of the approximation are taken, until the required degree of accuracy is achieved.

This process yields a number of possibilities $\alpha_1, \ldots, \alpha_r$ (roots of a polynomial) for the key. But the correct α can be determined by comparing (C_1, M_1) with respect to the possibilities $\alpha_1, \ldots, \alpha_r$. Once the correct α is known we can obtain an approximation for $\hat{\beta}$ from the pair (C_1, M_1). Now we use the KLL algorithm to determine β from $\hat{\beta}$.

4.2 Encryption based on Class 2 numbers

For Class 2 numbers we can without the loss of generality, use composite exponentiation of the form $\alpha_1^{\beta_1} \alpha_2^{\beta_2}$. The secret key is $K = (\alpha_1, \alpha_2, \beta_1, \beta_2)$. Encryption is defined by

$$C \stackrel{s}{=} \alpha_1^{\beta_1} \times (\alpha_2 + M)^{\beta_2}. \tag{2}$$

Decryption firstly involves the division of the cryptogram C by $\alpha_1^{\beta_1}$, further processing is the same as in the previous case.

Since $\alpha_1^{\beta_1}$ is fixed, the generator is subject to the attacks outlined in Section 4.1. In particular, observing a long sequence of cryptograms allows the enemy to choose cryptograms that correspond to consecutive plaintexts. In this case, the following equations have to be solved

$$C_1 \stackrel{s}{=} K(M_1 + \alpha)^\beta .$$
$$C_2 \stackrel{s}{=} K(M_1 + 1 + \alpha)^\beta ,$$
$$C_3 \stackrel{s}{=} K(M_1 + 2 + \alpha)^\beta ,$$
$$C_4 \stackrel{s}{=} K(M_1 + 3 + \alpha)^\beta .$$

The above two schemes have a common characteristic – they are subject to the 'cluster' attack. This follows from the property that if the distance $d(M, M')$ is small, then for the corresponding cryptograms, $d(C, C')$ tends to be small. The same observation is applicable for the key. To fix this problem, we can apply several encryption iterations each of the form of either (1) or (2). As a matter of fact, two iterations are enough.

There is however a better solution, which we describe below.

4.3 Better encryption algorithm

For an irrational $\gamma = b_1 b_2 b_3 \ldots$, we denote

$$\gamma \mid_t = b_1 b_2 \ldots b_t.$$

$\gamma \mid_t$ is a rational (or integer if the parameter t is smaller than the number of digits in the integer part of γ) which is created from the irrational γ by truncating all decimal digits after the t-th position. For an irrational $\gamma \in \mathcal{I}$ and any pair of positive integers t, s $(t < s)$, we define $\gamma \mid_t^s$ as

$$\gamma \mid_t^s = \gamma \mid_s - \gamma \mid_t .$$

For example, if $\gamma = a_1 a_2 \ldots$ then

$$\gamma \mid_2^4 = a_1 a_2 a_3 a_4 - a_1 a_2 = a_3 a_4,$$

$\gamma \mid_{2}^{4}$ holds the digits of γ in positions 3 and 4. We consider $\gamma \mid_{t}^{s}$ as subsequence of digits from $t + 1$ to s. It can also be treated as an integer or a rational (the decimal point can be placed arbitrarily).

A third alternative for the encryption algorithm, is to encode the message in a one-to-one way into the minimal polynomial of a random quadratic irrational. Denote this as β_M; so as long as M is not a square and odd integer, β_M is a root of $x^2 - M = 0$. Next select a sufficiently long subsequence $\beta_M \mid_{t_2}^{t_1}$ from β_M. The cryptogram is

$$C \stackrel{s}{=} \alpha^{\beta_M \mid_{t_2}^{t_1}}. \tag{3}$$

For decryption, the receiver (who knows α) retrieves the sequence $\beta_M \mid_{t_2}^{t_1}$ and uses the KLL algorithm to find the minimal polynomial and the message M. This method of encryption creates many different cryptograms for the same message if the subsequence $\beta_M \mid_{t_2}^{t_1}$ is selected from different places, so the cluster and approximation attacks do not apply.

5 Conclusion

We have described a pseudo random sequence generator based on the binary expansion of a class of transcendental numbers which appears to resist the KLL attack. We have described an encryption scheme based on this generator and have made a preliminary security analysis of this scheme. We have described possible attacks on one version of the encryption scheme based on Class 1 numbers. We have observed that cryptanalysis of encryption schemes based on Class 2 transcendentals seems to be more difficult than Class 1 schemes.

ACKNOWLEDGMENTS

Authors thank Arjen Lenstra for providing [11].

References

1. A. Baker. *Transcendental Number Theory*. Cambridge University Press, 1975.
2. M. Blum and S. Micali. How to generate cryptographically strong sequences of pseudo-random bits. *SIAM Journal on Computing*, 13:850–864, November 1984.
3. E. Borel. Lecons sur la Theorie des Fonctions. 2nd Ed, 1914,pp.182-216.
4. R.L. Burden and J.D. Faires. *Numerical Analysis*. PWS Publishing Company, Boston, 1993.
5. H. Feistel. Cryptography and computer privacy. *Scientific American*, 228(5):15–23, 1973.
6. M. R. Garey and D. S. Johnson. *Computers and Intractability A Guide to the Theory of NP-completeness*. W.H. Freeman and Company, 1979.

7. R. Kannan, A.K. Lenstra, and L. Lovász. Polynomial factorization and nonrandomness of bits of algebraic and some transcendental numbers. *Mathematics of Computation*, 50:235–250, 1988.

8. D.E. Knuth. *The Art of Computer Programming, Volume 2*. Addison-Wesley, 1981.

9. H.W. Lenstra, A.K. Lenstra, and L. Lovász. Factoring polynomials with rational coefficients. *Math. Ann.*, 261:513–534, 1982.

10. R.L. Rivest Cryptography *Handbook of Theoretical Computer Science, Chap. 13* Edited by J. van Leeuwen Elsevier Science Publishers, 1990.

11. A. Schönhage. The fundamental theorem of algebra in terms of computational complexity. Preliminary report, Mathematisches Institut der Universität Tübigen, August 1982.

12. A. Shamir. On the generation of cryptographically strong pseudo-random sequences. *ACM Transactions on Computer Systems*, 1(1):38–44, 1983.

13. J.S. Vandergraft. *Introduction to Numerical Computations*. Academic Press, New York, 1983.

14. Andrew C. Yao. Theory and application of trapdoor functions. In *Proceedings of the 23rd IEEE Symposium on Foundation of Computer Science*, pages 80–91, New York, 1982. IEEE.

Breakthroughs in
Standardisation of IT Security Criteria

Eugene F. Troy
Project Manager, IT Security Criteria & Evaluations
National Institute of Standards and Technology
Gaithersburg, Maryland, USA

Background - Why the Common Criteria:

There are four main Driving Factors affecting the current activity in IT security criteria and product evaluations. These factors are all closely inter-related:

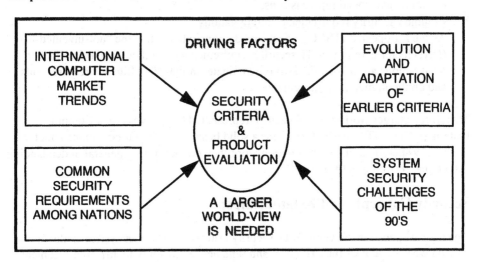

Figure 1: Rationale for Evolving to the Common Criteria

- International Computer Market trends.
  The notion of a "national IT product manufacturer" is no longer useful. Almost all manufacturers in all countries desire to sell into the international market and are typically based multi-nationally too. These manufacturers understandably have no particular desire to develop and sell numerous variants of their popular products to meet various national security restrictions or demands.
- The need for evolution and adaptation of earlier criteria.
  In the US, the Trusted Computer Security Evaluation Criteria (TCSEC), the so-called "Orange Book" was published by NSA in the early 1980's to establish the US government's military IT security requirements. The TCSEC has only six requirement sets, all for stand-alone operating systems, of which only four have been used to any great extent. None of them, when closely examined, addresses good commercial IT security requirements or the age of connectivity. The TCSEC

worked well to describe security requirements for stand-alone mainframes, and it has been difficult to translate into network and database terms. The Information Technology Security Evaluation Criteria (ITSEC), published in the early 1990's by the European Commission, is similarly limited in that it doesn't address the security functions needed, containing only requirements for the assurance aspect. The function sets mentioned in the ITSEC are mainly those from the old TCSEC, and are considered "examples".

- <u>System security challenges of the 90's.</u>
  This brings us to the challenges of the "real world" of today, that of distributed systems, the World Wide Web, the internet and intranet -- widespread connectivity and routine trans-national information flows. Distributed access, dispersed and cooperative work-patterns, the need for routine protection of information in transit -- all of these are today's IT security problems and are crucial to global society. Unsecured, they represent serious avenues for attack that can have widespread economic and social repercussions.

- <u>Common security requirements among nations.</u>
  As the Europeans' ITSEC project first demonstrated, when the "not-invented-here" (NIH) factor is removed, IT security requirements are rather standard no matter what nations are involved. Two predominate: military/intelligence requirements and civil/commercial requirements.

Therefore, a larger world-view is needed, represented by the work on a Common Criteria (CC) for IT Security that forms a solid basis for trust among nations about the IT security specification and evaluation work they do, permitting general understanding and mutual recognition of these efforts.

Security Concepts and Relationships:

One may ask, what is the value of IT security criteria and product/system evaluation -- why do we need to bear the extra cost and time needed to apply them? The argument that forms the basis for security criteria and evaluation is generally as follows:

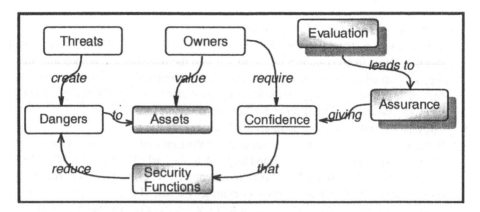

Figure 2: Security Concepts and Relationships

- Owners place some value on IT-based *assets*.
- *Threats* in the operating environment create *dangers* to these IT-based assets.
- Security features are applied to the operating environment and to the IT to reduce these known dangers (risk) to the assets. Those features which are incorporated into the IT are called *security functions* of the IT.
- However, owners expect and need to have some *confidence* that these embedded security functions in fact do the job needed and do it predictably. Otherwise, why bother with the added overhead and expense of including them?
- *Evaluation* of the security functions in the IT against accepted criteria leads to *assurance*, which gives the needed confidence in two ways: first that the security features are the right ones to meet the threats, and second that these security features are implemented appropriately, i.e., work predictably to do their job.

Twofold Purpose of IT Security Criteria:

IT security criteria help to provide the following two major benefits:

First, security criteria give a well-understood and common vocabulary and syntax for describing IT security requirements for product and systems. This requirements language can be viewed on two levels, as shown in the Common Criteria:
- The Protection Profile and Security Target constructs which first identify the relevant factors forming the basis for the IT security requirements, and then state those requirements in a standardised way that can be generally understood by both users and vendors.
- A catalogue of functional requirements that are complete enough to be useful in specifying security features for IT products and systems and are well-enough understood to be evaluable.

Second, security criteria provide a solid technical basis for deciding to trust (i.e., have assurance about, have confidence in) the security functions in IT products and systems. This trust basis comes from performing a well-understood process of evaluating the IT product against a set of factors that are well-known to help provide this trust. These trust factors are expressed in the form of:
- A series of evaluation assurance levels, increasingly stringent 'packages' containing assurance requirements of various types which are known to work together in a mutually supportive way.
- A catalogue of all those individual assurance requirements comprising the assurance levels, plus others which could be specified additionally to help provide extra assurance as needed.

Context of IT Security Evaluations:

The complete context of IT security evaluation is represented by a number of factors related to product development.

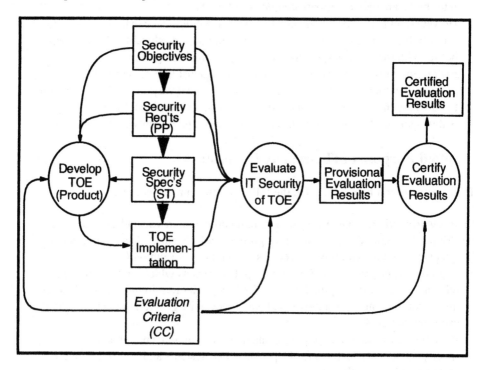

Figure 3: Context of IT Security Evaluations

During development, there is the increasingly specific instantiation of a product with its security features, moving from security objectives to requirements, on to security specifications, and then to the implementation of the security features along with the rest of the product. The security requirements are stated as much as possible in the context of known security functional requirements from the criteria, amplified as necessary to be product implementation specific. There are a variety of assurance criteria-driven deliverables produced during the development process that help provide confidence in the correctness of product implementation against the functional requirements.

After development, the completed IT security product is subjected to a series of evaluator actions, also specified in the criteria, to validate the correctness of implementation and to determine the effectiveness of the product in meeting the security threats and policies that are the basis for the requirements.

Finally, there is a series of actions by some authoritative body (currently governmental) to review the case made by the evaluators that the product indeed meets a valid set of requirements and implementation specifications. This process is called 'certification of results' and is generally followed by entry of the product onto an approved list of evaluted products available for user procurement guidance.

A Brief History of IT Security Criteria:

The history of IT security criteria is rather complex (see Figure 4). The salient elements relate first to the growth in national initiatives, followed by growing recognition of the inutility of individual national action, which was then succeeded by a number of joint efforts that have culminated in the Common Criteria and its acceptance into the process of becoming an International Standard.

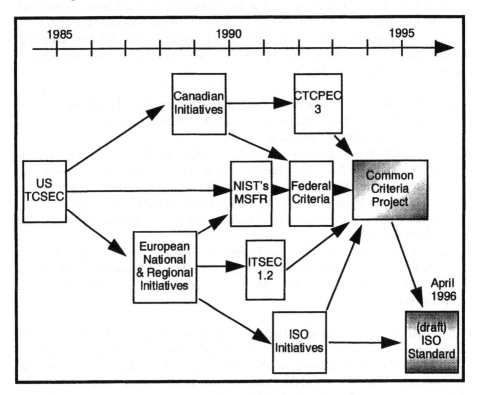

Figure 4: Brief History of IT Security Criteria

1. The first IT security criteria was the TCSEC, the fabled 'Orange Book', first published by the US National Security Agency (NSA) in 1993 based on earlier work done in conjunction with the National Bureau of Standards. This volume was fixed in form and application, and oriented towards multi-user operating systems without external connectivity. The TCSEC was very good for its time, and despite its requirement-set inflexibility, its fundamental technical requirements continue to be used in later criteria, and have been carried forward into the CC. The TCSEC was

subsequently "interpreted" for both networks and databases. It has formed the basis for NSA product evaluations to the present time.

2. Owing to the inflexibility of the TCSEC and to the need to set up their own trusted product evaluation programs, several European nations and Canada began their own criteria development efforts in the late 1980's. The Europeans rather quickly pooled their efforts after a number of unilateral forays. Initial versions of the Canadian Trusted Computer Product Evaluation Criteria (CTCPEC) and the European Community's Information Technology Security Evaluation Criteria (ITSEC) came out in 1990.

3. The ITSEC was the initial impetus for the search for a truly international standard which was begun in late 1990 by a working group of the International Standards Organisation (ISO). That group is called Working Group 3 of ISO's Subcommittee 27, and has been led from the beginning by Professor Svein Knapskog of the Norwegian University for Science and Technology.

4. In the United States, NSA and NIST agreed to jointly re-work the Orange Book to bring it up to date technically and to make its security requirement sets more broadly applicable to non-military IT products.
- The first US effort was the Minimum Security Functional Requirements (MSFR), an update of the TCSEC's C2 requirements set with the goal of being more useful to private industry and civil government bodies. The MSFR was heavily influenced by the ITSEC's Security Target philosophy, which separated functional and assurance requirements and justified each against expected threats in the intended environment of use.
- The second US effort was the draft Federal Criteria (FC) version 1, published in early 1993. The FC was in turn influenced strongly by the MSFR work and by the CTCPEC. One of the Canadian authors of the latter was an active member of the FC working group.

In 1993, the US and Canada agreed to harmonise their criteria, based on the draft FC and the newest version of the CTCPEC. They jointly announced these plans to the European Community, a decision was then made to pool North American and European criteria development efforts, and the Common Criteria (CC) effort was thereby born. That agreement was the first of the breakthroughs referred to in the title of this paper. This new project held promise to lead to the greater breakthrough everyone was hoping for -- the collapse of all ongoing criteria efforts into a single international criteria. The goal was to harmonise the several criteria into one, which would then be turned over to ISO as a contribution to the international standard. In large part, that objective has been achieved in April 1996, when ISO/SC27/WG3 accepted Parts 1 through 3 of the CC trial version 1.0 as the basis for its further work. This occurrence is indeed the second and key breakthrough that was long sought by the IT security community.

Common Criteria Project Participants:

The four national security agencies of France, Germany, the Netherlands and the United Kingdom which authored the ITSEC joined with Canada, NSA and NIST to form the Common Criteria Editorial Board in mid-1993. Initial plans were highly optimistic, envisioning that the several criteria involved could be 'aligned' in six months of hard work. In fact, it has taken over four times that long, due to the work required to resolve many fundamental differences in viewpoint. Trial-use version 1.0 of the Common Criteria was published in late January 1996, after two previous widely-reviewed draft versions. It is envisioned that after one year of application by the project participants and others, a new version 2 will be completed and handed over to ISO, and the criteria part of the project will be completed. There are other aspects of the CC Project which will continue on; these will be addressed shortly.

Overview of the Common Criteria Structure:

The CC consists of three major parts, following the original ISO criteria structure.

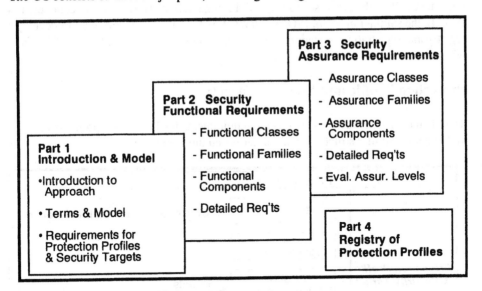

Figure 5: Common Criteria Structure

Part 1 consists of the introduction and presentation of the general model and concepts of IT security evaluation. Additionally, Part 1 includes normative requirements for the structure and content of the two constructs for stating requirements specified in the CC. These are the Protection Profile and Security Target, which will be described more fully later.

Part 2 is the catalogue of functional requirements. The attempt here has been, in the words of one of the CC principal authors, to write down everything we know about IT security functions and can evaluate. These security functions are grouped at a very high level into ten broad classes, each of which contains a number of families of related indivisible functional requirement components. The notion is that there should be very few unique and typically new security requirements at the level of "function" (more abstract than product-specific implementing mechanism) which are not covered by the catalogue, although some slow evolution is anticipated. It is expected that most seemingly unique security requirements will in fact be variants of the known functional requirement components in Part 2, and they can be stated through refinement of those component requirements to be more specific or detailed as needed.

Part 3 is the catalogue of assurance requirements, consisting of a set of discrete assurance components similarly to Part 2, plus a grouping of selected components into a series of seven increasingly rigorous packages called Evaluation Assurance Levels (EALs). The source criteria all have used variants of these levels in order to gauge the amount of assurance to be provided about an IT security product. Part 3 also contains evaluation criteria for Protection Profiles (PPs) and Security Targets (STs).

A new Part 4 is the initial registry of predefined PPs. It is anticipated that this document will be the precursor for a wider PP registration effort, possibly conducted by ISO. In the summer of 1996, ISO/SC27/WG3 undertook a new work item to develop a registry and registration procedures for PPs, that is expected to pave the way for this wider effort.

Protection Profile and Security Target:

The PP and ST constructs for specifying requirements for IT security products or systems have similar structures and numerous common elements.

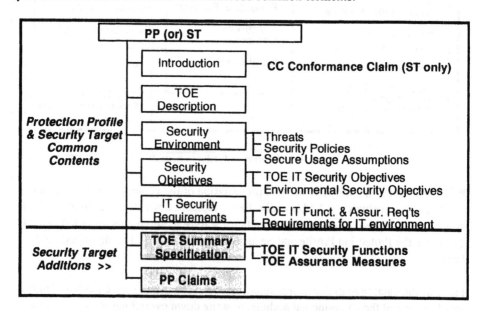

Figure 6: Protection Profile and Security Target Structure

Although they have much commonality, the PP and ST differ in two important respects:

- General versus specific product/system orientation -- The PP is intended to state a more general set of requirements that any of a number of IT security products or systems might fulfill. The ST is a specific requirement set for a single product or system (called Target of Evaluation or TOE in the CC).

- Detail of structure -- The PP and ST have common content down through the level of security requirements which are based on security objectives, which in turn are derived from statements about threats and security policies. The ST, covering a specific product or TOE, will go further. It is anticipated that STs typically will be formed from known and evaluated PPs in order to meet market requirements, but will amplify the requirements by containing detailed specifications for one single TOE or product. In such cases, the ST will also provide a claim of conformance of its TOE to a certain PP along with statements justifying that conformance.

It is expected that both the PP and ST will be formally evaluated, applying the criteria for them given in Part 3. This evaluation will make sure that security objectives to be met flow logically from the stated threats and security policies to be addressed, and that security requirements (both functional and assurance) fully cover the objectives. The PP is to be evaluated prior to its being registered for general use. The ST is to be evaluated in the first phase of the evaluation of the TOE which it describes and specifies.

Part 2 Functional Security Requirements - Classes:

Part 2 contains nine agreed classes or major groupings of specifiable IT security functional requirement components.

FAU -- Security Audit (35)

FCO -- Communication (Non-Repudiation) (4)

FCS -- Cryptographic Support *(in version 2)* (40)

FDP -- User Data Protection (46)

FIA -- Identification & Authentication (27)

FPR -- Privacy (Anonymity, etc.) (8)

FPT -- Protection of Trusted Security Functions (43)

FRU -- Resource Utilisation (8)

FTA -- TOE Access (11)

FTP -- Trusted Path (2)

Figure 7: Classes of CC Security Functional Requirements

These classes cover the known range of IT security functions. They are further subdivided into 75 families of 184 related components. Additionally, there is one draft Cryptographic Support class consisting of 15 families and 40 components that has been proposed and has been published with the CC version 1, but has not yet been agreed for inclusion in the CC.

The CC and Crypto

Because of the extreme sensitivity of the topic of crypto-algorithmic strength estimations, the CC project participants decided not to include requirements for assessment of the mathematical properties of algorithms in the criteria itself. It is assumed in CC evaluations that, when required for a particular product, some other entity will make the appropriate algorithmic assessment and then provide the results to the evaluators expressed in terms of "strength of function".

It is recognised and agreed by the CC project participants that cryptographic administration and application need to be covered in the CC, especially to cover the security challenge posed by distributed systems and networks in general. There is a general understanding, however, that crypto is fundamentally a mechanism, or implementation detail of more general functional requirements. The CC requirements in Parts 2 and 3 are not intended to go down to the level of mechanism implementation -- these details can be specified in a Security Target as detailed refinements of existing requirements. (For example, a requirement for authentication by smart-card algorithm is seen as simply a more-specific implementation of several security functional component requirements in the family FIA_UAU, User Authentication.) There is a counter-argument that cryptographic support is such an important topic in today's environment that it should be covered in the CC anyway. During the trial-use period a joint working group of the CC project organisations will resolve this question for CC version 2.

Late in the development of the CC, too late for the participants to come to agreement on it, a set of draft material on crypto based on the first view was presented for inclusion. It was agreed that the material was to be saved for possible future use in version 2 by being placed into a Technical Report (TR) to accompany the CCv1.0 during the trial use and comment period. This TR covers various aspects of cryptographic implementation, administration and key management. The crypto TR is generally available with the CC for public review and comment.

Assurance Requirements -- Classes:

Part 3 contains nine classes of specifiable assurance components covering both the correctness of TOE development and implementation and the effectiveness of the TOE in meeting its stated security objectives.

```
ACM - Configuration Management
ADV - Development
ATE - Tests
AVA - Vulnerability Assessment
ADO - Delivery and Operation
AGD - Guidance Documents
ALC - Life-cycle Support
-------------------------------------------------------------
APE - Protection Profile Evaluation
ASE - Security Target Evaluation
```

Figure 8: Classes of CC Assurance Requirements

Useful combinations of assurance components are combined into the seven Evaluation Assurance Levels (EALs) included in Part 3, as discussed next. Individual assurance components are specifiable when needed to augment these EALs for particular product needs.

Evaluation Assurance Levels:

The seven EALs are increasingly-strong packages of mutually supportive components covering requirements from each of the classes which have been developed for normal use in PPs and STs.

```
Level EAL1 - (new)
The lowest level which should be considered for purposes of evaluation

Level EAL2 - (like C1 - E1)
Best that can be achieved without imposing some additional tasks on a
developer

Level EAL3 - (like C2 - E2)
Allows a conscientious developer to benefit from positive security engineering
design without alteration of existing reasonably sound development practices

Level EAL4 - (like B1 - E3)
The best that can be achieved without significant alteration of current good
development practices.

Level EAL5 - (like B2 - E4)
The best achievable via pre-planned, good quality, careful security-aware
development without unduly expensive practices.

Level EAL6 - (like B3 - E5)
A "high tech" level for (mainly military) use in environments with *significant*
threats and moderately valued assets.

Level EAL7 - (like A1 - E6)
The greatest amount of evaluation assurance attainable whilst remaining in the
real world for real products.  EAL7 is at the limits of the current technology.
```

Figure 9: CC Evaluation Assurance Levels (EALs)

The EALs cover a broad range of assurance, from simple verification of minimal development requirements (EAL1) to the full formalisation of theorem-proving applied against mathematical models of the TOE's functions (EAL7). As might be anticipated, the normal range of assurance for commercial IT security products is in the lower middle (EAL3 and EAL4), which are achievable by a conscientious developer using sound engineering and good commercial development practices. EAL3 is intended to be comparable to the assurance requirements in the TCSEC's C2.

Part 4 -- Registry of Predefined Protection Profiles:

A central part of the CC notion is the Protection Profile. Well-crafted PPs for products with wide usefulness, expressed in terms of known and widely accepted functional and assurance requirement statements in the CC, are the goal. Part 4 is intended primarily as an initial set of PP examples and a stimulus for development of other PPs. The current version of Part 4 includes three PPs, two commercially-oriented ones from previous criteria (CS1/C2 and CS3) and a new low-end firewall (filtering packet router). Ultimately, it is envisioned that there will be a living catalog or registry of PPs. To that end, there were already at least 20 projects underway by Summer 1996 to develop various kinds of PPs against the CC.

The Future: Trial Use Period and Follow-On Tasks:

Now that CC version 1 is available to all and its ISO acceptance is a reality, it will be useful to look to the follow-on phases of the project, most of which have already begun. The project sponsors have entered into a one-year trial-use period for CCv1, in which it is being tested and built upon for practical use. The following implementation activities are under way:

- A significant number of trial evaluations are either planned or already initiated in the participating nations. These early evaluations are mainly comparisons, in which a product is evaluated against both the existing criteria and the new CC. The evaluators are expected thereby to grow in confidence that a CC-based evaluation will produce a predictably acceptable result.
- A major part of the current activities is dedicated to developing a common evaluation methods manual. This is a major activity which is one of the key underpinnings for all later CC-based work, especially the following.
- Preliminary work has begun on exploring the basis for mutual recognition of evaluations between North America and Europe. The three legs supporting mutual recognition are a stable and common criteria, common methods for their application, and mutual expectations of similar competence of the evaluators.
- The project sponsors are actively soliciting comments from the international IT security community on the CC version 1. Comments will be received until the end of October 1996. Directions for formatting and submitting comments are in each of the CC volumes. The CC has been made available both electronically on project participant internet websites (visit NIST's site: http://csrc.nist.gov/nistpubs/cc) and on CD-ROM that is free upon request from any of the participants.

- It is anticipated that CC version 2 will be developed during 1997, based on all the feedback gathered from the trials, the comments and other related work going on during this year. This definitive version will represent the end of the criteria-development phase of the project. At that point, the CC will be fully relinquished to ISO for completion as an international standard.
- Later on, it is envisioned that implementing guidance, like that published to accompany the Orange Book, will be needed. There is already a proposal in ISO to begin work on the first volume, guidance on preparation of Protection Profiles and Security Targets.
- As development and evaluation of globally-acceptable Protection Profiles is a major goal of the project, procedures for their international registration are required. At the present time, ISO's new work item holds significant promise to develop those registration procedures.

Summary:

The results of the Common Criteria project can indeed be viewed as a major breakthrough in the field of IT security. For the first time, six nations, representing both the military interests as well as civil government and private industry, have not only sat down at the table to iron out their philosophical differences in IT security but have achieved a great measure of accord. Admittedly, this accord not been won easily; it has come at a significant expense of both time and energy. Notwithstanding, the result is a very flexible and extensible approach that is designed to meet the needs of today and tomorrow; indeed the CC is the next generation criteria.

In doing so, the developers of the new CC have been careful to protect the fundamental technical principles of IT security, such as the Trusted Computing Base and Reference Mediation on the one side and effectiveness and correctness on the other. The resulting approach represented by the CC version 1 is a major contribution to international harmonisation. The fact that it has already been accepted by ISO as the basis for further work towards an international standard is indicative of the success of the project.

The desired end-state is now in sight -- a level playing field for IT security products world-wide, where it should make no difference to the consumer where a product is manufactured or evaluated. The degree of trust to be placed in a product's secure and predictable operation will be known and accepted.

Tailoring Authentication Protocols to Match Underlying Mechanisms*

Liqun Chen, Dieter Gollmann and Christopher J. Mitchell

Information Security Group
Royal Holloway, University of London
Egham, Surrey TW20 0EX, UK
E-mail:{liqun,dieter,cjm}@dcs.rhbnc.ac.uk

Abstract. Authentication protocols are constructed using certain fundamental security mechanisms. This paper discusses how the properties of the underlying mechanisms affect the design of authentication protocols. We firstly illustrate factors affecting the selection of protocols generally. These factors include the properties of the environment for authentication protocols and the resources of the authenticating entities. We then consider a number of authentication protocols which are based on mechanisms satisfying different conditions than those required for the ISO/IEC 9798 protocols, in particular the use of non-random nonces and the provision of identity privacy for the communicating parties.

1 Introduction

This paper is concerned with the fundamental security mechanisms which are used to construct entity authentication protocols, and how the strength of these mechanisms affects protocol design. For the purposes of this paper, the mechanisms are divided into two categories, cryptographic mechanisms and time variant parameters (TVPs). The first category includes symmetric encryption, digital signatures, cryptographic check functions (CCFs) and zero knowledge techniques. The second includes clocks for timestamping, nonce generators and sequence numbers. An authentication protocol is typically constructed using at least one mechanism from both categories.

When a new protocol is needed in a particular environment, the designer must first discover what mechanisms are available. For implementation reasons there may be limits on available mechanisms, because every mechanism has particular properties corresponding to the particular environment in which the protocol works. In such a case, the protocol must meet the needs of the environment and be tailored to the 'strength' of the available mechanisms.

A variety of entity authentication protocols[2] have recently been standardised by the International Organization for Standardization (ISO) and the International Electrotechnical Commission (IEC). The four published parts of ISO/IEC

---

* This work has been jointly funded by the UK EPSRC under research grant GR/J17173 and the European Commission under ACTS project AC095 (ASPeCT).
[2] They are referred to as entity authentication mechanisms in ISO/IEC documents.

9798 cover the general model (9798-1, [1]), authentication protocols based on symmetric encipherment (9798-2, [3]), authentication protocols based on public key algorithms (9798-3, [2]), and authentication protocols based on CCFs (9798-4, [4]). A fifth part of ISO/IEC 9798, covering authentication protocols using zero knowledge techniques, is currently at CD stage [5]. The ISO/IEC 9798 protocols use TVPs, e.g. timestamps, nonces and sequence numbers, to prevent valid authentication information from being accepted at a later time.

The protocols proposed in ISO/IEC 9798 have all been designed to use mechanisms meeting rather stringent requirements, which may be difficult to meet in some practical environments. Hence in this paper we look at alternatives to the ISO/IEC 9798 protocols which are still sound even when the mechanisms do not meet such 'strong' conditions.

The discussion starts in Section 2 with an illustration of factors which affect protocol selection, including the properties of the environment for an authentication protocol, and the resources of the authenticating entities. We then look at the ISO/IEC 9798 protocols in the context of requirements on the mechanisms. In Sections 3, 4, 5, 6, 7 and 8, we consider possible alternatives to the ISO/IEC 9798 schemes, which do not put such strict conditions on the mechanisms. In the final section, we give a summary of the contributions of this paper.

2 Factors affecting protocol selection

We now discuss two important factors which affect protocol design, namely the properties of the protocol environment and the resources of the authenticating entities. As we show in subsequent sections, if the environment and entity resources for authentication satisfy less stringent conditions than those required for the ISO/IEC 9798 protocols, then the ISO/IEC 9798 protocols cannot be used, and a different protocol, tailored to match the properties of the underlying mechanisms, needs to be designed.

2.1 Properties of the environment for protocols

Before considering what underlying mechanisms are available and selecting a protocol which uses them, the designer must know the environment in which the protocol will work. A particular environment may impose stringent requirements on the mechanisms and the protocol itself. For the purposes of this paper we consider the following aspects of the environment of an authentication protocol.

Communications channel. Communications channels are used to transmit messages during the process of authentication. The major property of interest here is whether a channel is broadcast or point-to-point.

⋄ *Broadcast channel*, where there exist messages from a variety of senders and/or for a variety of receivers. Typically, to make a broadcast channel operate correctly, the sender's and/or receiver's names have to be sent across the channel.

◊ *Point-to-point channel*, where the channel is reserved for a particular sender and receiver who both know the initiator and terminator of each message, so that their names do not need to be sent across the channel.

Other properties of the communications channel which affect the design of authentication protocol include whether or not interceptors can modify messages and/or introduce totally spurious messages. However we do not address these issues in detail here.

Entity identifier. The major property of interest here is which entities are authorised to know a particular entity identifier during the process of authentication. There are two main cases of interest, namely that the identifier of an entity is allowed to be transferred in clear text (e.g., ISO/IEC 9798 parts 2, 3 and 4) or that it is only allowed to be known to particular entities (e.g., [6, 13, 17]).

Trust relationship. The trust relationship describes whether one entity believes that the other entities will follow the protocol specifications correctly (e.g. [14]). The trust relationship of particular interest here is between authentication servers and clients. For instance, a client may not trust an individual server [9].

2.2 Resources of authenticating entities

We consider the following aspects of the resources of the authenticating entities.

Knowledge. An entity might have different kinds of knowledge at the start of authentication, namely shared secret information with another entity, private information (e.g. the private part of the entity's own asymmetric key pair), reliable public information (e.g. the public part of another entity's asymmetric key pair), or no knowledge of another entity.

Computational ability. The entities may or may not have the computational ability to perform certain operations, namely computation of complex cryptographic algorithms (e.g. a digital signature algorithm), random and secure generation of a key or an offset for a key, and generation of predictable or unpredictable nonces.

Time synchronisation. The entities may or may not have access to securely synchronised clocks or synchronised sequence numbers.

3 Absence of trust in individual third parties

We consider a situation where two entities, who share no secret based on symmetric cryptography or hold no public information based on asymmetric cryptography, want to complete unilateral or mutual authentication. Typically they will have to get assistance from a third party, referred to as an authentication server. However, in some environments, these clients have no reason to trust an individual server, and in such a case the ISO/IEC 9798 protocols cannot be used directly. In order to design a protocol which does not require trust in individual servers, a range of possible approaches can be followed.

Firstly, a client can choose which from a set of servers to trust, typically by applying a security policy or considering their history of performance and reliability. Yahalom et al. [21] proposed a protocol which allows a client or his agent to make such a choice. A limitation of this scheme is that a client may sometimes find it difficult to distinguish between trustworthy and untrustworthy servers.

Secondly, a set of moderately trusted servers who are trusted by users collectively, but not individually, can be used. Gong [11] proposed a protocol with multiple servers such that a minority of malicious and colluding servers cannot compromise security or disrupt service. In that protocol two clients participate in choosing a session key, and each relevant server is responsible for converting and distributing a part of the session key. Two variations on this approach have been described in [9]. In both variants the servers each generate a part of a session key, which can be successfully established between a pair of entities as long as more than half the servers are trustworthy. Both schemes from [9] have the advantage of requiring considerably fewer messages than Gong's protocol.

A third approach, based on asymmetric cryptography, is to separate transfer of authentication information from issue of authentication information, i.e. to let the issuer be off-line. One instance of this approach is where a master server (sometimes called the certification authority) issues a certificate which is then held by another on-line secondary server. The certificate is valid for a period of time, during which there is no need to further contact the master server, since the certificate is available from the secondary server. The client does not need to trust the secondary server, but does need to trust the off-line master server.

4   Entity identity privacy

During authentication, the identities of the entities concerned may need to be sent across the communications channel, either embedded within or alongside the protocol messages. There are two main reasons why this may be necessary.

- Depending on the nature of the communications channel, all messages may need to have one or more addresses attached. More specifically, if a broadcast channel is being used, then, in order for recipients to detect the messages intended for them, a recipient address must be attached. In addition, many authentication protocols require the recipient of a protocol message to know the identity of the entity which (claims to have) originated it, so that it can be processed correctly (e.g. deciphered using the appropriate key). If this information is not available, as would typically be the case in a broadcast environment, then the originator address also needs to be attached.
- Certain authentication protocols, including some of those in ISO/IEC 9798, require the recipient's name to be included within the protected part of some of the messages, in order to protect the protocol against certain types of attack.

However, depending on the nature of the communications channel, communicating entities may require a level of anonymity, which would prevent their name and/or address being sent across the channel (except in enciphered form). For example, in a mobile telecommunications environment, it may be important for users that their identifiers are not sent in clear across the radio path, since that would reveal their location to an interceptor of the radio communications.

Of the two reasons listed for sending names across the channel, the second is usually simpler to deal with, since alternative protocols can typically be devised which do not require the inclusion of names in protocol messages; for example, as described in ISO/IEC 9798-4, 'unidirectional keys' can be employed. We therefore focus our attention on the anonymity problems arising from the use of authentication protocols in broadcast environments. It is important to note that these anonymity problems are not dealt with in any published part of ISO/IEC 9798.

There are two main approaches to providing entity anonymity in broadcast channels, i.e., the use of temporary identities which change at regular intervals, and the use of asymmetric (public key) encipherment techniques. We now discuss three examples which make use of these approaches. In each case we consider a 'many-to-one' broadcast scenario, where many mobile users communicate with a single 'base' entity. In this case anonymity is typically only required for the mobile users who can only receive from the base, and hence there may be no need for the base address to be sent across the channel.

First observe that in GSM (Global System for Mobile Telecommunication) [13], temporary identities (TMSIs) are transmitted over the air interface instead of permanent identities (IMSIs). TMSIs are changed on each location update and on certain other occasions as defined by the network. A mobile user identifies himself by sending the old TMSI during each location update process, which has to be sent before authentication takes place and must therefore be sent unencrypted. However, the new TMSI is returned after authentication has been completed and a new session key has been generated so that it can be, and is, transmitted encrypted. In certain exceptional cases, including initial location registration, the user has to identify itself using its IMSI. In these cases an intruder may be able to obtain the IMSI from the GSM air interface. Thus the GSM system only provides a limited level of anonymity for mobile users.

Second consider a mutual authentication protocol, also outlined in [17], which has been designed for possible adoption by two third generation mobile systems, namely UMTS (Universal Mobile Telecommunications System) and FPLMTS (Future Public Land Mobile Telecommunication Systems). Like the GSM solution, this scheme also uses temporary identities to provide identity and location privacy. However, in this protocol, temporary identities are used at every authentication exchange including the case of a new registration, so that permanent identities are never transmitted in clear text.

The principals involved in this protocol are one of number of users, A, a single 'base' entity B and an authentication server S. The protocol makes use of two types of temporary identities: S-identities shared by A and S, including a current I_A and a new I'_A, and B-identities shared by A and B, also including a current J_A and a new J'_A. There are two versions of the protocol, depending on whether or not A and B already share a valid temporary B-identity J_A and secret key K_{AB}. The protocol makes use of five cryptographic check functions $F1$ - $F5$, each of which takes a key and a data string as input. Note also that $\oplus$ denotes bit-wise exclusive-or.

Version 1: A and B share K_{AB} and J_A. Then the message exchanges are as follows (where $M_i : A \rightarrow B : x$ means that the ith exchanged message is sent from A to B and contains data x).

$$M_1 : A \rightarrow B : J_A, R_A$$
$$M_2 : B \rightarrow A : R_B, F4_{K_{AB}}(R_A) \oplus J'_A, F3_{K_{AB}}(R_B, R_A, J'_A)$$
$$M_3 : A \rightarrow B : F3_{K_{AB}}(R_A, R_B)$$

Version 2: A and B share no secret, A and S share K_{AS} and I_A, and B and S have a secure channel which is available for exchanging messages 2 and 3.

$$M_1 : A \rightarrow B : I_A, R_A$$
$$M_2 : B \rightarrow S : I_A, R_A$$
$$M_3 : S \rightarrow B : F1_{K_{AS}}(R_A) \oplus I'_A, O, K_{AB}, F3_{K_{AS}}(I'_A, R_A, O)$$
$$M_4 : B \rightarrow A : F1_{K_{AS}}(R_A) \oplus I'_A, O, F3_{K_{AS}}(I'_A, R_A, O), R_B,$$
$$F4_{K_{AB}}(R_B) \oplus J'_A, F3_{K_{AB}}(R_B, R_A, J'_A)$$
$$M_5 : A \rightarrow B : F3_{K_{AB}}(R_A, R_B)$$

where O is a key offset so that $K_{AB} = F2_{K_{AS}}(O, B)$. The resulting session key is $K'_{AB} = F5_{K_{AB}}(R_A, R_B, J'_A)$.

Our third example is based on the use of asymmetric encipherment. In this case, no temporary sender identity is required in the first message because the real sender identity can be encrypted using the public encipherment transformation of the receiver. In order to keep the receiver identity confidential, a temporary receiver identity is needed. In this example, A is one of a number of users of a single broadcast channel which is used for communicating with a single 'base' entity B, E_X denotes public key encipherment using the public key of entity X and H denotes a hash-function. The messages are:

$$M_1 : A \rightarrow B : E_B(A, R_A, R'_A)$$
$$M_2 : B \rightarrow A : R'_A, E_A(R_B, R_A, B)$$
$$M_3 : A \rightarrow B : H(R_A, R_B, A)$$

where the nonce R'_A is a temporary identifier for A, which should be unpredictable for any third parties. In this protocol, the nonces R_A and R_B are used to meet two requirements, namely to verify the freshness of messages by using a challenge-response scheme and, possibly, to establish a session key shared between A and B. These nonces have to be unpredictable as well. Note also that A and B must have reliable copies of each other's public keys before starting the protocol.

5 Avoiding abuse of digital signatures

During authentication based on digital signatures with 'unpredictable' nonces, entity A typically challenges entity B by sending a nonce R_A. B then sends A a signature-based message containing this nonce in reply to the challenge. By choosing the nonce appropriately A may be able to use B's signature for malicious purposes.

Here, as throughout this paper, a digital signature function is defined to include use of either a hash-function or a redundancy function to prevent an impersonator claiming that a randomly chosen value is actually a signature.

The authentication protocols of ISO/IEC 9798-3 [2] discuss means of dealing with this problem, and to help avoid the worst consequences a nonce chosen by the signer can also be included in the relevant signature. However, the same problem may still exist. We now consider a protocol given in Clause 5.2.2 of ISO/IEC 9798-3 (see also [18]).

$$M_1 : B \rightarrow A : R_B, D_1$$
$$M_2 : A \rightarrow B : CertA, R_A, R_B, B, D_3, S_A(R_A, R_B, B, D_2)$$
$$M_3 : B \rightarrow A : CertB, R_B, R_A, A, D_5, S_B(R_B, R_A, A, D_4)$$

Note that D_1 - D_5 are (application dependent) data strings, S_X denotes the signature function of entity X, R_A and R_B are nonces, and $CertX$ denotes the certificate of entity X. R_A is present in the signed part of M_2 to prevent B from obtaining the signature of A on data chosen by B. Similarly R_B is present in the signed part of M_3. However, this approach cannot completely avoid the abuse of signatures for the following two reasons.

1. Although both nonces are included in both signatures, B selects its nonce before A. This means that A is in a more favourable position than B to misuse the other party's signature. To prevent this, B can generate an extra nonce and add it into the signed message, e.g. in M_3 a nonce R'_B can be included in D_4 and D_5:

$$M_3 : B \rightarrow A : CertB, R_B, R_A, A, R'_B, D'_5, S_B(R_B, R_A, A, R'_B, D'_4).$$

2. It is possible for users of the signatures to 'bypass' some nonces involved in the protocol if other signatures in different contexts use nonces in the same way. For example, a different protocol might make use of a message $S_A(R, X, B, D_2)$ or $S_B(R, Y, A, D_4)$, where R is a random number and X or Y has a particular meaning. The signatures of the previous protocol could then potentially be successfully abused in this protocol.

The above discussion implies that changing protocol construction only makes abuse of digital signatures more difficult, and cannot protect against such attacks completely, because the protocol itself cannot detect the misuse of digital signatures involved. However, there exist means of avoiding these problems, such as the following.

- 'Key separation' is a well known and widely used technique, i.e. using different keys for different applications (see, e.g. [18]).

- Another approach is using sequence numbers rather than unpredictable nonces to control the uniqueness of authentication exchanges. Because the values of the sequence numbers are agreed by both the claimant and verifier, neither party to a protocol can be persuaded to sign information which has some 'hidden meaning'.

- Last, but not least, observe that zero knowledge protocols are specifically designed to prevent this type of attack because they do not generate digital signatures. However, as Bengio et al. pointed out, these protocols are still open to interactive abuse through middleperson attacks [8]. It is then left to the implementation to ensure that those attacks are not feasible in practice.

6   Predictable and unpredictable nonces

The nonce-based protocols specified in Parts 2, 3 and 4 of ISO/IEC 9798 all require the nonces used to be unpredictable, i.e. the nonces must be generated in such a way that intercepting third parties cannot guess future nonce values. However, in some circumstances it may be necessary to use nonces which are predictable, i.e. generated using a deterministic process known to the interceptor. For example, it may be difficult for an entity to generate random or unpredictable pseudo-random numbers, particularly if the entity is implemented in a portable device, such as a smart card, and use of a simple counter for nonce generation may be the only practical possibility.

We now consider how secure nonce-based authentication protocols can be devised even if the nonces are predictable (as long as they are still 'one time'). This can be achieved by cryptographically protecting all the messages in a protocol including the nonces.

Before proceeding we briefly distinguish between predictable nonces and sequence numbers, both of which are used to control the uniqueness of messages. They are both used only once within a valid period of time and there is the possibility that they can be predicted in advance by a third party. However, a predictable nonce is used as a challenge, so that the responder does not need to know it before receiving it and to record it after using it. On the other hand, a sequence number is agreed by the claimant and verifier beforehand according to some policy, and will be rejected if it is not in accordance with the agreed policy. Furthermore, use of sequence numbers may require additional 'book keeping' for both the claimant and verifier. Typically every entity will need to store a 'send' sequence number and a 'receive' sequence number for each other entity with which they communicate.

It has been observed in [12] that in a protocol using symmetric encryption with a nonce, if the nonce is unpredictable then either the challenge or the response can be transmitted unencrypted; however if the nonce is predictable then both the challenge and response have to be encrypted. Otherwise the protocol cannot protect against replay attacks.

We now illustrate that this logic also applies to protocols using digital signatures and CCFs. The following example shows how digital signatures can be used in conjunction with a predictable nonce to produce a secure authentication protocol. This is a modification of the protocol given in Clause 5.2.2 of [2] (the notation is as used previously).

$$M_1 : B \rightarrow A : CertB, R_B, S_B(R_B)$$
$$M_2 : A \rightarrow B : CertA, R_A, S_A(R_A, R_B, B)$$
$$M_3 : B \rightarrow A : S_B(R_B, R_A, A)$$

Another example, this time based on the use of a CCFF, is the following modification of the protocol given in Clause 5.2.2 of [4].

$$M_1 : B \rightarrow A : R_B, F_K(R_B)$$
$$M_2 : A \rightarrow B : R_A, F_K(R_A, R_B, B)$$
$$M_3 : B \rightarrow A : F_K(R_B, R_A)$$

The above analysis means that there is a good range of protocols available to support both unpredictable and predictable nonces. Note that when using a predictable nonce as a challenge, since a future challenge is predictable to the responder, the verifier of the challenge has to depend on the honesty and competence of the responder [12].

7 Disclosure of plaintext/ciphertext pairs

There are a number of different models for known plaintext attacks, chosen plaintext attacks and chosen ciphertext attacks on cipher systems (e.g. [7] and [10]). Although obtaining plaintext/ciphertext pairs far from guarantees that attacks will be successful, it is typically the necessary first step for an attacker. Whether or not the attacker has access to a plaintext/ciphertext pair during authentication depends on the nature of the communications channel, the authentication protocol, and the details of the cryptographic processing.

We use the term 'plaintext/ciphertext pair' rather loosely here, to cover matching pairs of input and output for a variety of cryptographic algorithms, including encipherment algorithms, digital signatures and cryptographic check functions. Whether or not disclosing a plaintext/ciphertext pair is a problem depends on the strength of the algorithm, and whether the same algorithm and key are used for applications other than the authentication exchange.

In some situations the disclosure of plaintext/ciphertext pairs is not a security problem and that is what ISO/IEC 9798 assumes. However we are concerned here with situations where disclosure of pairs may be a security threat, and we consider ways of avoiding the threat. The following unilateral authentication protocols are given in ISO/IEC 9798 parts 2, 3 and 4.

Example 1: Symmetric encryption with nonce [3]:

$$M_1 : B \rightarrow A : R_B, D_1$$
$$M_2 : A \rightarrow B : D_3, E_{K_{AB}}(R_B, B, D_2)$$

Example 2: Digital signature with timestamp [2]:

$$M_1 : A \rightarrow B : T_A, B, D_1, S_A(T_A, B, D_2)$$

Example 3: CCF with sequence number [4]:

$$M_1 : A \rightarrow B : N_A, B, D_1, F_{K_{AB}}(N_A, B, D_2)$$

The first protocol is based on symmetric encryption. It depends on the optional text field D_2 whether a plaintext/ciphertext pair can be obtained. If D_2 is predictable data (including a null string), the plaintext/ciphertext pair is exposed. If D_2 includes some unpredictable data not supplied in D_1, it depends only on the structure of the cryptographic operation whether the plaintext/ciphertext pair is exposed.

In both the second protocol based on digital signature and the third protocol based on CCF, D_2 cannot be unpredictable for B, otherwise B cannot verify the signature and the cryptographic check value respectively. Those two protocols expose plaintext/ciphertext pairs.

Observe different time variant parameters based authentication protocols without an unpredictable optional text field placed in the cryptographically protected part of the token. In timestamp or sequence number based protocols, it is rather difficult to avoid disclosure of plaintext/ciphertext pairs, since the intruder can choose when to start the protocol or what predictable number is used in the protocol. When using an unpredictable nonce, the nonce has to be cryptographically protected in order to prevent the protocol disclosing a plaintext/ciphertext pair. The following examples of unilateral authentication protocols, which do not disclose plaintext/ciphertext pairs, do not depend on the predictability of D_1, D_2 and D_3.

Example 4: Symmetric encryption with nonce:

$$M_1 : B \rightarrow A : E_{K_{AB}}(R_B, D_1)$$
$$M_2 : A \rightarrow B : D_3, E_{K_{AB}}(R_B, B, D_2)$$

Example 5: Digital signature and asymmetric encipherment with nonce:

$$M_1 : A \rightarrow B : E_B(R_A, B, D_1, S_A(R_A, B, D_2))$$

Example 6: CCF with nonces:

$$M_1 : A \rightarrow B : R_A, B, D_1, F_{K_{AB}}(R_A, B, D_2) \oplus R'_A, F_{K_{AB}}(R'_A, D_3)$$

Note that the first nonce R_A in Example 6 can be replaced by a timestamp T_A or sequence number N_A.

8 Using poorly synchronised clocks

The authentication protocols with timestamps specified in ISO/IEC 9798 require the communicating parties to have synchronised clocks. There are several approaches to achieving secure clock synchronisation and re-synchronisation (e.g., [19, 20, 16]). However, in some environments, time stamp based protocols need to be used although the parties do not have exactly synchronised clocks. For example, in a mobile telecommunications system, a mobile user may find it difficult to keep a clock synchronised with the clock of its service provider. The user and network may still wish to use a timestamp-based authentication protocol rather than a nonce-based one to reduce the number of messages exchanged, and to allow the detection of forced delays.

Two points must be considered when using a timestamp-based protocol in such an environment. Firstly the size of the acceptance window to match poorly synchronised clocks must be selected. The size of this window can be either fixed or dynamically changed depending on the environment, in particular on the delays in the communications channels and the quality of the clocks in use. Secondly all messages within the current acceptance window must be logged, and second and subsequent occurrences of identical messages within that window must be rejected (see Annex B of [3] and [15]).

9 Summary

This paper discusses how the properties of underlying mechanisms affect the design of authentication protocols and how to tailor authentication protocols to match underlying mechanisms. A number of alternatives to the ISO/IEC 9798 protocols, which do not put such strict conditions on the underlying mechanisms, have been proposed and analysed. We now summarise them as follows.

◇ If authentication servers are not trusted by clients individually, three possible approaches are: (1) allowing clients to choose trustworthy servers, (2) using a set of moderately trusted servers instead of a single one, (3) using off-line master servers.

◇ In order to preserve entity identity privacy, two possible methods are: (1) based on asymmetric cryptography the entity identity can be hidden by using the public part of the receiver's asymmetric key pair, (2) one or more temporary entity identities instead of the real entity identity can be transmitted unencrypted.

◇ In order to avoid abuse of digital signatures, four possible approaches are: (1) let the signer generate an unpredictable nonce and insert the nonce into the signed message, (2) use different keys for different applications, (3) use sequence numbers rather than unpredictable nonces to control the uniqueness of authentication exchanges, (4) use zero knowledge techniques.

◇ When using a predictable nonce as a challenge, all messages, both the challenge and response, have to be protected cryptographically. Some possible examples are: (1) in a symmetric encryption based protocol, the challenge and response are respectively a nonce and a function of the nonce encrypted by the shared key, (2) in a digital signature based protocol, the private parts of the challenger's and responder's asymmetric key pairs are used in the challenge and response respectively, (3) in a CCF based protocol, the challenge is a nonce concatenated with a CCF of the nonce, and the response is a CCF of a function of the nonce.

⋄ In unpredictable nonce based protocols without unpredictable optional text fields it is possible to avoid disclosing plaintext/ciphertext pairs by using cryptographic protection for every message. In timestamp or sequence number based protocols without unpredictable optional text fields, it appears to be rather difficult to avoid giving plaintext/ciphertext pairs.

⋄ When using poorly synchronised clocks in authentication protocols, one approach is to take the following steps: (1) select the size of the acceptance window to match the poorly synchronised clocks, (2) log all messages and reject the second and subsequent occurrences of identical messages within the acceptance window.

10   Acknowledgements

The authors would like to thank Volker Kessler and Günther Horn for a useful suggestion which improved the authentication protocol using asymmetric encipherment in Section 4, and thank anonymous referees and Anish Mathuria for careful comments on an earlier version of this paper.

References

1. ISO/IEC 9798-1: 1991. Information technology — Security techniques — Entity authentication mechanisms — Part 1: General model. September 1991.
2. ISO/IEC 9798-3: 1993. Information technology — Security techniques — Entity authentication mechanisms — Part 3: Entity authentication using a public key algorithm. November 1993.
3. ISO/IEC 9798-2: 1994. Information technology — Security techniques — Entity authentication — Part 2: Mechanisms using symmetric encipherment algorithms. December 1994.
4. ISO/IEC 9798-4: 1995. Information technology — Security techniques — Entity authentication — Part 4: Mechanisms using a cryptographic check function. March 1995.
5. ISO/IEC 2nd CD 9798-5: 1995. Information technology — Security techniques — Entity authentication — Part 5: Mechanisms using zero knowledge techniques. June 1996.
6. M.J. Beller, L. Chang, and Y. Yacobi. Privacy and authentication on a portable communications system. *IEEE Journal on Selected Areas in Communications*, 11:821–829, 1993.
7. S.M. Bellovin and M. Merritt. Limitations of the Kerberos authentication system. *Computer Communication Review*, 20(5):119–132, October 1990.

8. S. Bengio, G. Brassard, Y.G. Desmedt, C. Goutier, and J. Quisquater. Secure implementation of identification systems. *Journal of Cryptology*, 4:175–183, 1991.

9. L. Chen, D. Gollmann, and C. Mitchell. Key distribution without individual trusted authentication servers. In *Proceedings: the 8th IEEE Computer Security Foundations Workshop*, pages 30–36. IEEE Computer Society Press, Los Alamitos, California, June 1995.

10. I. Damgard. Towards practical public key systems secure against chosen ciphertext attacks. *Lecture Notes in Computer Science 576, Advances in Cryptology - CRYPTO '91*, pages 445–456, 1991.

11. L. Gong. Increasing availability and security of an authentication service. *IEEE Journal on Selected Areas in Communications*, 11:657–662, 1993.

12. L. Gong. Variations on the themes of message freshness and replay. In *Proceedings: the Computer Security Foundations Workshop VI*, pages 131–136. IEEE Computer Society Press, Los Alamitos, California, June 1993.

13. ETSI/PT12 GSM-03.20. Security related network functions. August 1992.

14. B. Klein, M. Otten, and T. Beth. Conference key distribution protocols in distributed systems. In P. G. Farrell, editor, *Codes and Cyphers, Proceedings of the Fourth IMA Conference on Cryptography and Coding*, pages 225–241. Formara Limited. Southend-on-sea. Essex, 1995.

15. K-Y. Lam. Building an authentication service for distributed systems. *Journal of Computer Security*, 2:73–84, 1993.

16. K.Y. Lam and T. Beth. Timely authentication in distributed systems. In *Lecture Notes in Computer Science 648, Advances in European Symposium on Research in Computer Security*, pages 293–303. Springer-Verlag, 1992.

17. C. Mitchell. Security in future mobile networks. The 2nd International Workshop on Mobile Multi-Media Communications (MoMuC-2), Bristol University, April 11th-13th 1995.

18. C. Mitchell and A. Thomas. Standardising authentication protocols based on public key techniques. *Journal of Computer Security*, 2:23–36, 1993.

19. M. Reiter, K. Birman, and R. van Renesse. Fault-tolerant key distribution. Technical Report 93-1325, Department of Computer Science, Cornell University, Ithaca, New York, January 1993.

20. B. Simons, J.L. Welch, and N. Lynch. An overview of clock synchronization. In *Lecture Notes in Computer Science 448, Advances in Fault-Tolerant Distributed Computing*. Springer-Verlag, 1990.

21. R. Yahalom, B. Klein, and T. Beth. Trust-based navigation in distributed systems. European Institute for System Security, Karlsruhe University, Technical Report 93/4, 1993.

On the Design of Security Protocols for Mobile Communications

Yi Mu and Vijay Varadharajan

Department of Computing, University of Western Sydney,
Nepean, Kingswood, NSW 2747, Australia
Email: {yimu, vijay}@st.nepean.uws.edu.au

Abstract. Use of mobile personal computers in open networked environment is revolutionalising the way we use computers. Mobile networked computing is raising important information security and privacy issues. This paper is concerned with the design of authentication protocols for a mobile computing environment. The paper first analyses the authentication initiator protocols proposed by Beller,Chang and Yacobi (BCY) and the modifications considered by Carlsen and points out some weaknesses. The paper then suggests improvements to these protocols. The paper proposes secure end-to-end protocols between mobile users using both symmetric and public key based systems. These protocols enable mutual authentication and establish a shared secret key between mobile users. Furthermore, these protocols provide a certain degree of anonymity of the communicating users to be achieved vis-a-vis other system users.

1 Introduction

The rapid progress of portable communications technology is making mobile personal computing increasing popular [4]. In the near future the use of mobile computers combined with wireless access will be a common way of interaction between users. Mobile communications is more vulnerable to security attacks such as interception and unauthorised access than fixed network communications [9, 6]. Hence services for securing mobile communications are vital in guaranteeing authentication and privacy of legitimate users. [9] considers the provision of key security services for the portable mobile computing environment. These include authentication service to counteract masquerading, confidentiality and integrity services to protect information against unauthorised eavesdropping and modification, authorisation to control resource access and usage, and non-repudiation and auditing services.

Beller, Chang, Yacobi (BCY) [1, 2] have recently proposed a set of authentication and key distribution protocols, specficaly aimed at implementation in a low-cost portable telephone. In principle, these protocols are also applicable to portable mobile computers. Carlsen proposed some improvements to these protocols in [3]. In this paper, we focus on the design of authentication and key distribution protocols for mobile computing environment. We consider the interception attacks against the BCY's and Carlsen's authentication initiator protocols. We also consider ways of providing anonymity of legitimate mobile users participating in a communication to other users in the system. We will see that Carlsen modified BCY protocols have some vulnerabilities when it comes to interception attacks and furthermore they do not provide an appropriate level of anonymity of legitimate users. This paper provides improved solutions to these problems.

The paper is organised as follows: Section 2 outlines the mobile computing environment and describes the the interception attacks and the anonymity concerns in this environment. Section 3 considers the interception attacks against the BCY and Carlsen's protocols, and presents improvements to counteract these vulnerabilities. Section 4 proposes new end-to-end protocols which are secure against interception attacks as well as providing the desired level of user anonymity. Finally, Section 5 gives our conclusions.

2 Mobile Computing Environment

A simple mobile computing environment is shown in Figure 1. Mobile Computing Stations (MS) access the mobile network via a mobile network system. For instance, the network system may consist of Base Stations, Location Register, and Mobile Switching Component. The Location Register contains information related to the location and the subscription of the users in its domain. We will assume that an Authentication Server is present in every domain. This is logically a distinct entity; in practice, it may be co-located with the Location Register. The Authentication Servers store confidential information such as keys and are assumed to be physically protected.

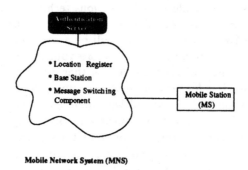

Fig. 1. Mobile Networked Computing Environment.

The mobile stations can move from one place to another, either within its domain or between domains. The usual place of the user is referred to as the "home" domain and the visiting place is referred to as the "visiting" domain. The base stations are controlled and managed by entities called Home Location Register (HLR) or Visiting Location Register (VLR). A combined entity of BSs and the corresponding local HLR is sometimes called a Home Server

(HS). Similarly, we will refer to a Visiting Server (VS) in a visiting domain. A symmetric key based protocol uses an on-line Authentication Server (HAS or VAS). In some cases, the Authentication Servers may be part of the Location Registers, but logically they are two separate entities. The Authentication Servers are assumed to be physically protected and are trusted entities.

2.1 Security Issues

We identify two types of adversaries, namely "insider" and "outsider" , who can attack the system. The outsider is an attacker who can only mount attacks over the air interface (such as eavesdropping and modification) whereas the insider is an attacker who can also mount attacks within the system, such as system tampering and intrusion as well as illegal information modification. For instance, a malicious system operator is regarded as an insider. The mobile network and the systems should be secure against attacks that can be mounted by both insiders and outsiders.

Let us now consider the aspects of interception, anonymity, and authentication in this environment.

An *interception* attack involves an attacker gaining access to information for which s/he is not authorised. For instance, the information could be the key in a key agreement protocol. Consider the interception attack on a simple public key based system described below.

Assume that we have two legitimate communicating users Alice and Bob and an attacker Eve. (1) Alice sends Bob her public key. Eve intercepts the key and sends her own public key to Bob. Bob is not aware that he is receiving a fake key. (2) Bob sends Alice his public key. Eve intercepts it and sends instead her own public key to Alice. Alice is also not aware of this fraudulent process. (3) When Alice communicates with Bob, she uses Eve's public key instead of Bob's and vice versa. (4) Eve is able to decrypt Alice's message and then she encrypts it using Bob's public key and sends it to Bob. Similarly, Eve decrypts Bob's message and encrypts it using Alice's public key, and then sends it to Alice. This has led to two shared channels, one between Alice and Eve, and the other between Bob and Eve.

Another important concern in the mobile environment is the anonymity of users. We consider the following anonymity requirement. The identity of a communicating user is known only to the user herself, to the communicating partner and to the home mobile network service, HS and HAS. Other entities such as the VS in the visiting domain as well as all other users should not have access to the communicating users' identities. To address this issue, we introduce the notion of a subliminal identity, written as ID_{sub}. Each user is issued a subliminal identity; in a symmetric key system, this is issued by the HAS, whereas in a public key system HS issues this identity. We will see below how the subliminal identity is used in the authentication protocols.

The protocols described in this paper employ both symmetric as well as public key based systems. Any of the well known symmetric key algorithms that are available (such as DES) can be used. In the case of public key systems, in this paper, we will use the Modular Square Root (MSR) technique [7, 10], and the Diffie-Hellman (DH) public key algorithm [5]. Under the MSR, the modulus of the public key system is a product of two large prime numbers: $N = pq$, p and q are primes. If a user A wishes to send a message x to user B who knows the factorisation of N, A calculates $m \equiv x^2 \bmod N$ and sends it to B. B can decrypt by taking "Modular Square Root", which we denote $\sqrt{m} \bmod N$. In order to take modular square roots efficiently, it is necessary to know the prime factors of the modulus. MSR can also be used to provide digital signatures. If B knows the factorisation of N, then s/he can sign a message m by calculating $x \equiv \sqrt{m} \bmod N$. When A receives x, s/he can verify the signature by computing $m = x^2 \bmod N$.

3 BCY and Carlsen Protocols

In this section, we analyse the BCY [1, 2] and Carlsen [3] protocols and describes some of the problems associated with them.

3.1 Notation

The following notations are used in the description of the protocols.

- A, B, C, D: End-users.
- A_{sub}: Subliminal identity of end-user A.
- HS: Home Domain Server.
- VS: Visiting Domain Server.
- HAS: Home Authentication Server.
- VAS: Visiting Domain Authentication Server.
- TA: Trusted Certification Authority.
- $E_{key}(content)$: $content$ encrypted with a symmetric key
- $D_{key}(content)$: $content$ decrypted with symmetric key
- k_A: Private key of user A
- $k_{A,B}$ or K_{AB}: Shared symmetric key between A and B.
- K_A: Public key of user A.
- k_s: Session key.
- h: A strong one-way hash function.
- n_A: Nonce generated by user A.
- $Cert_A$: public key certificate signed by a TA.
- T_A : Timestamp generated by A.
- $A \rightarrow B$: $message$: A sends $message$ to B.

3.2 Symmetric Key Protocols

Symmetric key cryptography is particularly suitable for situations where minimal computer power and less computational time are required. These were the important considerations behind the choice of symmetric key based techniques in systems such as the GSM (Group Special Mobile of the European Telecommunications Standard Institute - ETSI), the DECT (Digital European Cordless Telephone), and the interim Standard IS-54 of the Telecommunications Industries Association (TIA) for U.S. Digital Cellular.

Carlsen improved BCY symmetric key based initiator protocol is given in Table 1. In the protocol, the initiator A sends her identity to HS which then passes it to HAS in cleartext. HAS generates a session key k_s and distributes it to HS and A (via HS). This process establishes a secure channel between HS and A.

Table 1. Carlsen improved BCY Initiator Protocol

$$
\begin{array}{ll}
1: & A \rightarrow HS: A, n_A \\
2: & HS \rightarrow HAS: A, n_A, HS, n_{HS} \\
3: & HAS \rightarrow HS: E_{k_{HS}}(k_s, n_{HS}, A), E_{k_A}(n_A, HS, k_s) \\
4: & HS \rightarrow A: E_{k_A}(n_A, HS, k_s), E_{k_s}(n_A), n'_{HS} \\
5: & A \rightarrow HS: E_{k_s}(n'_{HS})
\end{array}
$$

First let us consider the issue of user anonymity. Note that the identity of the user (initiator) is revealed to other users of the system in the above protocol, as any other user can obtain the cleartext user identity. We introduce the notion of a subliminal identity for the user for a particular communication session. The subliminal identity is issued to the user by the HS (in a public key system) or the HAS (in a symmetric key system). This quantity is composed of a number (e.g. a sequence number) along with a timestamp. This will allow the HS or the HAS to perform efficient search of the database when required to locate a specific subliminal identity. Only the HS and/or the HAS know the mapping between this subliminal identity and the real user identity. The use of subliminal identities help to conceal the real user identities to outsiders. It is updated at the end of each session.

Let us now consider the Improved Initiator Protocol (IIP) shown in Table 2. In Step 1, the user A sends her subliminal identity A_{sub}, which was issued previously by HAS. Note that HS is not aware of the mapping from A_{sub} to A[1]. In Step 3, HAS sends to HS the user A's identity along with the session key k_s encrypted using HS's secret key k_{HS}. Furthermore, HAS issues to A a new subliminal identity (A'_{sub}) for future use, along with the session key k_s encrypted under k_A. This leads to the establishment of a shared session key between A and HS.

First note that the use of the subliminal identity helps to conceal the real identity of the initiator to other system users. Second, there is an important

---

[1] We may assume that HS also knows this mapping; however in this paper, we will not consider this option.

Table 2. Improved Initiator Protocol (IIP)

1:	$A \rightarrow HS$	: $A_{sub}, HS, n_A, h_0,$
		where $h_0 = E_{k_A}(A, h(A_{sub}, HS, n_A))$
2:	$HS \rightarrow HAS$	: $HS, HAS, A_{sub}, n_A, n_{HS}, h_0, E_{k_{HS}}(h(HS, HAS, n_{HS}))$
3:	$HAS \rightarrow HS$	: $HAS, HS, n_{HS}, n_A, e_1, e_2, h_1, h_2,$
		where $e_1 = E_{k_A}(k_s, A'_{sub}),$
		$e_2 = E_{k_{HS}}(k_s, A),$
		$h_1 = E_{k_A}(h(A'_{sub}, HAS, n_A, k_s)),$
		$h_2 = E_{k_{HS}}(h(A, HS, HAS, n_{HS}, n_A, k_s))$
4:	$HS \rightarrow A$	: $HS, A_{sub}, n'_{HS}, e_1, h_1,$
		$E_{k_s}(h(A_{sub}, HS, n'_{HS}))$
5:	$A \rightarrow HS$	: $A_{sub}, HS, E_{k_s}(h(A_{sub}, HS \underline{n'_{HS} + 1})$

difference between this improved protocol and the earlier version. In our protocol, we have carefully separated the information which needs to be signed (for integrity and authentication) from that which needs to be encrypted (for confidentiality). It is particularly important to adhere to this principle in the design of protocols; mixing of these two aspects leads to lack of clarity in protocol design which is often an important source for protocol flaws. Furthermore this separation is useful when it comes to obtaining export licenses where it is necessary to justify to the authorities the functionality of the various cryptographic interfaces and their use.

After A and HS obtain the session key k_s, A can have a secure conversation with HS and request to connect to user B. We will consider this stage of the protocol when we discuss secure end-to-end protocols in Section 4.

3.3 Public Key Protocols

A disadvantage of any symmetric key based protocol is the need for an on-line Authentication Server; in a public key system, this requirement can be simplified and the authentication can be done in an offline manner[2]. On the other hand, public key systems are more computationally intensive compared to the symmetric key based ones. However, the public key technology employed by the BCY protocols is promising for mobile communications due to their reduced complexity.

Table 3 presents the public key protocol by BCY and subsequently improved by Carlsen [3]. It is based on the Modular Square Root (MSR) technique developed by Rabin [7] and the Diffie-Hellman public key system. We will refer to it as the MSR+DH protocol. This protocol requires an offline trusted authority TA which provides the certificates for users.

---

[2] However in some circumstances, it may be necessary to have some part of the server to be on-line for revocation purposes. We will not discuss this aspect in this paper.

Table 3. MSR+DH protocol.

1: $HS \rightarrow A : n_{HS}, N_{HS}, P_{HS}, T_{HS}, e_1$

where $e_1 = \sqrt{h(N_{HS}, P_{HS}, T_{HS})} \bmod N_u$ and $P_{HS} = \alpha^{k_{HS}} \bmod N$

2:     A : retrieves local time T_A

verifies $T_A \leq T_{HS}$

calculates $h(N_{HS}, P_{HS}, T_{HS})$

verifies $h(N_{HS}, P_{HS}, T_{HS}) = e_1^2 \bmod N_u$

3:     A : chooses random x

encrypts $e_2 = x^2 \bmod N_{HS}$

encrypts $e_3 = E_x(n_{HS}, A, P_A, e_4)$

where $e_4 = \sqrt{h(A, P_A)} \bmod N_u$ and $P_A = \alpha^{k_A} \bmod N$

4: $A \rightarrow HS : e_2, e_3$

5:     HS : decrypts $x = \sqrt{e_2} \bmod N_{HS}$

decrypts $D_x(e_3) = n_{HS}, A, P_A, e_4$

verifies n_{HS} and $h(A, P_A) = e_4^2 \bmod N_u$

6:     A : calculates $\eta = (P_{HS})^{k_A} \bmod N$ and $k_s = E_\eta(x)$

      HS : calculates $\eta = (P_A)^{k_{HS}} \bmod N$ and $k_s = E_\eta(x)$

Let $P_A = \alpha^{k_A} \bmod N$ be the public key of A and $P_{HS} = \alpha^{k_{HS}} \bmod N$ be the public key for HS. N provided by TA is a large integer composed of two large primes. In Step 1, HS sends the concatenation of n_B, N_{HS}, P_{HS}, T_{HS}, e_1 to A, where h is a publicly known strong one-way hash function, e_1 represents the certificate of HS, t_{HS} denotes the expiration time of the certificate and N_u is public information. The verification of $h(N_{HS}, P_{HS}, T_{HS}) = e_1^2 \bmod N_u$ in Step 2 ensures that HS is authenticated. The verification of n_{HS} and $h(A, P_A) = e_4^2 \bmod N_u$ in Step 5 ensures that A is authenticated. Finally, in Step 6, the session key k_s to be shared by HS and A is calculated using the DH algorithm.

The protocol as it stands is not secure against an interception attack. Consider an attacker E residing between A and HS intercepting the message flows. Intercepting the message in Step 1 can lead to the following attack: The attacker E intercepts the message from HS and sends a modified message (with $n'_{HS}, N'_{HS}, P'_{HS}, t'_{HS}, e'_1$) to A. As E can be a legal user of the system, he has his own certificate e'_1 (signed by TA). When this attack is successful, E can further impersonate the communications between HS and A.

1: $HS \rightarrow E : n_A, N_{HS}, P_{HS}, t_{HS}, e_1$

$(e_1 = \sqrt{h(N_{HS}, P_{HS}, t_{HS})} \bmod N_u$ and $P_{HS} = \alpha^{k_{HS}} \bmod N)$

2: $E \rightarrow A : n'_{HS}, N'_{HS}, P'_{HS}, t'_{HS}, e'_1$

$(e'_1 = \sqrt{h(N'_{HS}, P'_{HS}, t'_{HS})} \bmod N_u$ and $P'_{HS} = \alpha^{k_E} \bmod N)$

The reason for this weakness in the protocol is due to the fact that the user's identity is not included in the signed certificate. Therefore this problem can be solved by including the user identity within the signed e_1.

$$e_1 = \sqrt{h(HS, N_{HS}, P_{HS}, t_{HS})} \bmod N_u.$$

Furthermore, it will also be useful to modify e_4 in the protocol by including an expiration time T_A for P_A.

4 Secure End-to-End Protocols

So far we have been considering security services from one user A to a mobile network service authority such as the HS or the HAS. From a user point of view, in a mobile computing environment, securing the end-to-end path from one mobile user to another is the primary concern. The end-to-end security service minimises the interferences from the operator controlled network components. In this section, we present secure end-to-end authentication and key distribution protocols between two mobile users. We consider both symmetric key and public key based approaches.

4.1 Symmetric Key Based End-to-End Protocol

We need to consider two scenarios: the intra-domain and the inter-domain. The intra-domain case considers the establishment of a secure communication between two mobile station users within a single domain. The inter-domain case addresses the situation where the two mobile station users wishing to have a secure communication reside in two different domains. In this paper, we only consider the intra-domain protocol here. The inter-domain protocols are described in [8]. Before describing the intra-domain protocol, let us briefly describe the assumptions and the parameters used in the protocol.

Assumptions and System Parameters:

Mobile Station Users A or B: Belong to HS.
   A has subliminal identity A_{sub} issued by HAS and a secret symmetric key k_A.
   B has subliminal identity B_{sub} issued by HAS and a secret symmetric key k_B.

Home Server HS: Does not know the secret keys of A and B.
   Does not have the mapping from the subliminal identities to real identities.
   Has its secret symmetric key k_{HS}.
Home Authentication Server HAS: Keeps the mapping of user subliminal identities to real user identities. Keeps the secret symmetric keys of the users in its domain. Keeps the secret symmetric key of its domain's HS.

The end-to-end intra-domain protocol is given in Table 4. The main objective of this protocol is to provide mutual authentication between mobile station users A and B, and to establish a secret shared conversation key k_{AB} between them.

In the first two steps, A authenticates herself to HAS through HS and requests a key to communicate with B. The nonce n_A is used by A to identify its request for communication with B. Upon verification of the request in Step 3, HAS carries out the following tasks: It generates a secret conversation key k_{AB}, which is encrypted under k_A and k_B separately for distribution to A and B respectively. HAS also generates a session key k_s which is encrypted under HS's secret key k_{HS}. The session key is used to provide a secure channel between A and HS for that particular session. HAS also generates new subliminal identity A'_{sub} which will be used by A in its next communication.

In Step 4, HS passes the secret conversation key k_{AB} and A'_{sub} encrypted under A's key to A. Note that HS is not able to read either of them. Step 5 is a response from A to HS acknowledging that she has received the message in Step 4.

In Step 6, HS sends the secret conversation key k_{AB} encrypted under B's key (sent by HAS) and B_{sub} to B. Step 7 is a response from B to HS acknowledging that he has received the message in Step 6.

Message in Steps 8, 9, and 10 conclude the setup phase. In step 8, HS informs A that B has successfully received the message which contains the conversation key k_{AB}. After receiving the response from A in Step 9, HS confirms to B in Step 10 that A has also obtained k_{AB}.

End-to-end symmetric key based protocol allowing the establishment of a secret conversation key between two users in different domains is given in [8]. We have considered two cases. The first case describes the situation whereby a user A residing in domain HS wishes to communicate with a user B residing in domain VS. The second case considers the situation where the user A in domain HS travels to domain $VS1$ and wishes to communicate with a user C in domain $VS2$.

4.2 Public Key Based End-to-End Protocol

Let us now consider the intra-domain end-to-end authentication protocol using a public key approach. Once again let us start by outlining the system parameters and the assumptions.

Assumptions and System parameters:

TA: is an offline Trusted Certification Authority
- Provides a reliable guarantee of the mapping between a user and his public key, by producing a signed certificate using the MSR algorithm.

A, B: End users
- Each user has a public key certificate signed by TA. For instance, A's certificate, $Cert_A$ is given by $h(K_A, T_A, A)$ signed using k_{TA}, where $K_A = g^{k_A} \bmod N$ is A's Diffie-Hellman public key and k_A is the corresponding private key, k_{TA} is the private key of TA, and g and N are publicly known parameters. T_A is the expiry time of the public key of A.
- Each user has TA's public key, K_{TA}.

Table 4. Symmetric Key End-to-End Protocol within the Home Domain

---

Initiator ($A \leftrightarrow HS$):

1:    $A \rightarrow HS : A_{sub}, HS, e_1, n_A, h_1,$

       where $e_1 = E_{k_A}(A, B)$, $h_1 = E_{k_A}(h(A, B, A_{sub}, HS, n_A))$

2: $HS \rightarrow HAS : HS, HAS, A_{sub}, e_1, n_A, n_{HS}, h_1,$

       $E_{k_{HS}}(h(HS, HAS, A_{sub}, n_{HS}))$

3: $HAS \rightarrow HS : HAS, HS, e_2, e_3, e_4, h_2, h_3, h_4, n_A, n_{HS}$

       where $e_2 = E_{k_{HS}}(A, B, B_{sub}, k_s)$,

       $e_3 = E_{k_A}(A'_{sub}, B, k_s, k_{AB})$,

       $e_4 = E_{k_B}(A, k_s, k_{AB})$,

       $h_2 = E_{k_{HS}}(h(A, B, B_{sub}, HAS, HS, n_{HS}, k_s))$

       $h_3 = E_{k_A}(h(HAS, A'_{sub}, B, n_A, k_s, k_{AB}))$,

       $h_4 = E_{k_B}(h(HAS, A, n_A, k_s, k_{AB}))$

4:    $HS \rightarrow A : HS, A_{sub}, e_3, n_A, n'_{HS}, h_3,$

       $E_{k_s}(h(HS, A_{sub}, n_A, n'_{HS}))$

5:    $A \rightarrow HS : A_{sub}, HS, E_{k_s}(h(A_{sub}, HS, n'_{HS}))$

Response ($HS \leftrightarrow B$):

6:    $HS \rightarrow B : HS, B_{sub}, e_4, h_4, n_A, n''_{HS}, E_{k_s}(h(HS, B_{sub}, n''_{HS}))$

7:    $B \rightarrow HS : B_{sub}, HS, n_A, n_B, n''_{HS},$

       $E_{k_s}(h(B_{sub}, HS, n''_{HS}))$, $E_{k_{AB}}(h(A, B, n_A, n_B))$

Finish-off: ($A \leftrightarrow B$):

8:    $HS \rightarrow A : HS, A_{sub}, n_A, n_B, n'''_{HS}, E_{k_s}(h(HS, A_{sub}, n_A, n_B, n'''_{HS}))$,

       $E_{k_{AB}}(h(A, B, n_A, n_B))$

9:    $A \rightarrow HS : A_{sub}, HS, n_B, n'''_{HS}, E_{k_{AB}}(h(B, n_B)), E_{k_s}(h(A_{sub}, HS, n'''_{HS}))$

10: $HS \rightarrow B: HS, B_{sub}, n^*, E_{k_{AB}}(h(B, n_B)), E_{k_s}(h(HS, B_{sub}, n^*))$

- Each user has a subliminal identity, A_{sub}, which was issued by the HS. This could have been issued during the last connection or obtained at the time of initial registration.

HS: Home Server,

    - Has a public key certificate signed by TA: $Cert_{HS} : h(K_{HS}, T_{HS}, HS)$ signed using k_{TA}, where $K_{HS} = g^{k_{HS}} \bmod N$ is the Diffie-Hellman public key of HS and k_{HS} is the corresponding private key. T_{HS} is the expiry time of the public key of HS.

    - Has the user identities to subliminal identities mappings.

    - Has TA's public key, K_{TA}.

The protocol is shown in Table 5. In Phase one, A and HS establish a session key and A informs HS her wish to communicate with B. In Step 1, the initiator A sends HS her public key certificate, K_A, T_A and a nonce n_A. Upon verification of the public key K_A using the certificate, HS calculates the session key $k_{A,HS} = K_A^{k_{HS}} \bmod N$, which will be shared by A and HS. In Step

Table 5. Public Key based End-to-End Protocol with A as the Initiator

Phase one: $A \leftrightarrow HS$

1:          $A \rightarrow HS$: $A_{sub}, HS, Cert_A, n_A, K_A, T_A, Ek_A(h(A_{sub}, HS, n_A))$

2:          $HS \rightarrow A$: $HS, A_{sub}, Cert_{HS}, K_{HS}, T_{HS}, n_{HS}, Ek_{HS}(h(HS, A_{sub}, n_A, n_{HS}))$

3:          $A \rightarrow HS$: $A_{sub}, HS, E k_{A,HS}(h(A_{sub}, B, HS, n_{HS})), E k_{A,HS}(B)$

Phase two: $HS \leftrightarrow B$

4:          $HS \rightarrow B$: $HS, B_{sub}, Cert_{HS}, K_{HS}, T_{HS}, Cert_A, K_A, T_A, n'_{HS}, n_A$

                       $E k_{HS}(h(HS, B_{sub}, n'_{HS}, n_A))$

5:          $B \rightarrow HS$: $B_{sub}, HS, n_A, n_B, E k_{B,HS}(h(B_{sub}, HS, n_A, n_B, n'_{HS})),$

                       $E k_{A,B}(h(A, B, n_A, n_B))$

Phase three: $A \leftrightarrow B$

6:          $HS \rightarrow A$: $HS, A_{sub}, Cert_B, K_B, T_B, n_A, n_B, n''_{HS}, E k_{A,HS}(A'_{sub}),$

                       $E k_{A,HS}(h(HS, A_{sub}, A'_{sub}, n_A, n_B, n''_{HS})), E k_{A,B}(h(A, B, n_A, n_B))$

7:          $A \rightarrow B$: $n_B, E k_{A,B}(A, B, n_B + 1)$

2, HS sends its certificate and its signature including the nonce n_A and n_{HS}. Upon receiving this message, A calculates the session key $k_{A,HS} = K_{HS}^{k_A} \bmod N$. Using $k_{A,HS}$, A then sends a message to HS requesting communication with B.

In Phase two, HS and B establish a session key and B is informed of the request from A to communicate with him. In Step 4, HS sends both its certificate as well as A's certificate, along with the nonces n_{HS} and n_A. Upon verification of the public key of HS using the certificate, B calculates the session key $k_{B,HS} = K_{HS}^{k_B} \bmod N$. B also calculates the session key $k_{A,B}$ which will be shared between A and B. B uses $k_{A,B}$ in Step 5 to confirm to HS that it accepts A's request for communication.

In Phase three, the setup process is concluded. In Step 6, HS passes to A the part of the message in Step 5 which contains B's confirmation using $k_{A,B}$. HS also sends A the certificate of B and issues a new subliminal identity A'_{sub} for A. In Step 7, A communicates with B by sending back B's nonce suitably modified and encrypted using the shared key $k_{A,B}$.

5 Concluding Remarks

This paper has addressed the design of security protocols for mobile computing environment. First, the paper analysed Beller, Chang, Yacobi's and Carlsen's authentication initiator protocols and highlighted some of their weaknesses. The paper then suggested improvements to these protocols to counteract interception attacks and to preserve the anonymity of users against outside attackers. The paper then proposed secure end-to-end protocols between mobile users in an intra-domain environment. It described the design of such protocols using both symmetric and public key based systems. These protocols enable mutual authentication as well as the establishment of a shared secret key between mobile users. Furthermore, these protocols provide a certain degree of anonymity of the communicating users to other system users. This has been done by introducing the notion of subliminal identities. The inter-domain situations have been considered in [8].

References

1. Beller, M. J., Chang, L.-F., Yacobi, Y.: Privacy and authentication on a portable communications system. IEEE Journal on Selected Areas in Communications **11** (1993) 821–829.
2. Beller, M. J., Yacobi, Y.: Fully-fledged two-way public key authentication and key agreement for low-cost terminals. Electronics Letters **29** (1993) 999–1001.
3. Carlsen, U.: Optimal privacy and authentication on a portable communications system. ACM Operating Systems Review **28** (1994) 16–23.
4. Cox, D. C.: Portable digital radio communication - an approach to tetherless access. IEEE Communications Magazine **27** (1990).
5. Diffie, W., Hellman, M.: New directions in cryptography. IEEE Trans. Info. Theory **22** (1976) 644–654.
6. Hardjono, T., Seberry, J.:. Information security issues in mobile computing. In Proceedings of IFIP Sec'95 (1995) pp. 143–155.
7. Rabin, M. O.:. Digitalized signatures and public-key functions as intractable as factorization. Tech. Rep. TR 212 MIT Lab. Computer Science, Cambridge., Mass. Jan. 1979.
8. Varadharajan, V., Mu, Y.:. Authentication of mobile communications systems. In Proceedings of the 1996 IFIP Mobile Communication Conference (Sept 1996).
9. Varadharajian, V.:. Security for personal mobile networked computing. In Proceedings of the International Conference on Mobile and Personal Communications Systems (Australia, April 1995).
10. Williams, H. C.: A modification of RAS public-key encryption. IEEE Trans. Info. Theory **IT-26** (1980) 726–729.

A Framework for Design of Key Establishment Protocols

Colin Boyd

Information Security Research Centre
School of Data Communications
Queensland University of Technology
Brisbane Q4001
AUSTRALIA
boyd@fit.qut.edu.au

Abstract. A framework is described in which designs for key establishment protocols may be described at a high level of abstraction. This enables protocols to be designed without concern for implementation details, while emphasising the fundamentals elements required. Consideration of the natural alternatives leads to the definition of four fundamental classes of key establishment protocols into one of which most published protocols can be placed. Abstract and concrete protocols are described in each class, including some novel concrete designs.

1 Introduction

Key establishment protocols are typically used at the start of a communications session to establish a session key which is used during the session to provide secure communications. Recent research has led to a number of techniques to analyse such protocols [4, 10]. However, although these techniques have been successful in identifying problems, it is doubtful whether any protocol which successfully passes analysis can be confidently pronounced secure. For this reason it is important to make decisions during the design phase of protocols which minimise, or preferably eliminate, choices which will lead to problems. As understanding of cryptographic protocols has advanced, various design principles have been suggested [1, 11] while Gong and Syverson [7] have considered a method to simplify design and analysis; but none of these can be thought of as a methodology for design.

The aim of this paper is to describe a framework in which designs for key establishment protocols can be made. The general approach is to derive abstractly both the fundamental security properties that such protocols should have and the fundamental cryptographic properties that must be used in order to achieve these properties. This enables a classification of key establishment protocols into four fundamental classes. By instantiating the classes with differing choices of cryptographic mechanisms a variety of concrete protocols can be designed. The result can be seen as a restricted *subclass* of all possible protocols. Although no proof of security is supplied, the abstract nature of the specifications means that the required security properties can be verified in a very transparent way.

In section 2 the model of security is explained. By considering the abstract security properties and cryptographic mechanisms four protocol classes naturally emerge. The framework may be used as a practical way to design protocols that are simple to understand and therefore easy to analyse. Section 3 is dedicated to giving examples of protocols that fit into the framework. These are mainly given at an abstract level which is independent of the particular type of cryptosystems used.

1.1 Protocol Aims

In a key establishment protocol the aim is for two or more users to share a key for a subsequent session. It is possible for a trusted server to generate the key, for one or more of the users to do so, or for a combination of principals to be responsible for key generation. Whichever of these is the case, keying information needs to be passed between users over insecure physical channels; cryptography is used to provide secure logical channels. In this paper an assumption is made, which is true in all practical cases, that the session key will be a shared key for use with a symmetric cryptosystem. However the methods used to exchange the keying information may be either symmetric or asymmetric techniques. Whenever such an exchange takes place the following two conditions must hold.

1. The key must not be allowed to become known to any users apart from those participating in the protocol or trusted servers.
2. Each recipient must be sure that the key is a new key for use with the stated users.

These conditions are uncontroversial and form the basis of what is meant in this paper by a secure key exchange protocol. In practice further aims are often desired, for example that each user has confirmation that the other users possess the new key. These additional requirements can always be achieved by additional steps independent of the actual key exchange and are not addressed further in this paper.

1.2 Secure Channels

Key establishment protocols make use of cryptography to ensure that the distributed keys are delivered to their destination in a manner that maintains confidentiality and ensures integrity. There are a variety of different cryptographic functions available and it is important to select amongst them such that the correct security services are provided. Many cryptographic protocols use either conventional symmetric cryptographic algorithms or public key cryptosystems, while others make use of key agreement procedures, such as those originally conceived by Diffie and Hellman [5]. Cryptographic security properties can broadly be classified into two types.

- **Confidentiality** allows the sender of a message to decide those users who will be able to receive it. These recipients will be defined by their possession of the required decryption key.

- **Authentication** allows the recipient of a message to decide those users who may have sent it. These senders will be defined by their possession of the required encryption key. Although not in general a part of the definition, it will be assumed in this paper that all authentication algorithms are *irreversible*; that is, it is not possible for any entity to recover any message sent only on an authentication channel.

Abstract security channels may be defined, of confidentiality or authentication type, between pairs of users in possession of keys which allow cryptographic confidentiality or authentication to take place between them. These channels can be used to provide the security services needed to transport keying material; furthermore, since these are defined in an abstract manner the particular cryptographic functions used can be left unspecified. The notation

$$S \xrightarrow{c} A : m$$

means that S sends the message m on a confidentiality channel to A. This is an action performed by S and forms part of the definition of the process which represents S; it does not imply that A will receive any such message, but does imply that no entity except A will receive it.

$$S \xrightarrow{a} A : m$$

means that A receives the message m on an authentication channel from S. This is an action performed by A and forms part of its process definition.

1.3 Protocol Specification

The common informal way to express protocols uses the notation

$$A \to B : m$$

to express that the message m is transferred from principal A to principal B. Often the complete protocol is presented merely as a series of such exchanges. This notation appears easy to understand but has pitfalls in that it tends to give a false impression of what occurs, or may potentially occur, at individual processes representing principals.

It is *a priori* completely unknown to A who will receive m, or to B who has sent m. (The situation is quite different from that of the use of logical secure channels described above where confidentiality to the recipient and/or authentication from the sender is assured.) Furthermore, the actions of principals are often critical in deciding the security of a protocol. Ideally, each principal should be fully specified as an individual process.

In this paper two different notations are used. The first notation uses the abstract security channels and takes a single user viewpoint. In other words it only specifies the messages to and from a single user, A (Alice), along abstract secure channels. Because this notation is abstract it can be used only to describe classes of protocols in which many concrete protocols will lie. In order to aid understanding we also use the notation

$$A \rightarrow * : M$$

to denote that Alice sends the message M unencrypted (the symbol '$*$' is intended to invoke the idea that plaintext messages are essentially broadcast to all principals).

In order to give examples of specific protocols which may be concretely implemented, the second notation shows all messages in a successful run of the protocol. This is close to the established notation mentioned above with the difference that the abstract cryptographic properties are shown via the notation introduced by Mao and Boyd [8]. Thus a message M may be transformed in three different ways.

- $[M]_K$ denotes a string that cannot be calculated from M without knowledge of K. In particular it shows that the principal which formed the string was in possession of K. Such a transformation is used to implement an authentication channel so, as mentioned above, it is also required that M cannot be recovered from $[M]_K$ (perhaps not even given K). A typical implementation of such a transformation would be a cryptographic check sum.
- $\langle M \rangle_K$ denotes a string that can be used to recover M only with possession of K. Such a transformation is used to implement a confidentiality channel. A typical implementation would be any encryption algorithm.
- $\{M\}_K$ denotes that both the above properties are provided. Such a transformation would implement a combined authentication and confidentiality channel. A typical implementation would be an encryption algorithm with a one-way hash function included.

Nearly all the examples given in this paper have the typical scenario of two users A and B who wish to share a new session with the help of a trusted server S (although the framework is by no means restricted to this case). The following notation is used uniformly in this paper.

K_{AB} is the value of the new session key.
K_{AS} is a long-term key initially shared by A and S.
K_{BS} is the corresponding long-term key shared by B and S.
N_X is a random value (nonce) chosen by any principal X for this run of the protocol.

2 Model of Security

The model of security is very simple. Each system user may be considered as holding a database of mappings each of which associates a key value to a set of one or more users. In the case of session keys this set of users are those with whom the session key is shared. From a strict security viewpoint it is not important if a user *overestimates* who may be in possession of a particular key. However if he should *underestimate* there could be serious consequences since when using that key he may allow access to more users than intended when sending a message (loss of confidentiality) or be unsure as to the sender when receiving a message (loss of authentication). Therefore one critical aspect of a secure protocol which results in the update of the user's database is that he should know the identity of all users who will, or might, gain access to that key and adds them to the database mapping for that key.

The only other important issue is to appreciate that session keys *age*. Cryptographic keys which are used for securing general purpose sessions are viewed as *short-term* keys and in usual practice are discarded at the finish of a session (or even periodically during it). Therefore in the model each user's key database should only contain session keys that are new. For this reason a secure protocol must ensure that if a session key is to be added to the database there must be some assurance that the mapping is a new one. To make a protocol practical it is, of course, necessary to add that each user efficiently learns the value of the session key. The following definition of a secure protocol is thus arrived at.

Definition 1. A protocol to establish a session key is **secure** if it is secure for all users involved. A protocol is secure for Alice if:

- Alice has acceptable assurance of who may have the key value.
- Alice has acceptable assurance that the key is fresh.

The main idea of this paper is to observe that there are two ways that each of the above security properties can be achieved, depending on whether the user *sends* a message to achieve it, or *receives* a message to achieve it. It follows that Alice has four flavours of protocol that can be used to derive a new secure session key, depending on how she derives the names of the other key recipients and how she derives key freshness.

- Alice may derive the names of the other recipients by choosing which users receive a message containing keying material sent on a confidentiality channel, which will be termed **recipients by choice (RC)**. Alternatively she may be told the names in a message received on an authentication channel from a trusted party, which will be termed **recipients by imposition (RI)**.
- Alice may derive key freshness either by choosing a fresh input to the key, which will be termed **freshness by input (FI)**, or by receiving a fresh item bound to a component of the key, which will be termed **freshness by receipt (FR)**.

This leads to the classification of four classes of protocol as shown in table 1. Two of these types are well known already. The classic Needham-Schroeder protocol [9] (in its repaired form) and its many variants lie in the class RI/FR. Even in this class a considerable variety of abstract protocols are possible within the framework. The class of protocols in RC/FI includes a number of published examples of *key agreement* protocols[1]. The two other classes do not appear to have been explored widely in the literature. A related paper [3] examines in detail the class RI/FI which has a number of unusual and attractive properties.

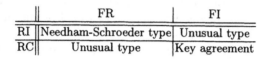

	FR	FI
RI	Needham-Schroeder type	Unusual type
RC	Unusual type	Key agreement

Table 1. Classification of Key Establishment Protocols

Notice that table 1 classifies protocols from the viewpoint of a single user; different users may perceive the same protocol to be in different classes — examples of such protocols are given below. Furthermore, it is possible that Alice could achieve assurance of recipient names and/or freshness in more than one way, but for simplicity such protocols are not considered here. Since there is no obvious alternative method of achieving the goals of a key establishment protocol the following conjecture is tempting to make. Of course, to prove such a conjecture formal definitions are necessary.

Conjecture 1 *Every secure key establishment protocol which provides assurance of the freshness of the session key and assurance of which other users know the key falls into (at least) one of the classes of table 1.*

3 Some Abstract Protocol Designs

In general the session key will be the result of multiple inputs by different users. Sometimes there will be only one input (such as in the Needham-Schroeder type protocol) but key agreement protocols typically have keys depending on more than one input.

Definition 2. Suppose that the session key K is defined as

$$K = f(i_1, i_2, \ldots, i_n)$$

The j'th component of the function f is **essential** if it is infeasible to find K without knowledge of the value of that component. The j'th component is **defining** if f is *collision free* in that component; in other words, for a given session key K it is infeasible to find two different values of the j'th component and any values of the other components such that f maps both values to K.

---

[1] In order to include the classic Diffie-Hellman protocol [5] in this class it is necessary to flexible about interpreting the way that recipients of the key are chosen.

Although these are natural properties of key-defining functions it is easy to show that they are independent by constructing (pathological) examples which satisfy one but not the other. For example, if $f(x_1, x_2) = g(x_1 \oplus x_2)$ for a collision resistant function g, then f is essential in both components but defining in neither. On the other hand, if $f(x_1, x_2) = g(x_1|x_2)$ (here '|' denotes concatenation of strings) where g is collision resistant, then f is defining in both components but if the first component is very short (in the extreme case just one bit long) then the output can be guessed with high probability without knowledge of its value so that the first component is not essential. Of course when f is the identity function (such as in the Needham-Schroeder class) its only component is both essential and defining. Ways to construct practical functions with these properties are mentioned in the examples.

The framework can be used to design specific protocols by ensuring that the following properties are held for any specific principal Alice. It follows from Definition 2 that protocols which satisfy these properties are secure in the sense of Definition 1.

Confidentiality Alice must **either** receive a message binding an essential component of the session key to the names of the other users who are sent the key (authentically) from a trusted user **or** send an essential component of the key (confidentially) to the other users who will have the key.

Freshness Alice must **either** receive a defining component of the session key bound to an item Alice knows to be fresh (authentically) from a trusted user **or** choose freshly a defining component of the key.

Value Alice must be able to efficiently calculate the session key value.

In the rest of this section a number of protocol designs using the framework are considered. In order to illustrate how the protocols which result are related to existing published protocols attention is concentrated on the most common situation of interest, namely where two nodes in a network wish to establish a session key, but initially share secure channels only with a mutually trusted third party. Each of the four classes is considered in turn.

3.1 Recipients by Imposition and Freshness by Receipt

This is by far the most commonly considered situation in published protocols with this architecture, starting with the seminal paper of Needham and Schroeder [9]. The abstract protocol which defines this class shows only the messages sent over secure channels to and from Alice.

- $A \rightarrow * : N_A$
- $S \xrightarrow{c} A : K_{AB}$
- $S \xrightarrow{a} A : K_{AB}, A, B, N_A$

Many protocols in this class have been published before. To give a slightly unfamiliar example now, a variant is presented where the key value is sent on one (confidentiality) channel from the server, while the freshness and recipients of the session key are sent on a separate (authentication) channel. Confidentiality may be obtained by any particular encryption algorithm such as DES. Authentication may be achieved by a key dependent message authentication code (MAC) which could also be based on a block cipher or could use a one-way hash function.

1. $A \rightarrow B$: A, N_A
2. $B \rightarrow S$: A, B, N_A, N_B
3. $S \rightarrow B$: $\langle K_{AB} \rangle_{K_{BS}}, [K_{AB}, A, B, N_B]_{K_{BS}},$
   $\langle K_{AB} \rangle_{K_{AS}}, [K_{AB}, A, B, N_A]_{K_{AS}}$
4. $B \rightarrow A$: $\langle K_{AB} \rangle_{K_{AS}}, [K_{AB}, A, B, N_A]_{K_{AS}}$

3.2 Recipients by Imposition and Freshness by Input

In this class of protocols Alice receives information from the server on who else knows the session key, but provides her own input to the protocol (for Bob) which will be part of the input to the function determining K_{AB}. Many of the resulting protocols appear to be novel and have a number of interesting features. The abstract protocol defining this class is as follows.

- $A \rightarrow * : N_A$
- $S \xrightarrow{c} A : KC_S$
- $S \xrightarrow{a} A : KC_S, A, B$

where KC_S is an essential component of the session key. This interesting class of protocols is explored further a related paper [3]; here one particular example is given.

1. $A \rightarrow S : A, B, N_A$
2. $S \rightarrow B : \langle K_S \rangle_{K_{AS}}, [A, B, K_S]_{K_{AS}}, \langle K_S \rangle_{K_{BS}}, [A, B, K_S]_{K_{BS}}, N_A$
3. $B \rightarrow A : \langle K_S \rangle_{K_{AS}}, [A, B, K_S]_{K_{AS}}, [N_A]_{K_{AB}}, N_B$
4. $A \rightarrow B : [N_B]_{K_{AB}}$

In this description K_S is a special 'token key' chosen randomly by S for this run only of the protocol. The session key K_{AB} is a result of the function

$$K_{AB} = f(K_S, N_A, N_B)$$

where the first component of f is essential and the second and third components are defining. An example of such a function is $f(k, x, y) = h(k, x|y)$ where h is a keyed hash function [2] and $x|y$ denotes concatenation of the strings x and y.

A major advantage of this type of protocol is that the number of messages and the number of rounds required are the same as that normally required for protocols using timestamps [6], but without the problems associated with timestamps, such as synchronisation of clocks. Another useful property is that the messages from the server can be cached as *renewable tokens* for use later. The price for these advantages is that revocation lists of 'blacklisted' tokens must be maintained in order to protect against the compromise of the token key K_S.

3.3 Protocol Symmetry

In the first two classes when the server chooses the key it is natural to expect both Alice and Bob to have symmetrical views of the protocol. In the other two classes, considered in the two next subsections, Alice obtains recipient information by directing her own key component through an appropriate confidentiality channel. Unless Bob does the same, then the protocol will naturally be asymmetrical. We note, however, that even in the classes already considered Alice and Bob could achieve freshness in different ways. For example in the following protocol Alice achieves freshness by input from the server, but Bob achieves it using his broadcast nonce.

1. $A \rightarrow S : A, B, N_A$
2. $A \rightarrow B : A$
3. $S \rightarrow A : \{K_S, A, B, N_A\}_{K_{AS}}$
4. $S \rightarrow B : \{K_S, A, B\}_{K_{BS}}$
5. $B \rightarrow A : N_B$

where the session key is $K_{AB} = f(K_S, N_B)$. In order that both A and B can obtain freshness of K_{AB} both the components of f should be defining and also its first component must be essential. A suitable function would be $f(x, y) = h(x) \oplus h(y)$ where h is a collision resistant function. Notice that this protocol may be completed in two rounds since the first two and last three messages can be run in parallel.

3.4 Recipients by Choice and Freshness by Input

Protocols where Alice and Bob both participate in choosing the session key are often known as *key agreement* protocols. These usually employ public key cryptography and do not require a (real-time) server. An example is the following.

1. $A \rightarrow B : \langle N_A \rangle_{K_B}$
2. $B \rightarrow A : \langle N_B \rangle_{K_A}$

and $K_{AB} = f(N_A, N_B)$ where f has both components as defining and at least one essential, and K_A and K_B are A and B's public keys, respectively. An abstract protocol for this class is thus simply the following.

$$- A \xrightarrow{c} B : KC_A$$

where KC_A is a defining and essential component of the session key. (Alternatively KC_A could be only a essential component and a separate defining component N_A could be broadcast. It seems natural to combine these.) protocol with the same results can be achieved with symmetric cryptography and the server acting as a key translation server.

1. $A \rightarrow S : \{A, B, N_A\}_{K_{AS}}$

2. $B \rightarrow S : \{A, B, N_B\}_{K_{BS}}$
3. $S \rightarrow B : \{A, B, N_A\}_{K_{BS}}$
4. $S \rightarrow A : \{A, B, N_B\}_{K_{AS}}$

As above $K_{AB} = f(N_A, N_B)$ where f has both components as defining and at least one essential. An asymmetric version, in which only Alice has input to the session key (she simply chooses the value for K_{AB}) is as follows.

1. $B \rightarrow A : B, N_B$
2. $A \rightarrow S : \{A, B, K_{AB}, N_B\}_{K_{AS}}$
3. $S \rightarrow B : \{A, B, K_{AB}, N_B\}_{K_{BS}}$

Note that in this protocol B derives freshness by receipt and recipients by imposition so the protocol appears from his viewpoint to lie in the first class. This point emphasises the importance of looking at a protocol from the viewpoint of each individual process involved.

3.5 Recipients by Choice and Freshness by Receipt

This final class has abstract protocol as follows.

- $A \rightarrow * : N_A$
- $A \xrightarrow{c} B : KC_A$
- $S \xrightarrow{a} A : KC_B, N_A$

where KC_A is an essential component of the session key and KC_B is a defining component.

When Alice chooses an essential component of the key randomly, she also gains freshness of the key if the component is also defining. For this reason it appears at first that the protocols in this class are not of interest. However, if Alice's chosen component is re-used, as in the example in subsection 3.2, there may be useful examples. Consider the following protocol.

1. $A \rightarrow S : \{A, B, K'\}_{K_{AS}}$
2. $S \rightarrow B : \{A, B, K'\}_{K_{BS}}$
3. $A \rightarrow B : N_A$
4. $B \rightarrow S : N_A, [K', N_A]_{K_{BS}}$
5. $S \rightarrow A : N_A, [k', N_A]_{K_{AS}}$

The session key is $K_{AB} = f(K', N_A)$ where the first component of f is essential and the second is defining. Note that the last three messages in the protocol can be exchanged at any time following the first two, in particular at the start of a different session. Similarly the last two messages can be used to renew the session key. The protocol suffers the same limitations following compromise of K' as explained in section 3.2.

4 Conclusion

The framework described in this paper allows protocol designers to consider the necessary elements in a protocol and choose between a variety of different ways in which to include these. Selection of protocol methods may be made in accordance with the scenario in which the protocol will be applied. Protocols may be designed within the framework at an abstract level allowing flexibility in implementation.

Protocols that conform to the framework will satisfy the security requirements in a particularly transparent way. Readers are encouraged to consider any attacks on the protocol examples included in this paper. Ultimately the strength or weakness of the framework must rest on the security of the protocols which are designed within it.

In this paper the framework has been used mainly as a classification tool for key establishment protocols. However, There are a number of other useful purposes for the framework.

- By observing abstractly what are the minimum fundamental requirements for key exchange new protocols may be found, with different properties that may be of practical importance. A number of such examples are given in this paper and further ones may be revealed by more detailed study.
- Designers are able to recognise protocols which are of the same type as those that they require, enabling them to consider exisiting proved designs, or incorporating mechanisms used before.
- Analysts are able to recognise similar protocols which have been successfully or unsuccessfully analysed before, providing pointers to possible weaknesses or suggesting repairs to broken protocols.

Acknowledgements

I am grateful to an anonymous referee for a careful reading of this paper and pointing out a number of errors and omissions.

References

1. M. Abadi and R. Needham, "Prudent Engineering Practice for Cryptographic Protocols", DEC SRC Research Report 125, June 1994.
2. S. Bakhtiari, R. Safavi-Naini and J. Pieprzyk, "Keyed Hash Functions", *Cryptography: Policy and Algorithms*, Springer-Verlag, 1996, pp.201-214.
3. C. Boyd, "A Class of Flexible and Efficient Key Management Protocol", IEEE Security Foundations Workshop, IEEE Press, 1996, pp.2-8.
4. M. Burrows, M. Abadi, and R.M. Needham, "A Logic of Authentication", *ACM Transactions on Computer Systems*, Vol 8,1, February 1990, pp 18-36.
5. W. Diffie and M.E. Hellman, "'New Directions in Cryptography", *IEEE Transaction on Information Theory*, IT-22, pp.644-654, 1976.

6. L. Gong, "Efficient Network Authentication Protocols: Lower Bounds and Optimal Implementations", *Distributed Computing*, 9(3), 1995.

7. L. Gong and P. Syverson, "Fail-Stop Protocols: An Approach to Designing Secure Protocols", *Proceedings of IFIP DCCA-5, Illinois, September 1995.* (Available as `http://www.csl.sri.com/` ~`gong/failstop-protocols.ps.gz`.)

8. W. Mao and C. Boyd, "Methodical Use of Cryptographic Transformations in Authentication Protocols", *IEE Proceedings - Computers and Digital Techniques*, 142, 4, July 1995. pp.272-278.

9. R. M. Needham & M. D. Schroeder, "Using Encryption for Authentication in Large Networks of Computers", *Communications of the ACM*, 21,12, December 1978, 993-999.

10. R. Kemmerer, C. Meadows and J. Millen, "Three Systems for Cryptographic Protocols Analysis", *Journal of Cryptology*, 7,2, Spring 1994, pp.79-130.

11. Thomas Y.C. Woo and Simon S. Lam, "A Lesson on Authentication Protocol Design", *ACM Operating Systems Review*, July 1994.

On Period of Multiplexed Sequences

Jovan Dj. Golić

Information Security Research Centre, Queensland University of Technology
GPO Box 2434, Brisbane Q 4001, Australia
School of Electrical Engineering, University of Belgrade
Email: golic@fit.qut.edu.au

Abstract. Multiplexed and generalized multiplexed sequences for cryptographic and spread spectrum applications are introduced and their periods determined by using a recent result on the period of nonuniformly decimated sequences. Several published results are thus strenghtened and/or generalized. In particular, the period of the well-known multiplexed sequences is derived without the constraints assumed in the literature. The period of the so-called MEM-BSG sequences is also obtained.

1 Introduction

A class of binary pseudorandom sequences for cryptographic and spread spectrum applications known as the multiplexed sequences was proposed and analyzed in [12, 13] and widely popularized in [2]. Their use has also been recommended in an EBU standard for video encryption for pay-TV [18]. Multiplexed sequences are generated by a simple and fast scheme consisting of two linear feedback shift registers and a multiplexer whose address is controlled by one of the shift registers and whose inputs are taken from the other. They were shown [12, 13, 14] to possess good standard cryptographic properties such as long period, high linear complexity, and low out-of-phase autocorrelation. On the other hand, they were also shown to exhibit certain cryptographic weaknesses such as the collision test [1], the linear consistency test [16], the resynchronization weakness [6], and the autocorrelation and correlation weaknesses [9].

In this paper we study the periods of various generalizations of multiplexed binary sequences. Note that deriving the period of a pseudorandom sequence is generally a difficult algebraic problem which seems to be tractable only for relatively simple sequences and under special constraints. Our main objective is to obtain general results and thus show that the period can be controlled without the constraints usually assumed in the literature. We will employ the technique [10] based on interleaving and will also introduce some new classes of pseudorandom sequences which may be interesting for stream cipher and spread spectrum applications.

In Section 2, some basic definitions and results [10] regarding the periods of interleaved and nonuniformly decimated sequences are briefly reviewed. Multiplexed and generalized multiplexed sequences, defined as appropriate generalizations of the class of binary sequences introduced by Jennings [12], are analyzed in Sections 3 and 4, respectively. By deriving some additional results on decimated sequences, the periods of multiplexed and generalized multiplexed sequences are determined and the corresponding result of Jennings [12, 13] is thus generalized. In Section 5, the so-called MEM-BSG sequences based on the variable-memory binary sequence generator [8] are introduced. Their period is obtained and the corresponding result from [8] is thus strengthened and generalized. The results are summarized in Section 6.

2 Preliminaries

In this section we will give necessary definitions and results regarding the period of interleaved and nonuniformly decimated integer sequences. For a set of K periodic integer sequences $a_i = \{a_i(t)\}_{t=0}^{\infty}$ with periods P_i, $0 \leq i \leq K - 1$, respectively, the interleaved sequence b is defined by

$$b(i + Kt) = a_i(t), \quad 0 \leq i \leq K - 1, \quad t \geq 0. \tag{1}$$

It is clear that b is periodic with period $P_b \mid PK$, where $P = $ l.c.m. $(P_i : 0 \leq i \leq K - 1)$. The following lemma, proved in [10], specifies P_b more precisely.

Lemma 1. The period P_b of the interleaved sequence b satisfies

$$P_b = P(P_b, K) \tag{2}$$

where $(\cdot, \cdot)$ denotes the greatest common divisor.

An exact characterization of P_b in terms of the constituent sequences a_i, $0 \leq i \leq K - 1$, was also derived in [10], and then used as a basic result to determine the period of nonuniformly decimated sequences and thus generalize an old result of Blakley and Purdy [3]. Namely, let $a = \{a(t)\}_{t=0}^{\infty}$ be a periodic integer sequence with period P_a. Let $D = \{D(t)\}_{t=0}^{\infty}$ be a decimation sequence defined recursively by $D(t + 1) = D(t) + d(t)$, $t \geq 0$, where $D(0) = 0$ and the difference decimation sequence $d = \{d(t)\}_{t=0}^{\infty}$ is a periodic nonnegative integer sequence with period M such that $0 \leq d(t) \leq P_a - 1$, $t \geq 0$. The decimated sequence b is then defined by

$$b(t) = a(D(t)) = a\left(\sum_{i=0}^{t-1} d(i)\right), \quad t \geq 0, \tag{3}$$

see [3], [4], assuming that $\sum_i^j(\cdot) = 0$ if $j < i$. In a special case when $d(t) = d$, $t \geq 0$, the decimation is called uniform. Letting $N = D(M) \bmod P_a$, it is well known [4], [5], [7] that the decimated sequence b can be regarded as an interleaved sequence such that

$$b(i + Mt) = a(D(i) + Nt), \quad 0 \leq i \leq M - 1, \quad t \geq 0 \tag{4}$$

meaning that $\{b(i + Mt)\}_{t=0}^{\infty}$ is a decimated sequence obtained from the uniform decimation by N of $\{a(D(i) + t)\}_{t=0}^{\infty}$, $0 \leq i \leq M - 1$. Let P_b denote the period of

b. It clearly follows that $P_b \mid MP_a/(N, P_a)$. Blakley and Purdy [3] proved that $P_b = MP_a$ if $(N, P_a) = 1$. The period P_b of b in a general case when $(N, P_a) > 1$ was determined in [10] by using the established characterization of the period of interleaved sequences and an auxiliary lemma. The lemma and the theorem yielding the period are given below.

Lemma 2. Let a positive integer N, $1 \leq N \leq P_a - 1$, be such that the decimated sequences $\{a(i + Nt)\}_{t=0}^{\infty}$, $0 \leq i \leq P_a - 1$, are distinct, given a periodic integer sequence a with period P_a. Then for each i, $0 \leq i \leq P_a - 1$, the period P_i of $\{a(i + Nt)\}_{t=0}^{\infty}$ is equal to $P_a/(N, P_a)$.

Theorem 1. Let the decimated sequences $\{a(i + Nt)\}_{t=0}^{\infty}$, $0 \leq i \leq P_a - 1$, be distinct. Then the period P_b of the decimated sequence b is given by

$$P_b = M \frac{P_a}{(N, P_a)}. \tag{5}$$

If the uniformly decimated sequences are not distinct, then the period of the decimated sequence may be equal to or smaller than (5). They are distinct if $(N, P_a) = 1$, due to the inverse decimation property [11] of the proper decimation, that is, $a(i + t) = a(i + N(N^{-1}t))$, $t \geq 0$, where N^{-1} is the multiplicative inverse of N modulo P_a. Thus Theorem 1 implies the result of Blakley and Purdy [3]. It is interesting to find other, more general sufficient conditions for the decimated sequences to be distinct. To this end, first note that any periodic integer sequence can be regarded as a sequence over an appropriate sufficiently large finite field. Second, according to the fact that every periodic finite field sequence a is linear recurring, the *minimum polynomial* of a is defined as the unique monic polynomial of lowest degree being the characteristic polynomial of a linear recurrence satisfied by a, see [17]. The *linear complexity* of a is then defined as the degree of its minimum polynomial. Also, given a prime power q and a positive integer n, a q-ary maximum-length finite field sequence is a q-ary finite field sequence of linear complexity n and period $q^n - 1$, see [11]. Its minimum polynomial is primitive. Maximum-length sequences have wide applications due to their large period, low out-of-phase autocorrelation, and good distribution properties, see [11]. The following two propositions [10] give some interesting sufficient conditions for the decimated sequences to be distinct.

Proposition 1. Let $a = \{a(t)\}_{t=0}^{\infty}$ be a periodic integer sequence with period P_a and let N be a positive integer such that $1 \leq N \leq P_a - 1$. Then the uniformly decimated sequences $\{a(i + Nt)\}_{t=0}^{\infty}$, $0 \leq i \leq P_a - 1$, are distinct if the linear complexities of a and $a^{(N)} = \{a(Nt)\}_{t=0}^{\infty}$, regarded as finite field sequences, are equal. In particular, the linear complexities of a and $a^{(N)}$ are equal if their periods are equal, that is, if $(N, P_a) = 1$.

Proposition 2. Let a be a q-ary finite field sequence of period P_a and let N

be a positive integer such that $1 \leq N \leq P_a - 1$. Let the minimum polynomial of a be an irreducible polynomial of degree n. Then the linear complexities of a and $a^{(N)}$ are equal if any of the two following conditions is satisfied:

$1°$ $q = 2$ and n is a prime.
$2°$ $P_a = q^n - 1$ and $N = q^m - 1$ mod P_a, where n and m are positive integers.

Note that the second condition implies that a is a q-ary maximum-length sequence, which means that the minimum polynomial is primitive, and also that $N \geq 1 \Leftrightarrow n \nmid m$. Theorem 1 along with Proposition 1 and the second condition in Proposition 2 will be used in the following sections to analyze the period of the multiplexed sequences [12] and their various generalizations.

We conclude this section by a result, not given in [10], which provides an interpretation of the right-hand side of equation (5) in Theorem 1 which will be used in the following section.

Proposition 3. Given a decimation sequence D, define the reduced decimation sequence $\hat{D}$ as $\hat{D}(t) = D(t)$ mod P_a, $t \geq 0$. Then the period $\hat{M}$ of $\hat{D}$ is given by

$$\hat{M} = M \frac{P_a}{(N, P_a)} \tag{6}$$

where M is the period of the difference decimation sequence d and $N = \hat{D}(M)$.

Proof. It follows that $\hat{D}$ satisfies the recursion $\hat{D}(t+1) = (\hat{D}(t) + d(t))$ mod P_a, $t \geq 0$, with $\hat{D}(0) = 0$. We first prove that $M \mid k$, for a positive integer k, if and only if

$$\hat{D}(t + k) = \hat{D}(t) + r \pmod{P_a}, \quad t \geq 0 \tag{7}$$

for some integer r, $0 \leq r \leq P_a - 1$. Namely, $M \mid k$ implies (7) with $r = Nk/M$, whereas (7) implies that $d(t + k) = d(t)$, $t \geq 0$, and hence $M \mid k$. In particular, since (7) holds with $r = 0$ if $k = \hat{M}$, we get that $M \mid \hat{M}$. Let $\hat{M} = sM$. But then (7) also holds with $r = sN$, so that $sN = 0 \pmod{P_a}$. Therefore, s is the minimal positive integer such that $sN = 0 \pmod{P_a}$, that is, s equals $P_a/(N, P_a)$ and hence (6) is true.

3 Multiplexed Sequences

In this section we study the period of multiplexed sequences which are defined as an appropriate generalization of the class of binary sequences introduced and analyzed in [12, 13]. The multiplexed sequences are widely popularized by Beker and Piper in [2] and are referred to in [15] as well. Let $a = \{a(t)\}_{t=0}^{\infty}$ be a binary maximum-length sequence of period $P_a = 2^n - 1$, $n \geq 1$. Then the multiplexed sequence b is defined by

$$b(t) = a(t + \gamma(X(t))), \quad t \geq 0 \tag{8}$$

where $X = \{X(t)\}_{t=0}^{\infty}$ is a periodic integer sequence with period $P = 2^m - 1, m \geq 1$, such that $0 \leq X(t) \leq K, t \geq 0$, and γ is an injective mapping $\{0, \ldots, K\} \rightarrow \{0, \ldots, L\}$ with $0 \leq L \leq 2^{n-1} - 1$. Clearly, the multiplexed sequence b can be generated by a multiplexer-like scheme with X defining the addresses and $\gamma(X)$ defining the phase shifts of a forming the inputs to the multiplexer.

It follows that the multiplexed sequence b is a decimated sequence obtained from the phase shift $\{a(t + \gamma(X(0)))\}_{t=0}^{\infty}$ of a by the decimation sequence D corresponding to the difference decimation sequence d given by

$$d(t) = (1 + \gamma(X(t+1)) - \gamma(X(t))) \bmod (2^n - 1), \quad t \geq 0. \tag{9}$$

Accordingly, the period of the multiplexed sequence can be analyzed by applying Theorem 1 and Propositions 1 and 2. It remains to derive the period M of d. This is accomplished in Theorem 2 by using the following two lemmas.

Lemma 3. Let d_1 and d_2 be arbitrary difference decimation sequences with periods M_1 and M_2, respectively, where $0 \leq d_1, d_2 \leq P_a - 1$. Let $d = (d_1 + d_2) \bmod P_a$. Then the period M of d satisfies

$$\frac{\text{l.c.m.}(M_1, M_2)}{(M_1, M_2)} \mid M \mid \text{l.c.m.}(M_1, M_2). \tag{10}$$

In particular, if $(M_1, M_2) = 1$, then $M = M_1 M_2$.

Proof. First, it is clear that $M \mid \text{l.c.m.}(M_1, M_2)$. Second, from $d_i = d - d_j$ (mod P_a), we also have that $M_i \mid \text{l.c.m.}(M, M_j), i \neq j$. Therefore, $\text{l.c.m.}(M_1, M_2) \mid \text{l.c.m.}(M, (M_1, M_2))$, which implies the lower bound in (10).

Lemma 4. Let a difference decimation sequence d be given by $d(t) = (\delta(t+1) - \delta(t)) \bmod P_a$ where δ is a periodic integer sequence such that $0 \leq \delta(t) \leq L$, $t \geq 0$, and $P_a > 2L$. Then the period of d is equal to the period of δ.

Proof. Define the sequence δ' as $\delta'(t) = \delta(t+1) - \delta(t), t \geq 0$. If δ is periodic with period $\hat{M}$, then δ' is periodic with period M such that $M \mid \hat{M}$. Since $0 \leq \delta(t) \leq L, t \geq 0$, and $P_a > 2L$, it follows that $d(t_1) = d(t_2)$ if and only if $\delta'(t_1) = \delta'(t_2)$, for any $t_1, t_2 \geq 0$. Hence the period of d is equal to the period M of δ'. It remains to prove that $M = \hat{M}$. This can be done either directly or by invoking Proposition 3, which is at our disposal already. Namely, it is readily checked that the reduced decimation sequence $\hat{D}$ corresponding to d is given by $\hat{D}(t) = (\delta(t) - \delta(0)) \bmod P_a, t \geq 0$. In view of $P_a > 2L$, $\hat{D}$ is periodic with the same period $\hat{M}$ as δ. Proposition 3 then yields that $\hat{M} = M P_a / (\hat{D}(M), P_a)$

$= M P_a / (\delta(M) - \delta(0), P_a)$. Since $\hat{M}$ and M are both independent of P_a, it then follows that $\delta(M) = \delta(0)$ and hence $\hat{M} = M$.

Theorem 2. If $n \nmid m$, then the period of the multiplexed sequence is equal to l.c.m.$(2^m - 1, 2^n - 1)$.

Proof. The period of d is equal to the period of d' defined by $d'(t) = (\gamma(X(t + 1)) - \gamma(X(t)))$ mod $(2^n - 1)$, $t \geq 0$, see Lemma 3. Since $0 \leq \gamma(X(t)) \leq L$, $t \geq 0$, where $P_a = 2^n - 1 > 2L$, Lemma 4 then yields that the period of d' equals the period of $\gamma(X)$ which is itself equal to $P = 2^m - 1$, because γ is an injection. Hence, $M = 2^m - 1$ and $N = (2^m - 1 + \gamma(X(2^m - 1)) - \gamma(X(0)))$ mod $(2^n - 1) = (2^m - 1)$ mod $(2^n - 1)$. If in addition $n \nmid m$, then by Propositions 2 and 1 we have that the period P_b of b is determined by Theorem 1. Therefore, $P_b = (2^m - 1)(2^n - 1)/(2^{(m,n)} - 1) = $ l.c.m.$(2^m - 1, 2^n - 1)$ if $n \nmid m$.

If $n \mid m$, then the period may be smaller than $2^m - 1$. Theorem 2 generalizes the result of Jennings [12, Theorem 2.4.3] which besides $n \nmid m$ assumes that $L \leq n - 1$ and also requires that $(2^m - 1, (2^n - 1)/(2^{(m,n)} - 1)) = 1$, which is restrictive.

4   Generalized Multiplexed Sequences

We now generalize the definition of multiplexed sequences by introducing a time-variable function γ. Let a be a binary maximum-length sequence of period $P_a = 2^n - 1$, $n \geq 1$. The generalized multiplexed sequence b is defined by

$$b(t) = a(t + \gamma_t(X(t))), \quad t \geq 0 \tag{11}$$

where X is a periodic integer sequence with period $P_1 = 2^m - 1$, $m \geq 1$, such that $0 \leq X(t) \leq K$, $t \geq 0$, and $\gamma = \{\gamma_t\}_{t=0}^{\infty}$ is a periodic sequence of injective mappings $\{0, \ldots, K\} \rightarrow \{0, \ldots, L\}$, $0 \leq L \leq 2^{n-1} - 1$, with period P_2. In particular, γ may be defined as the composition $\gamma_t = \gamma_2 \gamma_{1,t}$ where γ_2 is a fixed injection $\{0, \ldots, K\} \rightarrow \{0, \ldots, L\}$ and $\gamma_{1,t}$ is a time-variable bijection $\{0, \ldots, K\} \rightarrow \{0, \ldots, K\}$, $t \geq 0$. It can be realized by a multiplexer-like scheme whose address is transformed by a table implementing a set of bijections $\{0, \ldots, K\} \rightarrow \{0, \ldots, K\}$.

In order to prove Theorem 3 specifying the period of b, we need the following lemma.

Lemma 5. Let $(P_1, P_2) = 1$ and $\{X(t) : t \geq 0\} = \{0, \ldots, K\}$. Then the period P of the integer sequence $\gamma(X) = \{\gamma_t(X(t))\}_{t=0}^{\infty}$ is equal to $P_1 P_2$.

Proof. It is clear that $P \mid P_1 P_2$. We first prove that $P_1 \mid P$. The sequence $\gamma(X)$ can be regarded as the interleaved sequence formed from the sequences $a_i = \{\gamma_{i+P_2 t}(X(i + P_2 t))\}_{t=0}^{\infty}$, $0 \leq i \leq P_2 - 1$. Since P_2 is the period of γ, it follows that $a_i = \{\gamma_i(X(i + P_2 t))\}_{t=0}^{\infty}$. For each $0 \leq i \leq P_2 - 1$, the period of a_i equals the period of $\{X(i + P_2 t)\}_{t=0}^{\infty}$, because γ_i is injective, and the period of $\{X(i + P_2 t)\}_{t=0}^{\infty}$ equals P_1, in view of the inverse decimation property holding for $(P_1, P_2) = 1$. Lemma 1 then implies that $P_1 \mid P$.

Second, we prove that $P_2 \mid P$. The sequence $\gamma(X)$ can also be regarded as the interleaved sequence formed from the sequences $b_i = \{\gamma_{i+P_1 t}(X(i + P_1 t))\}_{t=0}^{\infty}$. Since P_1 is the period of X it follows that $b_i = \{\gamma_{i+P_1 t}(X(i))\}_{t=0}^{\infty}$, $0 \leq i \leq P_1 - 1$. Clearly, the period P_2 of γ is equal to l.c.m. $(p_j : 0 \leq j \leq K)$ where p_j denotes the period of $\{\gamma_t(j)\}_{t=0}^{\infty}$, $0 \leq j \leq K$. Since $(P_1, P_2) = 1$, then $(P_1, p_j) = 1$ as well, for each $0 \leq j \leq K$. Due to the inverse decimation property, the period of b_i is then equal to $p_{X(i)}$, $0 \leq i \leq P_1 - 1$. Now, since by supposition X assumes all the values from $\{0, \ldots, K\}$, we then get l.c.m. $(p_{X(i)} : 0 \leq i \leq P_1 - 1) = P_2$, which by virtue of Lemma 1 implies that $P_2 \mid P$. Hence $P_1 P_2 \mid P$ as well, which together with $P \mid P_1 P_2$ proves the lemma.

Theorem 3. Let $n \nmid m$, $(P_2, 2^m - 1) = 1$, $(P_2, 2^n - 1) = 1$, and let X assume all the values from $\{0, \ldots, K\}$. Then the period of the generalized multiplexed sequence is equal to P_2 l.c.m.$(2^m - 1, 2^n - 1)$.

Proof. First, by using Lemmas 4 and 5 one easily proves that the period M of the difference decimation sequence d corresponding to (11) is equal to $P_2(2^m - 1)$, see the proof of Theorem 2. Hence, $M = P_2(2^m - 1)$ and $N = P_2(2^m - 1) \bmod (2^n - 1)$.

In order to apply Theorem 1, one should then check the conditions from Propositions 1 and 2. We prove that the linear complexity condition from Proposition 1 is satisfied as a combination of the condition $2°$ from Proposition 2 and the period condition from Proposition 1. Namely, $a(Nt) = a'(P_2 t)$, where $a'(t) = a((2^m - 1)t)$, $t \geq 0$. Since $n \nmid m$, the decimation of a by $2^m - 1$ satisfies the condition $2°$ from Proposition 2 and hence the linear complexity condition from Proposition 1 as well. Accordingly, the period of a' equals $(2^n - 1)/(2^{(m,n)} - 1) = P'$ by Lemma 2. Further, since $(P_2, P') = 1$, the decimation of a' by P_2 satisfies the period condition from Proposition 1 and hence the linear complexity condition as well. Consequenly, the decimation of a by N satisfies the linear complexity condition from Proposition 1. Therefore, the period P_b of b is determined by Theorem 1, which means that

$$P_b = P_2 \frac{(2^m - 1)(2^n - 1)}{(2^n - 1, P_2(2^m - 1))} = P_2 \text{ l.c.m.}(2^m - 1, 2^n - 1), \qquad (12)$$

because $(P_2, 2^n - 1) = 1$.

If any of the conditions from Theorem 3 is not satisfied, then the period may in general be smaller than l.c.m.$(P_2, 2^m - 1, 2^n - 1)$.

We now analyze the period of a broad class of binary sequences obtained by combining the decimation and multiplexing operations. Let $a = \{a(t)\}_{t=0}^{\infty}$ be a binary maximum-length sequence of period $P_a = 2^n - 1$, $n \geq 1$. The decimated and multiplexed sequence b is defined by

$$b(t) = a(D(t) + \gamma_t(X(t))), \quad t \geq 0 \qquad (13)$$

where the sequences $X = \{X(t)\}_{t=0}^{\infty}$, with period $P_1 = 2^m - 1$, $m \geq 1$, and $\gamma = \{\gamma_t\}_{t=0}^{\infty}$, with period P_2, are defined in the same way as for the generalized multiplexed sequences, and the decimation sequence D is defined in Section 2 in terms of the difference decimation sequence d with period P_3. The decimated and multiplexed sequence can be generated by a multiplexer-like scheme and a clock-controlled linear feedback shift register. Clearly, b can be regarded as a decimated sequence with a difference decimation sequence d' being the sum modulo P_a of d and d'', with d and d'' corresponding to D and $\gamma(X)$, respectively. In view of Lemma 3, in a similar way as for Theorem 3 it is then readily checked that the period P_b of b is determined by the following theorem.

Theorem 4. Let $n \nmid m$, let $2^m - 1$, P_2, and P_3 be pairwise coprime, let $(P_2, 2^n - 1) = 1$ and $(D(P_3), (2^n - 1)/(2^{(m,n)} - 1)) = 1$, and let X assume all the values from $\{0, \ldots, K\}$. Then the period of the decimated and multiplexed sequence is equal to $P_2 P_3$ l.c.m.$(2^m - 1, 2^n - 1)$.

If any of the conditions from Theorem 4 is not fulfilled, then the period may in general be smaller than l.c.m.$(P_2, P_3, 2^m - 1, 2^n - 1)$.

5 MEM-BSG Sequences

In this section we study the period of the variable-memory binary sequence generator (MEM-BSG) proposed and analyzed in [8]. We achieve this by showing that the output sequences of a MEM-BSG belong to the class of generalized multiplexed sequences introduced in Section 4.

We first define a class of the so-called MEM-BSG sequences as a generalization of the class of output sequences of a MEM-BSG [8]. Let $a = \{a(t)\}_{t=0}^{\infty}$ be a binary maximum-length sequence of period $P_a = 2^n - 1$, $n \geq 1$. Let $X = \{X(t)\}_{t=0}^{\infty}$ be a periodic integer sequence with period $P_1 = 2^m - 1$, $m \geq 1$, and the range of values $\{X(t) : t \geq 0\} = \{0, \ldots, K\}$. Let $Y = \{Y(t)\}_{t=0}^{\infty}$ be a periodic integer sequence with period P_2 and the range of values $\{Y(t) : t \geq 0\} = \{0, \ldots, K\}$. Also, assume that for each $0 \leq j \leq K$ and every $t_1 < t_2$ such that $Y(t_1) = Y(t_2) = j$ and $Y(t) \neq j$, $t_1 < t < t_2$, it is true that $t_2 - t_1 \leq L + 1$, where $0 \leq L \leq 2^{n-1} - 1$. It follows that $K \leq L \leq P_2 - 1$. From the practical point of view, the MEM-BSG sequence b can then be generated by a time-varying memory of size $K + 1$ with $X(t)$ defining the read address at time t and $Y(t + L + 1)$ defining the write address of a memory location into which a bit $a(t + L + 1)$ is being written at time t, $t \geq 0$, assuming that the read operation precedes the write one. The initial content of the memory is completely specified by letting the write operation be carried out by using $Y(t)$ and $a(t)$ for $0 \leq t \leq L$, before the first bit of b is generated. More precisely, starting from Y, for each $0 \leq j \leq K$ form the integer sequence $f(j) = \{f(j,t)\}_{i=0}^{\infty}$ so that $f(j,t)$ is defined as the greatest t', $t' < t$, such that $Y(t') = j$; if such t' does not exist, then put $f(j,t) = f(j,P_2) - P_2$. The MEM-BSG sequence b is then defined by

$$b(t) = a(f(X(t),t+L+1)), \quad t \geq 0 \tag{14}$$

noting that $f(j,t) \geq 0$ for every $t \geq L+1$, for each $0 \leq j \leq K$. Equivalently, $b(t)$ is equal to the value of a at the last time t' before $t+L+1$ such that $Y(t') = X(t)$, $t \geq 0$.

In order to study the period of the MEM-BSG sequence b, we put (14) into a more convenient form. For each $0 \leq j \leq K$, introduce the integer sequence $g(j) = \{g(j,t)\}_{i=0}^{\infty}$ by $g(j,t) = f(j,t+L+1) - t$, $t \geq 0$. It follows that $0 \leq g(j,t) \leq L$ and that for each $t \geq 0$ the values $g(j,t)$, $0 \leq j \leq K$, are all distinct. Therefore, (14) reduces to

$$b(t) = a(t + g(X(t),t)) = a(t + \gamma_t(X(t))), \quad t \geq 0 \tag{15}$$

where $\gamma = \{\gamma_t\}_{i=0}^{\infty}$ is a sequence of injective mappings $\{0,\ldots,K\} \to \{0,\ldots,L\}$ defined by $\gamma_t(j) = g(j,t)$, $0 \leq j \leq K$, $t \geq 0$. Moreover, it is easy to check that γ is periodic with period P_2. Consequently, we have shown that every MEM-BSG sequence is a generalized multiplexed sequence, see (11). Theorem 3 then reduces to the following one.

Theorem 5. Let $n \not| m$, $(P_2, 2^m - 1) = 1$, and $(P_2, 2^n - 1) = 1$. Then the period of the MEM-BSG sequence is equal to P_2 l.c.m.$(2^m - 1, 2^n - 1)$.

If any of the conditions from Theorem 5 is not satisfied, then the period may in general be smaller than l.c.m.$(P_2, 2^m - 1, 2^n - 1)$. Theorem 5 strengthens and generalizes the result from [8] regarding the period of a particular MEM-BSG sequence. Namely, for a MEM-BSG with X and Y formed in a special way by using the phase shifts of two maximum-length sequences of periods $P_1 = 2^m - 1$ and $P_2 = 2^l - 1$, respectively, under additional conditions that $P_2 \leq n$ and that m, n, and l are pairwise coprime, it is proved in [8] that the period P_b of the MEM-BSG sequence satisfies $(2^m - 1)(2^n - 1) \mid P_b \mid (2^m - 1)(2^n - 1)(2^l - 1)$. Theorem 5 shows that the period P_b is in fact equal to this upper bound.

6 Conclusion

In this paper we study the periods of various generalizations of the well-known class of multiplexed binary sequences introduced in [12] for cryptographic applications. Multiplexed and generalized multiplexed sequences are defined as appropriate generalizations of the sequences from [12]. Their periods are derived by employing the results from [10] regarding the period of nonuniformly decimated sequences. The corresponding result of Jennings [12, 13] is thus generalized. A broad class of the so-called decimated and multiplexed sequences is also defined and their period is obtained by combining the results on decimated and generalized multiplexed sequences. They may be interesting for cryptographic and spread spectrum applications.

In addition, the so-called MEM-BSG sequences based on the variable-memory binary sequence generator [8] are then proposed. Their period is determined by showing that they belong to the class of generalized multiplexed sequences. The corresponding result from [8] is thus strengthened and generalized.

References

1. R. J. Anderson, "Solving a class of stream ciphers," *Cryptologia*, vol. 14(3), pp. 285-288, 1990.
2. H. Beker and F. Piper, *Cipher Systems: The Protection of Communications.* London: Northwood Publications, 1982.
3. G. R. Blakley and G. B. Purdy, "A necessary and sufficient condition for fundamental periods of cascade machines to be products of the fundamental periods of their constituent finite state machines," *Information Sciences*, vol. 24, pp. 71-91, 1981.
4. W. G. Chambers and S. M. Jennings, "Linear equivalence of certain BRM shift-register sequences," *Electron. Lett.*, vol. 20, pp. 1018-1019, Nov. 1984.
5. W. G. Chambers, "Clock-controlled shift registers in binary sequence generators," *IEE Proc. E*, vol. 135, pp. 17-24, 1988.
6. J. Daemen, R. Govaerts, and J. Vandewalle, "Cryptanalysis of MUX-LFSR based scramblers," *Proc. SPRC '93*, Rome, Italy, pp. 55-61, 1993.
7. J. Dj. Golić and M. V. Živković, "On the linear complexity of nonuniformly decimated *PN*-sequences," *IEEE Trans. Inform. Theory*, vol. IT-34, pp. 1077-1079, Sep. 1988.
8. J. Dj. Golić and M. J. Mihaljević, "Minimal linear equivalent analysis of a variable-memory binary sequence generator," *IEEE Trans. Inform. Theory*, vol. IT-36, pp. 190-192, Jan. 1990.
9. J. Dj. Golić, M. Salmasizadeh, and E. Dawson, "Autocorrelation weakness of multiplexed sequences," *Proceedings of International Symposium on Information Theory and Its Applications ISITA '94*, Sydney, Australia, pp. 983-987, 1994.
10. J. Dj. Golić, "A note on nonuniform decimation of periodic sequences," Cryptography: Policy and Algorithms - Brisbane '95, *Lecture Notes in Computer Science*, vol. 1029, E. Dawson and J. Golić eds., Springer-Verlag, pp. 125-131, 1996.
11. S. W. Golomb. *Shift Register Sequences.* San Francisco: Holden-Day, 1967.
12. S. M. Jennings, "A special class of binary sequences," Ph.D. thesis, University of London, 1980.
13. S. M. Jennings, "Multiplexed sequences: some properties of the minimum polynomial," Proc. Workshop on Cryptography, Burg Feuerstein, 1982, *Lecture Notes in Computer Science*, vol. 149, T. Beth ed., Springer-Verlag, pp. 189-206, 1983.
14. S. M. Jennings, "Autocorrelation function of the multiplexed sequences," *IEE Proc. F*, vol. 131, pp. 169-172, April 1984.
15. R. Lidl and H. Niederreiter, *Introduction to finite fields and their applications.* Cambridge: Cambridge University Press, 1986.
16. K. C. Zeng, C. H. Yang, and T. R. N. Rao, "On the linear consistency test (LCT) in cryptanalysis and its applications," Advances in Cryptology - CRYPTO '89,

Lecture Notes in Computer Science, vol. 435, G. Brassard ed., Springer-Verlag, pp. 164-174, 1990.

17. N. Zierler, "Linear recurring sequences," *J. Soc. Indust. Appl. Math.*, vol. 7, pp. 31-48, 1959.

18. Specification of the systems of the MAC/packet family. EBU Technical Document 3258-E, Oct. 1986.

Edit Distance Correlation Attacks on Clock-Controlled Combiners with Memory

Jovan Dj. Golić

Information Security Research Centre, Queensland University of Technology
GPO Box 2434, Brisbane Q 4001, Australia
School of Electrical Engineering, University of Belgrade
Email: golic@fit.qut.edu.au

Abstract. Edit distance based correlation attacks on binary keystream generators consisting of clock-controlled shift registers combined by a function with memory are introduced. Recursive algorithms for efficient computation of the proposed many-to-one string edit distances are derived for both the constrained and unconstrained irregular clocking. The distances are based on mutually correlated input and output feedforward linear transforms for regularly clocked combiners with memory. Linear transforms can also incorporate linear models of clock-controlled shift registers. In particular, linear transforms and the corresponding correlation coefficients are obtained for a special type of combiners with memory based on a time-varying memoryless function.

1 Introduction

Use of functions with memory in keystream generators was suggested in [16] to avoid the trade-off between the correlation immunity and the linear complexity shown in [18]. Irregular clocking in keystream generators is a well-known way of achieving long periods and high linear complexities [11] as well as the immunity to fast correlation attacks [13]. By combining the two principles, we obtain a clock-controlled combiner with memory as a binary keystream generator consisting of n clock-controlled shift registers, with not necessarily linear feedback, combined by a function with m bits of memory. When $m = 0$ the combiner is memoryless, see [19]. Concrete keystream generators of this type have not been proposed in the open literature, for example, see [17] and [11], although such a type can be regarded as a general model for any shift register based keystream generator. Input sequences to the combiner are decimated versions of the regularly clocked shift register sequences. Irregular clocking of the shift registers is controlled by n individual clock-generators in a constrained [3] or unconstrained [15], [1] way. In both the cases the number of clocks per output symbol is at least one, whereas in the constrained case this number is also upper-bounded by a positive integer, possibly different for different shift registers. It is assumed that the shift register feedback functions and the function with memory are known. The secret key controls the initial states of the shift registers, the initial state of the function with memory, and also the initial states and the structure of the clock-generators. The objective is to develop divide and conquer correlation attacks, that is, to reconstruct the initial contents of individual shift registers given a segment of the keystream sequence, based on the correlation between this sequence and a subset of the input shift register sequences. Since the clock-controlling sequences are not known, one should consider the correlation to regularly clocked shift register sequences. The main problems considered in this paper and posed in [6] are how to measure this correlation and how to achieve a divide and conquer effect.

When the shift registers are clocked regularly and the combining function is memoryless and zero-order correlation immune [18] (its output is correlated to at least one input), it is possible to apply the well-known correlation attack in its basic [19] or fast [13] form in order to reconstruct the initial state of the corresponding shift register. The attack [19] is based on the Hamming distance between the output and the shift register sequence. Fast correlation attacks are based on iterative probabilistic decoding procedures. Generalization of these attacks from a zero-order to a kth-order correlation immune combining function (its output is correlated to at least one set of $k+1$ inputs and is not correlated to any set of k inputs) is straightforward by using the fact that the output of a kth-order correlation immune function is correlated to at least one linear function of exactly $k+1$ input variables [20]. One should then consider the Hamming distance between the output sequence and a sum of the shift register sequences.

Correlation properties of regularly clocked combiners with one bit of memory are established in [14]. For a general combiner with m bits of memory, it is shown in [4], [9] that one should consider the Hamming distance based correlation between the linear functions of at most $m+1$ successive inputs and outputs, respectively. An efficient procedure, based on the linear sequential circuit approximation of functions with memory, for finding such pairs of mutually correlated input and output linear functions is also developed in [4], [9].

When the shift registers are clocked irregularly and the clock-controlling sequences are not known, the Hamming distance based correlations can not be used. However, in a particular case when the combining function is memoryless and zero-order correlation immune, one may apply the so-called generalized correlation attack based on the one-to-one string constrained Levenshtein-like distance [3] or on the constrained edit probability [5], in order to reconstruct the initial state of the corresponding shift register. The unconstrained case can be treated similarly, see [7].

Our main objective in this paper is to extend the results from [3] to arbitrary combining functions with or without memory, by introducing the appropriate edit distances incorporating the linear correlations [4], [9] of regularly clocked combiners with memory. Note that even the generalization to kth-order correlation immune memoryless functions is not trivial because the clock-controlling sequences need not be the same for different shift registers. We will thus define a many-to-one string edit distance as the minimum sum of elementary edit distances associated with the edit operations of deletion and substitution needed to transform a set of input strings into a single output string, where the transformation consists of the deletions of symbols in the input strings and the substitutions of symbols in the combination string obtained by a feedforward linear transform of the decimated input strings. The combinatorial optimization problem of computing this distance efficiently is solved by deriving a recursive algorithm for a suitably defined partial edit distance, for both the constrained and unconstrained case. The use of the proposed edit distances for a correlation attack on clock-controlled combiners with memory is then discussed. Its divide and conquer effect depends on a particular linearization of the function with memory. The feedforward linear transform to be used is the corresponding linear functior. of the input sequences that is found to be correlated to a linear function of the output sequence.

Moreover, the feedforward linear transform can also incorporate the linear models of clock-controlled shift registers [8], [10], which may considerably increase the divide and conquer effect. This is especially interesting for the stop-and-go shift registers [11], because the corresponding correlation coefficients are large. Accordingly, the combiners with memory may also involve the stop-and-go shift registers which are replaced by the corresponding linear models. The computational complexity to obtain the edit distances is exponential in the memory size of the feedforward linear transform which can be considerably smaller than the memory size of the combiner. This is demonstrated on a particular type of combiners with memory based on a time-varying memoryless function, see [12] or [2], for example. The combining function changes slowly in time under the control of another generator thus destroying the termwise correlation between the output and inputs. However, by the linear sequential circuit approximation method we show that the sum of the two consecutive output bits is then correlated to the sum of the two corresponding consecutive bits of any individual input or of any linear combination of the inputs. The value of the corresponding correlation coefficient is also determined. The feedforward linear transform with one bit of memory for the proposed edit distances is then the sum of the two consecutive bits in any single input sequence.

The rest of the paper is organized as follows. A many-to-one string edit distance based on linear approximations of functions with memory is defined in Section 2, along with a recursive algorithm for its efficient computation. The proof of the algorithm is given in the Appendix. The corresponding correlation attack is discussed in Section 3. The special case of a combiner with memory based on a time-varying memoryless function is analyzed in Section 4. Conclusions are presented in Section 5.

2 Edit Distances

Let $\mathbf{a}_k = (a_{k,t})_{t=1}^{N_k}$, $1 \leq k \leq K$, and $\mathbf{b} = (b_t)_{t=1}^{M}$ be $K + 1$ binary strings and let $L = \sum_{k=1}^{K} \sum_{j=0}^{m_k} l_{kj} x_{kj}$ be a linear boolean function such that $l_{k0} = l_{k,m_k} = 1$ for every $1 \leq k \leq K$. An edit transformation of the ordered set of input strings $(\mathbf{a}_1, \ldots, \mathbf{a}_K)$ into the output string $\mathbf{b}$ is defined in the following way: first, delete some symbols-bits from the input strings and thus obtain the decimated strings $\hat{\mathbf{a}}_k = (\hat{a}_{k,t})_{t=1}^{m_k+M}$, $1 \leq k \leq K$; second, apply the linear transform L in the

feedforward way to the decimated strings and thus generate a combination string
$\mathbf{c} = (c_t)_{t=1}^M$ by

$$c_t = \sum_{k=1}^K \sum_{j=0}^{m_k} l_{kj}\, \hat{a}_{k,t+j}, \quad 1 \le t \le M; \tag{1}$$

and, third, substitute some bits in the combination string $\mathbf{c}$ to obtain the output string $\mathbf{b}$. Let $m = \sum_{k=1}^K m_k$ denote the total memory size of the feedforward linear transform. When $m = 0$, the transform is memoryless. An edit transformation is called *constrained* or *unconstrained* with respect to the kth input string $\mathbf{a}_k$ depending on whether the number of consecutive deletions in that string is upper-bounded by a positive integer E_k or not. In both the cases $M \le N_k - m_k$ must hold, while in the constrained case we in addition have $N_k - m_k \le M + E_k(M + 1)$.

Elementary edit distances are generally defined as nonnegative real-valued functions:

- $d_k(x)$ is the elementary distance associated with deleting a bit x from the string $\mathbf{a}_k$, $1 \le k \le K$;
- $d(x, y)$ is the elementary distance associated with substituting a bit y for a bit x.

Edit distance D_L between an ordered set of input strings $(\mathbf{a}_1, \ldots, \mathbf{a}_K)$ and an output string $\mathbf{b}$ is defined as the minimum sum of elementary edit distances associated with the edit operations of deletion and substitution needed to transform $(\mathbf{a}_1, \ldots, \mathbf{a}_K)$ into $\mathbf{b}$. The edit distance is thus a solution to a combinatorial optimization problem over the set of the permitted edit transformations. To this end, we represent an edit transformation sequentially, as a sequence $\mathcal{E} = (\varepsilon_i)_{i=1}^T$, $T = \sum_{k=1}^K N_k - (K-1)M$, of deletions and substitutions. First note that until the initial conditions for the feedforward linear transform (1) are obtained, the first m_k substitutions in the kth input string are all dummy, for each $1 \le k \le K$, and only mean that the corresponding bits are not deleted. There are exactly $N_k - m_k - M$ deletions and m_k dummy substitutions in the kth input string and M substitutions in the combination string. Let for any $1 \le i \le T$, $\varepsilon_i = k$ and $\varepsilon_i = \tilde{k}$ denote the deletion and the dummy substitution in the kth input string, respectively, and let $\varepsilon_i = 0$ denote the substitution in the combination string $(c_t)_{t=1}^M$, which is formed according to (1). Both the deletions and substitutions take place immediately after the positions of the previously edited symbols, in each of the input strings. In order to ensure the unique sequential representation of edit transformations, it is also assumed that in any sequence of consecutive deletions one first deletes the symbols from $\mathbf{a}_1$, then from $\mathbf{a}_2$, and so on. Let $G(\mathbf{a}_1, \ldots, \mathbf{a}_K; \mathbf{b})$ denote the set of all the permitted edit sequences $\mathcal{E}$. Given $(\mathbf{a}_1, \ldots, \mathbf{a}_K)$, $\mathbf{b}$, and $\mathcal{E}$, the symbols to be deleted or substituted are uniquely determined. Consequently, for each edit operation ε_i one can define the corresponding elementary edit distance $\lambda(\varepsilon_i)$, where $\lambda(\varepsilon_i) = 0$ for the initial dummy substitutions in the input strings.

Example. Let $K = 1$, $m_1 = 1$, $L = x_{1,1} + x_{1,2}$, $E_1 = 1$, $N_1 = 9$, $M = 5$, $\mathbf{a}_1 = 100011101$, and $\mathbf{b} = 10110$. A permitted edit sequence is then $\mathcal{E} = \tilde{1}, 1, 0, 0, 1, 0, 1, 0, 0$, which gives rise to the decimated string $\hat{\mathbf{a}}_1 = 100101$ and the combination string $\mathbf{c} = 10111$. So, there are 3 deletions and 1 dummy substitution in the input string and 1 effective substitution in the combination string.

The edit distance can be expressed in terms of edit sequences by

$$D_L(\mathbf{a}_1, \ldots, \mathbf{a}_K; \mathbf{b}) = \min \left\{ \sum_{i=1}^{T} \lambda(\varepsilon_i) : \mathcal{E} \in G(\mathbf{a}_1, \ldots, \mathbf{a}_K; \mathbf{b}) \right\}. \quad (2)$$

The minimization by exhaustive search has exponential complexity and is not feasible except for very small string lengths. Using the one-to-one string approach from [3], we can define the *partial edit distance* $W(e_1, \ldots, e_K, s)$ as the edit distance $D_L(\mathbf{a}_1^{(m_1 + e_1 + s)}, \ldots, \mathbf{a}_K^{(m_K + e_K + s)}; \mathbf{b}^{(s)})$ between the ordered set of prefixes $(\mathbf{a}_1^{(m_1 + e_1 + s)}, \ldots, \mathbf{a}_K^{(m_K + e_K + s)})$ of lengths $m_1 + e_1 + s, \ldots, m_K + e_K + s$, respectively, and the prefix $\mathbf{b}^{(s)}$ of length s, where for an arbitrary string $\mathbf{a}$, $\mathbf{a}^{(n)}$ denotes the prefix of length n. It follows that

$$D_L(\mathbf{a}_1, \ldots, \mathbf{a}_K; \mathbf{b}) = W(N_1 - m_1 - M, \ldots, N_K - m_K - M, M). \quad (3)$$

However, it is impossible to derive a recursion for $W(e_1, \ldots, e_K, s)$ as in [3], except when the linear feedforward transform is memoryless. In order to solve the general problem, the main idea is to introduce a new *partial edit distance* as the edit distance between $(\mathbf{a}_1^{(m_1 + e_1 + s)}, \ldots, \mathbf{a}_K^{(m_K + e_K + s)})$ and $\mathbf{b}^{(s)}$ under an additional constraint that the last m_k undeleted symbols in the kth input prefix string are given for each $1 \leq k \leq K$, that is,

$$W(\underline{U}, \underline{e}, s) = \min \left\{ \sum_{i=1}^{m_1 + e_1 + \cdots + m_K + e_K + s} \lambda(\varepsilon_i) : \mathcal{E} \in G_{\underline{U}, \underline{e}, s} \right\} \quad (4)$$

where $\underline{e} = (e_1, \ldots, e_K)$, $\underline{U} = (U_1, \ldots, U_K)$, $U_k = (u_{kj})_{j=1}^{m_k}$ is an m_k-tuple of the last m_k undeleted bits in the kth input string, and $G_{\underline{U}, \underline{e}, s}(\mathbf{a}_1, \ldots, \mathbf{a}_K; \mathbf{b})$ or, simply, $G_{\underline{U}, \underline{e}, s}$ denotes the set of all the permitted edit sequences for the considered prefix strings. Then

$$W(\underline{e}, s) = \min\{W(\underline{U}, \underline{e}, s) : \underline{U} \in Z_2^m\} \quad (5)$$

where $Z_2 = \{0, 1\}$ and $m = \sum_{k=1}^{K} m_k$ is the memory size of the feedforward linear transform. The set of the permissible values of $(e_1, \ldots, e_K, s)$ is clearly

$$0 \leq s \leq M \quad (6)$$

$$0 \leq e_k \leq \min\{N_k - m_k - M, (m_k + s + 1)E_k\}, \quad 1 \leq k \leq K, \quad (7)$$

where $E_k = \infty$ in the unconstrained case. It is assumed that $W(\underline{U}, \underline{e}, 0) = \infty$ if for at least one k the m_k-tuple U_k can not be obtained by deleting e_k bits from the prefix $\mathbf{a}_k^{(m_k+e_k)}$.

Our objective is to establish a recursive property of $W(\underline{U}, \underline{e}, s)$ which will in view of (3) and (5) enable efficient computation of the constrained/unconstrained edit distance $D_L(\mathbf{a}_1, \ldots, \mathbf{a}_K; \mathbf{b})$. Assume for simplicity that the elementary edit distance associated with deletion is symbol-independent in each string $\mathbf{a}_k$, that is, $d_k(x) = d_k^*$, $x \in Z_2$, $1 \leq k \leq K$. The following theorem gives a recursion for computing the partial edit distance in the constrained case. Its proof is given in the Appendix. Let $\underline{U}' = (U_1', \ldots, U_K')$, $U_k' = (v_k, u_{k1}, u_{k2}, \ldots, u_{k,m_k-1})$, and $\underline{v} = (v_1, \ldots, v_K)$, where $\underline{U} = (U_1, \ldots, U_K)$, $U_k = (u_{kj})_{j=1}^{m_k}$, and all the variables are binary. Let $\underline{\delta} = (\delta_1, \ldots, \delta_K)$, $0 \leq \delta_k \leq E_k$, and let $\underline{e} - \underline{\delta} = (e_1 - \delta_1, \ldots, e_K - \delta_K)$. Since we need a different initial recursion for computing $W(\underline{U}, \underline{e}, 0)$ for $s = 0$, for any $U_k \in Z_2^{m_k}$, any $0 \leq n_k \leq m_k$, any permissible $0 \leq j_k \leq \min\{e_k, (n_k+1)E_k\}$, and each $1 \leq k \leq K$, we in addition introduce an auxiliary partial edit distance $W_k'(j_k, n_k)$, as the edit distance between the prefix $U_k^{(n_k)}$ and the prefix $\mathbf{a}_k^{(n_k+j_k)}$ where only deletions and dummy substitutions are allowed. The elementary edit distance $d'(x, y)$ associated with the dummy substitution is assumed to be zero if x and y are equal and infinite if they are different. This means that $W_k'(j_k, n_k)$ is infinite if $U_k^{(n_k)}$ can not be obtained as a decimated version of $\mathbf{a}_k^{(n_k+j_k)}$, under the given constraints. The partial edit distances are assumed to be infinite whenever the minimization is taken over an empty set or when the the argument values are not permissible.

Theorem 1. In the constrained case, the partial edit distance $W(\underline{U}, \underline{e}, s)$ satisfies the following recursion for $1 \leq s \leq M$

$$W(\underline{U}, \underline{e}, s) = \min_{\underline{\delta}, \underline{v}} \left\{ W(\underline{U}', \underline{e} - \underline{\delta}, s-1) + \sum_{k=1}^{K} \delta_k d_k^* + \right.$$

$$\left. d(\sum_{k=1}^{K} v_k + \sum_{k=1}^{K} \sum_{j=1}^{m_k} l_{kj} u_{kj}, b_s) : a_{k,m_k+e_k-\delta_k+s} = u_{k,m_k}, \quad 1 \leq k \leq K \right\} \quad (8)$$

and for $s = 0$ is given by

$$W(\underline{U}, \underline{e}, 0) = \sum_{k=1}^{K} W_k'(e_k, m_k) \quad (9)$$

where $W_k'(j_k, n_k)$ for $1 \leq n_k \leq m_k$ satisfies the recursion

$$W_k'(j_k, n_k) = \min_{\delta_k} \{W_k'(j_k - \delta_k, n_k - 1) + \delta_k d_k^* + d'(a_{k,n_k+j_k-\delta_k}, u_{k,n_k})\} \quad (10)$$

with the initial condition $W_k'(j_k, 0) = j_k d_k^*$, $1 \leq k \leq K$.

Clearly, one can combine Theorem 1 together with (3) and (5) in a recursive algorithm for the efficient computation of the edit distance. The computational complexity is determined by the number of argument values for which the computation of the partial edit distances is performed and by the cardinality of the set over which the minimizations in (8) and (10) are taken. Therefore, the time complexity is $2^m 2^K \prod_{k=1}^{K}(E_k + 1) \, O(M \prod_{k=1}^{K}(N_k - M + 1))$. As for the space complexity, it is not necessary to store the values of the partial edit distances for all the permissible s, but only for the two consecutive values of s at a time. Hence the space complexity is $2^{m+1} O(\prod_{k=1}^{K}(N_k - M + 1))$.

In the unconstrained case, the partial edit distance is determined by the following theorem, which can be proved by a similar technique as Theorem 1. Note that the time complexity is then reduced to $2^m (2^K + K) O(M \prod_{k=1}^{K}(N_k - M + 1))$. Let ε_k denote the binary K-tuple consisting of $K - 1$ zeros and a single one at the kth position, $1 \leq k \leq K$.

Theorem 2. In the unconstrained case, the partial edit distance $W(\underline{U}, \underline{e}, s)$ satisfies the following recursion for $1 \leq s \leq M$

$$W(\underline{U}, \underline{e}, s) = \min \left\{ \min_{1 \leq k \leq K} \{W(\underline{U}, \underline{e} - \varepsilon_k, s) + d_k^*\}, \right.$$

$$\min_{\underline{v}} \left\{ W(\underline{U}', \underline{e}, s - 1) + d(\sum_{k=1}^{K} v_k + \sum_{k=1}^{K} \sum_{j=1}^{m_k} l_{kj} u_{kj}, b_s) : \right.$$

$$\left. \left. a_{k,m_k + e_k + s} = u_{k,m_k}, \quad 1 \leq k \leq K \right\} \right\} \qquad (11)$$

and for $s = 0$ is given by (9) where $W_k'(j_k, n_k)$ for $0 \leq n_k \leq m_k$ satisfies the recursion

$$W_k'(j_k, n_k) = \min \{W_k'(j_k - 1, n_k) + d_k^*, \, W_k'(j_k, n_k - 1) + d'(a_{k,n_k + j_k}, u_{k,n_k})\} \qquad (12)$$

with the initial condition $W_k'(0,0) = 0, \, 1 \leq k \leq K$.

In correlation attacks, the edit sequences are random and so are the lengths of input strings given the length of the output one. It is then convenient to define a modified constrained edit distance for which an arbitrary number of deletions is allowed at the end of each input string. It follows that the modified edit distance is again given by (3) and (5) where the corresponding partial edit distance is determined by the recursion from Theorem 1 for $0 \leq s \leq M-1$, and for $s = M$ by the recursion from Theorem 2. In addition, by combining Theorems 1 and 2 one may obtain a general theorem for the combined constrained and unconstrained case.

3   Correlation Attack

Consider a binary keystream generator consisting of n clock-controlled shift registers combined by a function with m bits of memory, as described in Section 1. Irregular clocking is constrained, unconstrained, or combined. The objective of correlation attacks is to reconstruct the initial contents of individual shift registers given a segment of the keystream sequence. The starting point is a pair of mutually correlated feedforward linear transforms of the output and the input, respectively, in the same, but regularly clocked combiner. They could be obtained by the linear sequential circuit approximation method [4], [9] which works even for very large memory sizes. Let c denote the corresponding correlation coefficient different from zero, let $K, 1 \leq K \leq n$, be the number of individual inputs to the feedforward linear transform L of the input, and let m' be the memory size of L. Of course, the goal of a cryptanalyst is that c be large and that K and m' be small. In particular, if m is large, then m' should be smaller than m. Note that m' may be equal to zero not only for memoryless combiners, for which $m = 0$. The divide and conquer effect exists even if $K = n$ (for example, when the combiner is maximum-order correlation immune), because the shift register initial contents may then be reconstructed regardless of the unknown clock-control generators. Note that the divide and conquer effect can be considerably increased if the feedforward linear transforms make use of the linear models of the clock-controlled shift registers [8], [10]. This is especially effective for the stop-and-go shift registers [11] since the corresponding correlation coefficients are large.

The edit distances are then defined between the K regularly clocked shift register sequences (for the assumed initial contents) as the input strings and the linearly transformed keystream sequence (by the output feedforward linear transform) as the output string. Given a length M of the transformed keystream sequence, the lengths N_k, $1 \leq k \leq K$, of the input strings should be chosen to be larger than their expected values for random decimation sequences. In the constrained case, for the modified constrained edit distance, the lengths could be assumed to be maximum possible given M. For the edit distance based correlation attack, we can assume that the elementary edit distances associated with deletions and effective substitutions are all equal to one. The edit distances can then be computed by the recursive algorithms given in Sections 2. The time and space complexities for computing the edit distances are both exponential in m' and K.

With the edit distances so obtained, one should then apply essentially the same statistical procedure as the one proposed for a memoryless zero-order correlation immune combiner [3]. More precisely, compute an appropriate edit distance for every possible combination of the shift register initial contents and then choose the combination with the minimum distance. The underlying assumption for the correlation attack to be successful is that only the correct guess about the unknown shift register initial states, or just a few of its modifications, should yield the minimum value of the edit distance, provided that the observed segment of the keystream sequence is sufficiently long. To determine the required length theoretically, one has to solve the corresponding statistical problem, which for $K > 1$ appears to be different and more difficult than the particular instance corresponding to the one-to-one string memoryless case already discussed in [3]. The empirical results from [3] along with the theoretical results and observations from [7] for the one-to-one string case without effective substitutions ($c = \pm 1$) and with memoryless linear transform lead us to conjecture that the constrained edit distance correlation attack is successful for a sufficiently long observed keystream sequence provided that the effective substitution probability is different from one half ($c \neq 0$), see also [21]. A linear transform with memory and multidimensional input are not expected to make any essential difference in this respect. In view of [7], for the success of the unconstrained edit distance correlation attack it is necessary that the deletion rate (i.e., the relative expected number of deleted symbols) for each of the involved shift registers should be smaller than one half. If the deletion rate is bigger, then the use of appropriate edit probabilities instead of edit distances is required, but that is out of the scope of this paper. Under a reasonable assumption that the required keystream sequence length is linear in the lengths of the involved shift registers, this length can be obtained experimentally without having to do the exhaustive search over all the shift register initial contents.

4   Time-Varying Memoryless Combiner

A time-varying memoryless combiner is a binary keystream generator consisting of n clock-controlled shift registers combined by a memoryless function $f_t, t \geq 0$, that slowly changes in time under the control of another generator, see [12] or [2], for example. This means that only one, as in [12] or [2], or just a few, at most τ, bits in the truth table of f_t are changed per each output bit. This destroys the termwise correlation between the output and the inputs, because the positive and negative correlations cancel out in time for any given linear combination of the inputs. It turns out that the time-varying memoryless combiner is a combiner with memory whose state vector includes the 2^n bits specifying f_t and the internal state of the control generator. Our aim is to study the linear correlations between the output and the inputs by using the linear sequential circuit approximation method from [4], [9]. The method consists in finding the linear approximations of the output function and the component next-state functions of the combiner with memory, and then in solving the corresponding linear sequential circuit by the generating function technique, where the sequences produced by the correlation noise functions resulting from the linear approximations are regarded as the additional inputs.

When the 2^n state bits specifying f_t are approximated by the 2^n state bits specifying f_{t-1}, it follows that the the sum of two consecutive output bits at any time $t \geq 1$ is correlated to the sum of two consecutive bits of any input at the same time. The same holds for any linear combination of the inputs. Under an assumption that at any time on average $\tau/2$ bits are changed independently of the truth table, the correlation coefficient c is then equal to the product of

$1 - \tau/2^n$, which is close to 1, and the average value of the square of the correlation coefficient to the constant zero function of a randomly chosen boolean function of n variables which is easily seen to be 2^{-n}. A similar result is true even if one assumes that f_t is balanced for every $t \geq 0$. The corresponding input feedforward linear transform L has a single input, $K = 1$, and memory size only $m' = 1$. This demonstrates that the the memory size of L can be considerably smaller than the memory size of the whole combiner with memory. When the shift registers are clock-controlled, one may then apply the edit distance correlation attack described in Section 3.

5 Conclusion

In this paper we investigate the security of clock-controlled combiners with memory, consisting of clock-controlled shift registers combined by an arbitrary function with or without memory, with respect to correlation attacks. Irregular clocking may be constrained [3], unconstrained [15], [1], or combined. By using mutually correlated input and output feedforward linear transforms [4], [9] for regularly clocked combiners with memory, we develop correlation attacks on clock-controlled combiners with memory based on appropriately defined many-to-one string edit distances. The feedforward linear transforms can incorporate linear models of clock-controlled shift registers [8], [10]. This may considerably increase the divide and conquer effect, especially for keystream generators using the stop-and-go shift registers. Recursive algorithms for the efficient computation of the edit distances are derived for both the constrained and unconstrained case. The computational complexity is exponential in the memory size of the input feedforward linear transform which can be considerably smaller than the combiner memory size. This is demonstrated on a special type of combiners with memory based on a time-varying memoryless function.

The constrained Levenshtein-like distance correlation attack [3] on zero-order correlation immune memoryless combiners is thus extended to arbitrary combiners with memory. It remains to investigate whether similar results can be obtained for edit probabilities which would extend the constrained edit probability correlation attack [5]. The conditions for the proposed correlation attack to be successful depend on the solutions of the corresponding statistical problems, which appear to be different and more difficult than the particular yet unsolved one discussed in [3]. However, empirical results from [3] together with the results from [7] for the one-to-one string case without effective substitutions suggest that the constrained edit distance correlation attack is successful if the observed keystream sequence is sufficiently long. An analogous conclusion holds for the unconstrained edit distance correlation attack, with a difference that the deletion rate for each of the involved shift registers is required to be smaller than one half. The required keystream sequence length can be obtained by experiments.

Appendix

Proof of Theorem 1. For $s = 0$, (9) and (10) are direct consequences of the recursion from [3] for the one-to-one string edit distance corresponding to deletions and dummy substitutions. Assume now that $s \geq 1$. We start from the basic expression (4) for $W(\underline{U}, \underline{e}, s)$. The main idea is to partition the set $G_{\underline{U},\underline{e},s}$ of the permitted edit sequences according to the sequence of deletions *after the last substitution* and also according to the last m_k undeleted symbols in the kth input prefix string *before the last substitution* for each $1 \leq k \leq K$. Namely, let $G_{\underline{U},\underline{e},s}^{\underline{v},\underline{\delta}}$ denote the set of all the edit sequences in $G_{\underline{U},\underline{e},s}$ that end in a substitution followed by exactly δ_k deletions in $\mathbf{a}_k^{(m_k+e_k+s)}$ and that satisfy an additional condition that the last m_k undeleted symbols in the kth input string before the last substitution are given by $U_k' = (u_{kj}')_{j=1}^{m_k} = (v_k, u_{k1}, u_{k2}, \ldots, u_{k,m_k-1})$, where v_k is a binary variable, as well as that $a_{k,m_k+e_k-\delta_k+s} = u_{k,m_k}, 1 \leq k \leq K$. Rewriting (4) accordingly, we obtain

$$W(\underline{U}, \underline{e}, s) = \min_{\underline{v}} \left\{ \min_{\underline{\delta}} \left\{ \min \left\{ \sum_{i=1}^{m_1+e_1-\delta_1+\cdots+m_K+e_K-\delta_K+s-1} \lambda(\varepsilon_i) + \right. \right. \right.$$

$$\left. + d(\sum_{k=1}^{K} v_k + \sum_{k=1}^{K}\sum_{j=1}^{m_k} l_{kj}u_{kj}, b_s) + \sum_{k=1}^{K} \delta_k d_k^* : \right.$$

$$\left. \left. \left. \mathcal{E} \in G_{\underline{U},\underline{e},s}^{\underline{v},\underline{\delta}}, a_{k,m_k+e_k-\delta_k+s} = u_{k,m_k}, 1 \leq k \leq K \right\} \right\} \right\} \quad (13)$$

breaking the sum in (4) into the three parts corresponding to the sequence of edit operations before the last substitution, to the last substitution, and to the sequence of deletions after the last substitution, respectively. Given $\underline{v}$ and $\underline{\delta}$, the contributions of the second and the third part of the sum are constant and hence

$$W(\underline{U}, \underline{e}, s) = \min_{\underline{v}} \left\{ \min_{\underline{\delta}} \left\{ d(\sum_{k=1}^{K} v_k + \sum_{k=1}^{K}\sum_{j=1}^{m_k} l_{kj}u_{kj}, b_s) + \sum_{k=1}^{K} \delta_k d_k^* + \right. \right.$$

$$\min \left\{ \sum_{i=1}^{m_1+e_1-\delta_1+\cdots+m_K+e_K-\delta_K+s-1} \lambda(\varepsilon_i) : \mathcal{E} \in G_{\underline{U},\underline{e},s}^{\underline{v},\underline{\delta}} \right\} :$$

$$\left. \left. a_{k,m_k+e_k-\delta_k+s} = u_{k,m_k}, 1 \leq k \leq K \right\} \right\}. \quad (14)$$

Thus we have come to the critical point: the minimization over $G_{\underline{U},\underline{e},s}^{\underline{v},\underline{\delta}}$. Since every edit sequence in $G_{\underline{U},\underline{e},s}^{\underline{v},\underline{\delta}}$ ends in one substitution followed by exactly δ_k deletions in $\mathbf{a}_k^{(m_k+e_k+s)}$, $1 \leq k \leq K$, and the constraints relate to the number of consecutive deletions, it follows that an edit sequence $\mathcal{E}$ of length $m_1 + e_1 +$

$\cdots + m_K + e_K + s$ belongs to $G_{\underline{U},\underline{e},s}^{\underline{v},\underline{\delta}}$ if and only if its prefix of length $m_1 + e_1 - \delta_1 + \cdots + m_K + e_K - \delta_K + s - 1$ belongs to $G_{\underline{U}',\underline{e}-\underline{\delta},s-1}$. Therefore, by virtue of (4), we have

$$\min \left\{ \sum_{i=1}^{m_1+e_1-\delta_1+\cdots+m_K+e_K-\delta_K+s-1} \lambda(\varepsilon_i) : \mathcal{E} \in G_{\underline{U},\underline{e},s}^{\underline{v},\underline{\delta}} \right\} = W(\underline{U}',\underline{e}-\underline{\delta},s-1).$$

$$(15)$$

Consequently, (14) reduces to (8).

References

1. D. Coppersmith, H. Krawczyk, and Y. Mansour, "The shrinking generator," Advances in Cryptology - CRYPTO '93, *Lecture Notes in Computer Science*, vol. 773, D. R. Stinson ed., Springer-Verlag, pp. 22-39, 1994.

2. J. Dj. Golić and M. J. Mihaljević, "Minimal linear equivalent analysis of a variable-memory binary sequence generator," *IEEE Trans. Inform. Theory*, vol. IT-36, pp. 190-192, Jan. 1990.

3. J. Dj. Golić and M. J. Mihaljević, "A generalized correlation attack on a class of stream ciphers based on the Levenshtein distance," *J. Cryptology*, vol. 3(3), pp. 201-212, 1991.

4. J. Dj. Golić, "Correlation via linear sequential circuit approximation of combiners with memory," Advances in Cryptology - EUROCRYPT '92, *Lecture Notes in Computer Science*, vol. 658, R. A. Rueppel ed., Springer-Verlag, pp. 113-123, 1993.

5. J. Dj. Golić and S. V. Petrović, "A generalized correlation attack with a probabilistic constrained edit distance," Advances in Cryptology - EUROCRYPT '92, *Lecture Notes in Computer Science*, vol. 658, R. A. Rueppel ed., Springer-Verlag, pp. 472-476, 1993.

6. J. Dj. Golić, "On the security of shift register based keystream generators," Fast Software Encryption - Cambridge '93, *Lecture Notes in Computer Science*, vol. 809, R. J. Anderson ed., Springer-Verlag, pp. 90-100, 1994.

7. J. Dj. Golić and L. O'Connor, "Embedding and probabilistic correlation attacks on clock-controlled shift registers," Advances in Cryptology - EUROCRYPT '94, *Lecture Notes in Computer Science*, vol. 950, A. De Santis ed., Springer-Verlag, pp. 230-243, 1995.

8. J. Dj. Golić, "Intrinsic statistical weakness of keystream generators," Advances in Cryptology - ASIACRYPT '94, *Lecture Notes in Computer Science*, vol. 917, J. Pieprzyk and R. Safavi-Naini eds., Springer-Verlag, pp. 91-103, 1995.

9. J. Dj. Golić, "Correlation properties of a general binary combiner with memory," *J. Cryptology*, vol. 9(2), pp. 111-126, 1996.

10. J. Dj. Golić, "Linear models for keystream generators," *IEEE Trans. Comput.*, vol. C-45, pp. 41-49, 1996.

11. D. Gollmann and W. G. Chambers, "Clock-controlled shift registers: a review," *IEEE J. Select. Areas Commun.*, vol. 7(4), pp. 525-533, May 1989.

12. M. D. MacLaren and G. Marsaglia, "Uniform random number generators," *J. Ass. Comput. Machinery*, vol. 15, pp. 83-89, 1965.

13. W. Meier and O. Staffelbach, "Fast correlation attacks on certain stream ciphers," *J. Cryptology*, vol. 1(3), pp. 159-176, 1989.

14. W. Meier and O. Staffelbach, "Correlation properties of combiners with memory in stream ciphers," *J. Cryptology*, vol. 5(1), pp. 67-86, 1992.

15. M. J. Mihaljević, "An approach to the initial state reconstruction of a clock-controlled shift register based on a novel distance measure," Advances in Cryptology - AUSCRYPT '92, *Lecture Notes in Computer Science*, vol. 718, J. Seberry and Y. Zheng eds., Springer-Verlag, pp. 349-356, 1993.

16. R. A. Rueppel, "Correlation immunity and the summation generator," Advances in Cryptology - CRYPTO '85, *Lecture Notes in Computer Science*, vol. 218, H. C. Williams ed., Springer-Verlag, pp. 260-272, 1986.

17. R. A. Rueppel, "Stream ciphers," in *Contemporary Cryptology: The Science of Information Integrity*, G. Simmons ed., pp. 65-134. New York: IEEE Press, 1991.

18. T. Siegenthaler, "Correlation-immunity of nonlinear combining functions for cryptographic applications," *IEEE Trans. Inform. Theory*, vol. IT-30, pp. 776-780, Sep. 1984.

19. T. Siegenthaler, "Decrypting a class of stream ciphers using ciphertext only," *IEEE Trans. Comput.*, vol. C-34, pp. 81-85, Jan. 1985.

20. G. Z. Xiao and J. L. Massey, "A spectral characterization of correlation-immune combining functions," *IEEE Trans. Inform. Theory*, vol. IT-34, pp. 569-571, May 1988.

21. M. V. Živković, "An algorithm for the initial state reconstruction of the clock-controlled shift register," *IEEE Trans. Inform. Theory*, vol. IT-37, pp. 1488-1490, Sep. 1991.

A Faster Cryptanalysis of the Self-Shrinking Generator

Miodrag J. Mihaljević [*]

Institute of Applied Mathematics and Electronics
Institute of Mathematics, Academy of Arts and Sciences
mailing address: Šolina 4, 11040 Belgrade, Yugoslavia
e-mail: emihalje@ubbg.etf.bg.ac.yu

Abstract. A novel algorithm for cryptanalysis of the self-shrinking generator is presented and discussed, assuming that number of the shift register feedback taps is large and that length of the available generator output sequence is under a certain limit. It is based on a probabilistic approach and employes the generator output sequence not only for the hypothesis testing, but also for reducing the set of hypothesis which have to be tested. When the shift register length is L and the characteristic polynomial is known, the novel algorithm ensures the cryptanalysis with overall compexity $2^{L-\ell}$, $\ell \leq L/2$, assuming that required length of the generator output sequence is not greater than $\ell 2^{L/2} \binom{L/2}{\ell}^{-1}$. The proposed algorithm yields significant average gain of $2^{\ell-0.25L}$ in comparison with the best one published so far, which can work under the assumed conditions.

Key words: cryptology, binary sequence generators, shift registers, self-shrinking generator, cryptanalysis.

1 Introduction

A binary sequence generator, for stream ciphers cryptography [1], based on the shrinking principle (see [2], [4]) called self-shrinking generator was proposed and considered in [3]. The self-shrinking generator employes only one linear feedback shift register (LFSR) and the generator output is produced from the LFSR output sequence according to the following: If a pair happens to take the value 10 or 11, this pair is taken to produce the pseudo random bit 0 or 1, depending on the second bit of the pair. On the other hand, if a pair happens to be 01 or 00, it will be discarded. The self-shrinking generator can be considerd as an irregularly clocked LFSR whose clock is controlled in a certain manner that corresponds to the bit deletion with probability 0.75. The cryptographic key consists of the initial state of the LFSR and preferably also of the LFSR feedback logic. For

* This research was supported by the Science Fund, Grant No. 04M02.

practical applications it is assumed that the feedback connection is to produce maximal length LFSR sequence.

Recall that the shrinking generator [2] uses two binary LFSRs, sey LFSR 1 and LFSR 2, as basic components. The pseudo random bits are produced by shrinking the output sequence of LFSR 1 under the control of LFSR 2 as follows: The output bit of LFSR 1 is taken if the current output of LFSR 2 is 1, othervise it is discaded. It turns out that the self-shrinking generator and the shrinking generator are closely related to each other. In fact, it was shown that the self-shrinking generator can be implemented as a shrinking generator, and conversely, that the shrinking generator can be implemented as a self-shrinking generator. The latter implementation however cannot be accomplished with a maximum length LFSR. Thus the self-shrinking generator has its main interest in implementing the shrinking principle at lower hardware costs. Also recall that according to [2], the effective key size of the shrinking generator, measured in terms of the complexity of known cryptanalytic attacks, is roughly half of the maximum possible key size.

A security exmination of the self-shrinking generator is also precented in [3]. It is stated there that in view of the presently known cryptanalytic ataks the effective key size of the self-shrinking generator can be estimated to be, in average, 75% of the maximum possible value, assuming that the key determines the intial state of the LFSR and that the LFSR feedback logic is known.

The objective of this paper is examination of the previous statement.

Section 2 contains problem statement, summary of the previous relevant results and summary of the results achieved in this paper. The basic analysis of the self-shrinking generator relevant for the security examination is given in Section 3. A novel and faster algorithm for the self-shrinking generator cryptanalysis is proposed in Section 4. The main characteristic of the proposed algorithm are established in Section 5 including the gain over the algorithm [3]. The conclusions are given in Section 6.

2 Problem Statement and the Results

In this paper the self-shrinking generator security examination is considered under the following assumptions: the key determines the initial state of the LFSR only, the LFSR characteristic polynomial is known and has a large number of nonzero coefficients, and available length of the self-shrinking generator output sequence is under a certain limit (for example 10^{10}). In this case, according to the results published so far, the best one algorithm for cryptanalysis of the self-shrinking generator is given in [3].

The objective of this paper is to propose a novel algorithm for cryptanalysis of the self-shrinking generator faster than the procedure from [3], such that it does not depend on number of the LFSR feedback taps and can work assuming that length of the observable sequence is under the given limit.

Note that the general concepts for cryptanalysis of the clock-controlled shift registers proposed in [4] and [5] are inappropriate for the self-shrinking generator because they do not employ the fact that the LFSR is self-clocked. They are designed for the case when the irregularly clocked shift register is controlled by another shift register or by an autonomus binary sequences generator and they are based on the exhaustive search over all possible initial states of the irregularly clocked shift register. Recently, a concept of the fast attack on irregularly clocked shift registers is proposed in [6]. This concept promises to be a powerful one, but its performances highly depend on the deletion rate, number of the shift register feedback taps, and length of the available generator output sequence. So, when the deletion rate is high, the number of feedback taps is large, and the sequence length is limited, the procedure [6] is not appropriate.

The self-shrinking generator cryptanalysis [3] is based on the following. Assume that $(s_0, s_1, ...)$ is the known portion of the self-shrunken sequence. The bit s_0 is produced by a bit pair (a_j, a_{j+1}) of the original sequence where the index j is unknovn. Our aim is to reconstruct the original sequence in forward direction beginning with position j. As we know s_0 we conclude that $a_j = 1$ and $a_{j+1} = s_0$. For the next bit pair (a_{j+2}, a_{j+3}) there remain three possibilities, namely $a_{j+2} = 1$, $a_{j+3} = s_1$ if the bit pair was used to produce s_1, or the two alternatives $a_{j+2} = 0$, $a_{j+3} = 0$ and $a_{j+2} = 0$, $a_{j+3} = 1$ if the bit pair was discarded. For each of the three possibilities there are again three alternatives for the nxt bit pair. Therefore, for reconstructing n bit pairs, i.e., $N = 2n$ bits, we obtain a total of

$$S = 3^{n-1} = 3^{N/2} = 2^{((log_2 3)/2)N} = 2^{0.79N} \qquad (1)$$

possble solutions. However the solutions have different probabilities. We explain this fact by considering the above bit pair (a_{j+2}, a_{j+3}). Assuming that the original sequence is purely random, $a_{j+2} = 1$ with probability $1/2$. Hence the first alternative has probability $1/2$ and the other two cases have probability $1/4$. In turns of informaton theory the uncertainty about the bit pairs is

$$H = -(1/2)log_2(1/2) - (1/4)log_2(1/4) - (1/4)log_2(1/4) = 3/2 \qquad (2)$$

As for the reconstruction the individual bit pairs are supposed to be independent from each other, the total entropy for n bit pairs is $3n/2$. Therefore the optimum strategy for reconstructing N bits of the original sequence has average complexity $2^{3N/4}$. Following the previous consideration and the consequences when we take into account that the original sequence is produced by an LFSR assuming that the self-shrinking generator key determines the LFSR initial state only in [3] it is estimated the difficulty of finding the key. It is shown that, in general case, the cryptanalysis has average complexity $2^{0.75L}$, where L is the LFSR length, and when the LFSR has only few feedback taps and if they are concentrated around few locations a slightly faster attack are possible.

In this paper a novel and different approach to the LFSR initial state reconstructon is proposed and discussed. It is based on a probabilistic approach and employes the generator output sequence not only for the hypothesis testing, but also for reducing the set of hypothesis which have to be tested. It uses the following characteristic of the self-shrinking generator: The probability that a current LFSR state generates an output generator segment of length ℓ, $\ell < L/2$, is equal to the probability that there are exacty ℓ ones on odd positions in the considered LFSR state. Roughly speaking, the novel algorithm consists of repetition of the following steps until the true initial state is reached:

- suppose positions of ℓ ones on odd locations of a candidate for the LFSR state and set to zero the remained $L/2 - \ell$ odd locations, where $\ell \leq L/2$ is a suitably addopted parameter and L is an even number
- for each ℓ-dimensional segment of the available self-shrinking generator output sequence assume that the considered segment is generated by discarding of exactly $L - \ell$ bits from the corresponding LFSR state, and do the following:
  - according to the current assumption about the odd locations of the LFSR state candidate and the considered ℓ-dimensional output segment reconstruct values of certain ℓ even locations based on the self-shrinking rule
  - suppose values of remained $L/2 - \ell$ unknown locations in the LFSR state candidate
  - check the hypothesis whether the constructed candidate is the true one

The appropriate selection of the parameter ℓ in the first step could yield probabilistic reduction of the hypothesis which have to be tested in the second step.

Accordingly, in this paper a novel algorithm for cryptanalysis of the self-shrinking generator is proposed which ensures the cryptanalysis with overall compexity $2^{L-\ell}$, $\ell \leq L/2$, assuming that required length of the generator output sequence is not greater than $\ell 2^{L/2} \binom{L/2}{\ell}^{-1}$. So, the novel algorithm yields gain of $2^{\ell - 0.25L}$ in comparison with the algorithm [3].

3 Basic Analysis

Let a binary sequence $\{a_m\}_{m=1}^{M}$ be the L-length LFSR output (for simplicity L is an even number), and consider this sequence as a sequence of L-dimensional binary vectors $\mathbf{A}_k$,

$$\mathbf{A}_k = [a_{kL+1}, a_{kL+2}, ..., a_{kL+L}], \quad k = 0, 1, ..., \lfloor M/L \rfloor - 1, \tag{3}$$

where $\lfloor . \rfloor$ denotes the integer part, and for simplicity suppose that L is an even number. Obviously, each vector $\mathbf{A}_k$ is equal to an LFSR state, so that it determines whole $\{a_m\}_{m=1}^{M}$.

Each vector $\mathbf{A}_k$ produces a corresponding segment of the self-shrinking generator output sequence $\{s_n\}_{n=1}^{N}$, $N < M$. Denote these segments of different length as binary vectors $\mathbf{S}_k$, $k = 0, 1, ..., \lfloor M/L \rfloor - 1$. The vector $\mathbf{S}_k$ is a transformation of the vector $\mathbf{A}_k$ obtained by deletion certain bits from $\mathbf{A}_k$ according to the self-shrinking rule.

As the first, note that the following lemma holds.

Lemma 1. The probability that the L-dimensional vector $\mathbf{A}_k$ generates the corresponding vector $\mathbf{S}_k$ of length ℓ_k, $\ell_k < L/2$, is equal to the probability that there are exacty $L/2 - \ell_k$ zeros on odd positions in the vector $\mathbf{A}_k$.

Assuming that ℓ_k is a realization of an integer stochastic variable l_k we have
$$Pr(l_k = \ell_k) = 2^{-L/2} \binom{L/2}{\ell_k} \; , \quad \ell_k = 0, 1, ..., L/2 \; ,$$
where it is also assumed that in the statistical model the sequence $\{a_m\}$ can be considered as a realization of the sequence of independent stochastic binary variables which with equal probability take values 0 or 1.

Lemma 2. For an arbitrary ℓ-dimensional segment, $\ell \leq L/2$, of the sequence $\{s_m\}$ we can suppose that it is generated by deletion of $L - \ell$ bits from the corresponding LFSR state, and we can conclude that after examination of $2^{L/2} \binom{L/2}{\ell}^{-1}$ succesive nonoverlapping ℓ-dimensional segments we can expect that our assumption is correct in one case.

Accordingly a method for probabilistic reduction of the set of all possible hypothesis consists of the following two steps:

- Assume that considered segment of the output sequence, $\mathbf{S}_k$, is obtained by deletion a certain - suitable number of bits from the current shift register state, $\mathbf{A}_k$, and check this hypothesis in an appropriate way.
- Repeat the first step a number of times employing different segments of the available shift register output sequence, so that with probability close to 1 the first step assumption is correct at least in one case.

This probabilistic approach in a number of situations, with probability close to 1, yields significant reduction of the hypothesis which have to be tested.

4 Algorithm for Cryptanalysis of the Self-Shrinking Generator

The considerations in Section 3 are the basis for design of the following algorithm which, with probability close to 1, allows cryptanalysis of the self-shrinking generator with the complexity lower than the algorithm [3] complexity.

- *Input:* The LFSR characteristic polynomial and the generator output segment $\{s_n\}_{n=1}^{N}$.
- *Initialization:* Determine value of the algorithm parameter ℓ, $\ell \leq L/2$, so that ℓ has maximum possible value such that $\lfloor (N - L)/\ell \rfloor 2^{-L/2} \binom{L/2}{\ell} \geq 1$, and divide $\{s_n\}$ into $\lfloor (N - L)/\ell \rfloor$ ℓ-dimensional succesive nonoverlapping segments $\mathbf{S}_k$, $k = 0, 1, ..., \lfloor (N - L)/\ell \rfloor - 1$.

- For each of all $\binom{L/2}{\ell}$ different possibilities related to the Step 1, do Steps 1 - 5.

Step 1: Set to one ℓ previously unconsidered odd locations of the LFSR state, and set to zero the remained $L/2 - \ell$ odd locations.
- For each segment $\mathbf{S}_k = [s_{k\ell+i}]_{i=1}^{\ell}$, $k = 0, 1, ..., \lfloor (N - L)/\ell \rfloor - 1$, do Steps 2 - 5.

Step 2: Set to value of $s_{k\ell+i}$ the LFSR element on even location which left-side neighbour location contains i-th one in the subsequence of elements on the odd locations determined in the Step 1, $i = 1, 2, ..., \ell$.
- For each of all $2^{L/2-\ell}$ different possibilities for remained $L/2 - \ell$ zeros and ones of the LFSR state, do Steps 3 - 5.

Step 3: Set the remained $L/2 - \ell$ on even locations not touched in the Step 2 according to previously unused combination of zeros and ones.

Step 4: According to the current LFSR initial content determined by the Steps 1 - 3, generate $L^* \geq L$ output bits from the self-shrinking generator.

Step 5: If the L^*-length segment generated in the Step 4 is identical to the segment $\{s_{(k+1)\ell+i}\}_{i=1}^{L^*}$, $L^* \geq L$, conclude that the secret key is reconstructed and go to the Output. Othervise continue the searching procedure.
- *Output:* The reconstructed secret-key or the conclusion that the solution is not found because of the probabilistic nature of the algorithm.

5 Discussion

The proposed algorithm for cryptanalysis is based on a probabilistic principle, and employes the observed generator output sequence not only for the hypothesis checking, but also, for reducing the set of hypothesis which have to be tested.

Lemma 3. The probability p that the set of hypothesis, examinated by the algorithm for cryptanalysis, contains the true one about the secret-key is given by:

$$p = 1 - [1 - 2^{-L/2} \binom{L/2}{\ell}]^{\lfloor (N-L)/\ell \rfloor} . \tag{4}$$

Recall that L is the LFSR length, N is length of the observable output sequence, and ℓ, $1 < \ell \leq L/2$, is the algorithm parameter determined in the Initialization phase.

So, the proposed algorithm, with the probability close to 1, allows reconstruction of the secret-key, assuming sufficiently long obervable segment $\{s_n\}_{n=1}^{N}$.

On the other hand, according to the algorithm structure, the algorithm complexity is determined by the number of hypothesis which have to be tested, and it can be directly shown that the following holds.

Lemma 4. The number of hypothesis $\#H$ which have to be tested is upper bounded by:

$$\#H \leq \binom{L/2}{\ell} \lfloor (N-L)/\ell \rfloor 2^{L/2-\ell} . \tag{5}$$

Obviously, the algorithm parameter ℓ is a function of the parameters L and N. Note that Lemma 4, the parameter ℓ determination rule (see the Initialization phase), and the Algorithm structure yields the following lemma.

Lemma 5. The expected number of hypothesis $\bar{\#H}$ which have to be tested is upper bounded by

$$\bar{\#H} \leq 2^{L-\ell} . \tag{6}$$

Recall that the expected complexity of the cryptanalytic algorithm [3] is equal $2^{0.75L}$. Accordingly, the following lemma holds.

Lemma 6. The expected gain g obtained by the here proposed algorithm in comparison with the algorithm [3] is lower bounded by:

$$g > \frac{2^{0.75L}}{2^{L-\ell}} = 2^{\ell-0.25L} , \tag{7}$$

assuming that the required length N of the generator output is upper bounded by $\ell 2^{L/2} \binom{L/2}{\ell}^{-1}$.

The illustrative examples of the proposed algorithm complexity and the gain obtained by the proposed algoritm in comparison with the algorithm [3], assuming an arbitrary characteristic polynomial for the LFSR, are presented in Table 1.

Table 1: The upper bound on complexity of the proposed algorithm and expected lower bound on the gain g obtained by the proposed algorithm as the functions of the LFSR length L and the observed sequence length N.

L	N	upper bound on algorithm complexity	lower bound on gain
60	$1.88\ 10^5$	2^{35}	2^{10}
60	$6.96\ 10^7$	2^{32}	2^{13}
80	$5.85\ 10^7$	2^{45}	2^{15}
80	$4.11\ 10^9$	2^{43}	2^{17}
100	$4.86\ 10^8$	2^{57}	2^{18}
100	$3.12\ 10^9$	2^{56}	2^{19}

6 Conclusions

A novel and faster cryptanalytic attack on the self-shrinking generator is proposed and discussed. With probability close to 1, the algorithm yields significant gain in comparison with the algorithm [3]. The algorithm efficiency originates from the fact that it uses the observed sequence not only for the hypothesis testing purposes, but also, for reducing, in the probabilistic manner, the set of hypothesis which have to be tested. The algorithm performances does not depend on the number of the LFSR feedback taps and it employes the generator output sequence of length under the given limit.

References

1. R.A. Rueppel, "Stream ciphers" in G. Simmons, editor, *Contemporary cryptology, The Science of Information Protection.* IEEE Press, New York, 1992, pp. 65-134.
2. D. Coppersmith, H. Krawczyk, and Y. Mansour, "The shrinking generator", Advances in Cryptology - CRYPTO '93, *Lecture Notes in Computer Science*, vol. 773, pp. 22-39, 1994.
3. W. Meier and O. Staffelbach, "The self-shrinking generator", Advances in Cryptology - EUROCRYPT '94, *Lecture Notes in Computer Sciences*, vol. 950, pp. 205-214, 1995.
4. M.J. Mihaljević, "An approach to the initial state reconstruction of a clock-controlled shift register based on a novel distance measure", Advances in Cryptology - AUSCRYPT '92, *Lecture Notes in Computer Science*, vol. 718, pp. 349-356, 1993.
5. J.Dj. Golić and L. O'Connor, "Embeding and probabilistic correlation attacks on clock-controlled shift registers", Advances in Cryptology - EUROCRYPT '94, *Lecture Notes in Computer Science*, vol. 950, pp. 230-243, 1995.
6. J.Dj. Golić, "Towards fast correlation attacks on irregularly clocked shift registers", Advances in Cryptology - EUROCRYPT '95, *Lecture Notes in Computer Science*, vol. 921, pp. 248-261, 1995.

Modeling A Multi-level Secure Object-Oriented Database Using Views

Ahmad Baraani-Dastjerdi   *Josef Pieprzyk*   *Reihaneh Safavi-Naini*

Department of Computer Science
University of Wollongong
Wollongong, NSW 2522
AUSTRALIA

e-mail: [ahmadb, josef, rei]@cs.uow.edu.au

Abstract. In this paper, we employ the view model given by Bertino to propose a new design approach for a secure multi-level object-oriented database system. The central idea is to provide the user with a *multi-level view* derived from a single-level secure object-oriented database. Hence the database operations performed on the multi-level views are decomposed into a set of operations on the single-level objects which can be implemented on any conventional mandatory security kernel.

We show that this approach allows us to overcome the difficulties of handling content and context dependent classification, dynamic classification, and aggregation and inference problems in multi-level object-oriented databases.

1 Introduction

Recently, several secure models for object-oriented database system (OODBS) have been proposed. Some proposals consider *single-level objects.* That means that for every object, a unique security level is assigned and this level applies to the entire object [5, 15, 19]. This approach is attractive for its simplicity and its compatibility with the security kernel. Moreover we do not need to handle the multi-level update problem [19]. However in the real world, there are situations when it is necessary to classify instance variables of an object at different security levels. That is, the security model has to support *multi-level objects.* There has been also proposals that introduce a finer grain of classification by assigning the security level to each instance variable of an object [16, 13, 14]. However, these proposals allow us to model multi-level security in OODBS, but require trusted enforcement mechanisms on the object layer.

In order to maintain the security kernel compatibility of the single-level object and to overcome the difficulties of multi-level object, some researchers [15, 19, 12] proposed to design the schema which handles various security constraints that classify instance variables. For example, if we want the GAP instance variable of the class $STUDENT$[1] to be secret, we need to create a class $STUDENT\_GAP$

---

[1] The specification of STUDENT and EMPLOYEE have been shown in Figure 1 in Appendix A

with security level SECRET to be a subclass of the class *STUDENT* (see Figure 2 in Appendix B). Or if the security level of instance variable *address* depends on the value of instance variable *occupation* of the class *EMPLOYEE* (the security level of *address* is secret if the employee is a chancellor or vice-chancellor, and otherwise it is unclassified) then we shall create two classes *S_ADDRESS* and *U_ADDRESS* to be subclasses of *EMPLOYEE* such that either of them contains addresses related to secret or unclassified employees, respectively (see Figure 3 in Appendix B).

There are several problems with this approach. Firstly, if the value of an instance variable is changed dynamically, the schema evolution caused by that has to be handled. Secondly, because object instances do not have to be at the same security level with their class, it may happen that there are object instances at levels higher than the corresponding class. Therefore, certain object instances might end up in unexpected locations and be inaccessible to authorized users. For example, if a secret address object is created as an instance of *U_ADDRESS* by a secret user, no reference to it would appear in the instance variables in the *U_ADDRESS* object. As a result, secret subjects would fail to find it in the expected place under the secret subclass *S_ADDRESS*.

The view update problem disappears and view can almost freely be updated in an OODBSystem when a query language used for view definitions preserves object identities [17]. Recall that in relational databases, views containing the key of their (one) underlying base relation can be updated. We assume that the query language used for view definitions has object preservation property. The advantages in using the view approach are as follows.

1. View definitions can be regarded as subclasses, or superclasses of the base classes (virtual subclasses or superclasses) [1, 2, 17]. Views, therefore, provide the facility for a dynamic modification of the database schema but yet retaining its older versions. They also provide a means of handling various security constraints.

2. Views may be defined on arbitrary sets of classes and other views with different security levels. These views are called *multi-level views*. So by defining a multi-level view for unclassified and secret users, the possibility of storing certain data in an unexpected location which is not accessible to authorized users is eliminated.

3. A view definition can also be regarded as a constraint relating derived data to other data (stored or derived). It can be used to restrict the user access to the data that they actually need. Thus the view approach allows to handle inference and aggregation problems or at least minimize them.

4. View definitions are independent of the underlying data. If the database contents changes, it will not be necessary to reclassify the database because views will enforce the required classification rules. For example, if an unclassified employee becomes a vice-chancellor or chancellor, his related data must be reclassified to secret. Having declared views *U_ADDRESS* and *S_ADDRESS*, it is not necessary to reclassified the data.

The objective of this paper is to show how to use the view concept to implement a multi-level security policy on the top of a single-level OODBSystem. In Section 2, we describe the basic concept of multi-level secure databases. In Section 3, we discuss the essential features of the view model proposed in [4], and then extend the view model to incorporate the mandatory label-based security policy. We show how the multi-level view, the content and context-classification, and the dynamic classification can be supported by the model. In Section 4, we develop a security model for OODBS based on object views.

2 A Brief Concept of Multilevel secure Databases

A multi-level secure database contains information of varying security levels. The security levels may be assigned to the data depending on the content, context, aggregation, or time. An effective security policy for multi-level databases must ensure that users only access the information to which they are cleared. To fulfill this requirement, each entity is assigned a security attribute. Attributes associated with active entities (or subjects) are called *clearance levels* while attributes associated with passive entities (or so-called objects) are termed *security levels*. Subjects are not allowed to modify these attributes and/or their values. The modification of these attributes can only be done by system security officers. Moreover, the set of security and clearance levels form a partially ordered lattice with ordering relation "$\geq$" (for example, UNCLASSIFIED < CONFIDENTIAL < SECRET < TOP-SECRET). For security levels L_1 and L_2 , $L_1 \geq L_2$ means that security level L_1 *dominates* security level L_2 (if $L_1 > L_2$, it means that L_1 *strictly dominates* L_2).

In object-oriented databases, the encapsulation feature combined with the security labels provides a natural protection for objects. However after the value leaves the protection of the encapsulated object, the security may be compromised. To ensures that the security will not be compromised, the *flow of information* has to be restricted in some way. A number of models have been proposed, the earliest and the best known is the Bell-LaPadula model [3]. The Bell-LaPadula model is based on two properties: the *simple security property* and the **-property*. According to the simple security property, a subject is allowed to read information from an object (or a passive entity) if the clearance level of the subject dominates the security level of the object. The *-property requires that a subject has write access to an object if the subject's clearance level is dominated by the security level of the object. Informally, a subject can *read-down* (simple security property) and can *write-up* (*-property). A number of extensions to the Bell-LaPadula security model have been proposed [8, 10, 15, 20, 19, 16, 12, 13, 14]. These extensions address some specific problems related to database systems such as inferring unauthorized information from the legitimate responses that users have received from multi-level queries, and the information flow that occurs as a result of inheritance and the message passing in OODBS.

3   View Model

Views in an OODB model have been considered in a number of papers [1, 4, 7, 11, 17]. At present, there is no consensus on the view model for OODBS. We are considering the view model proposed by Bertino [4] and will show how security properties may be incorporated into such a model. Note that the conceptual Bertino's model is proposed for a context not relevant to security.

Smith in [18] identifies three "dimensions" of the protection:

1. *the data itself may be classified;*
2. *the existence of the data may be classified; and*
3. *the reason (or rule) for classifying the data may be classified.*

We define the notion of *access-view* to deal with the first two dimensions of the protection, The third security dimension is addressed by the introduction of *security-constraints*. The sets of access views and security constraints constitute meta-classes ACCESS-VIEWS and SECURITY-CONSTRAINTS, respectively. The meta-class ACCESS-VIEWS is used to enforce of the mandatory access control, and the meta-class SECURITY-CONSTRAINTS (which contains rules that the view must satisfy) is used to control the security levels of views.

An instance of the ACCESS-VIEWS has the following format:

> **access-view** av-name [parameters]
> [**properties** list_of_property_names_and_security_levels]
> [view-query]
> [**additional-properties** list_of_properties_and_security_levels]
> [**OID** true_or_false]
> [**identity-from** list_of_class_names]
> [**methods** list_of_method_specifications]
> [**superviews** list_of_views]
> [**av_level** view_security_level]
> [**av_range** upper_and_lower_security_level]
> [**mac-constraint set-of** <SECURITY-CONSTRAINTS> mac_names]

The *av-name* is the unique name of the access view, and must be different from the names of other access views, security constraints, and classes. Parameters are bound to the actual values at the time the view is evaluated. So when the value of the parameters are determined, corresponding objects are evaluated dynamically. A view is defined by a query (view-query) on an arbitrary set of classes and derived classes (or views). They are collectively called the *base class(es)*. With each view, there is an associated set of instance variables (and their security levels) that may be derived from the base classes (**properties**) or may be introduced (**additional-properties**). A view may also have methods associated with it. These methods might be inherited exactly from the base classes, be renamed or defined for the particular view (**methods**). In OODBS, when a query is executed on a view, the system returns a set of object identifiers of the view instances that satisfy the query. **OID** (which may be true or false) indicates whether or not persistent identifiers must be provided for the view instances (the default value is true). This allows users to send messages to the selected instances, and views may be used as classes. Moreover, the performance of database increases as view instances are not re-evaluated each time. If we want views to be evaluated dynamically, the **OID** must be false. The keyword **identity-from** is used for views whose view-queries contain joins. It determines a subset of the base classes which the view instances depend on for identity.

Two types of view hierarchies may be constructed. The first one is the *inheritance (or is-a) hierarchy* where a view has implicit or explicit super-views associated with it (**superviews**). The view inherits all of the instance variables, methods, and view constraints associated with its superclass (or super-view). For detailed discussion on inferred hierarchy for views, see [1, 2, 17].

The second type of view hierarchy is the *composite (or is-part-of) hierarchy*. Since an instance variable specification provided with the keyword **additional-properties** consists of the variable name and the domain where the domain can be a class or a view, then an instance of the view may be aggregation of a set objects, each of which belongs to some class or view. Such an aggregate view is sometime called a *composite object,* or here is called a *composite view.*

av_range imposes restrictions on the range of security levels of base classes (views) and properties used in the definition of a view. **av_level** indicates the security level of the name of the view. If the range is not indicated, then the view can contain the *single-level* properties and base classes (or views) that is indicated by **av_level**.

mac-constraint stands for mandatory access control constraints and consist of the set of security rules that the view must satisfy. Each element in this set is an instance of the class SECURITY-CONSTRAINTS.

The concepts are illustrated on two examples which use a university database shown in Figure 1 in Appendix A.

Example 1. Define the multi-level class STUDENT whose instance variable GAP is classified SECRET and others are UNCLASSIFIED. □

The class STUDENT can be represented as two views: a view *U_Student* with security level UNCLASSIFIED (containing the unclassified instance variables) and a view *S_Student* with security level SECRET containing the secret instance variable *GAP* and superview *U_Student* .

    access-view U_Student
    select S.idno, S.name, S.age, S.status, S.address, S.spouse, S.sex, S.subject, S.Start-Date, S.Graduate-Date, S.takes
    from S:STUDENT
    av_level UNCLASSIFIED

    access-view S_Student
    properties [Greatest_Point, SECRET]
    select S.GAP **from** S:STUDENT
    superview U_Student

av_level SECRET
av_range [UNCLASSIFIED, SECRET]

The access view *U_Student* is a single-level view, and contains all variable instances of the base class STUDENT except GAP. It is labeled UNCLASSIFIED. The access view *S_Student* which is defined on top of *U_Student* is a multi-level view because in addition to secret instance variable *Greatest_Point* it inherits all properties of superview *U_Student* which are unclassified.

Example 2. Define the class EMPLOYEE such that the instance variable "address" is classified SECRET if the value of the instance variable "occupation" is a "chancellor" or a "vice-chancellor" otherwise is UNCLASSIFIED. Other instance variables are UNCLASSIFIED. □

One possible solution is to create two views, *U_Employee* and *S_Employee*, labeled UNCLASSIFIED and SECRET, respectively.

access-view U_Employee
select E **from** E: EMPLOYEE
where not (E.occupation = "chancellor" **or** E.occupation = "vice-chancellor")
av_level UNCLASSIFIED

access-view S_Employee
select E **from** E: EMPLOYEE
where (E.occupation = "chancellor" **or** E.occupation = "vice-chancellor")
av_level SECRET

As shown in the above examples, the enforcement of security constraint has been done without the redesign of the schema.

There is a problem with the above examples. Because the security level of the view *U_Employee* is unclassified, unclassified users will know that there are expected employees who are classified at a higher level. In order to hide the classification rule but still enforce it (and also to simplify verification and assurance process), we introduce the meta-class SECURITY-CONSTRAINTS whose instances have the following format.

security-constraint sc-name [parameters]
[predicates] [property-name.**level** = property_security_level]
[predicates] [view-name.**level** = view_security_level]
[**sc_level** sc_security_level]

The *sc-name* is the unique name of the security constraint (must be different from the names of other security constraints, classes, and access views). Parameters are bound to the actual values at the time the view is evaluated. So when the value of the parameters are determined, the corresponding security rules the view must satisfy, are evaluated for the control of the security levels of the view and its properties.

sc_level indicates the security level of the security constraint. Note that the security level of properties and views can be indicated either in the access view definitions or in the security constraint definitions.

The types of constraints are *simple, content, and aggregate security constraints* [9]. A *simple constraint* classifies entire view property, or view, e.g., the view property *Greatest_point* in view *S_Student* is secret (i.e., all *Greatest_point* values will be secret). The view *S_Student* can also be defined as follows.

access-view S_Student	**security-constraint** Sc_S_Student
properties [Greatest_Point]	Greatest_Point.**level**= SECRET
select S.GAP	S_Student.**av_level** SECRET
from S:STUDENT	
superview U_Student	
mac-constraint set-of < SECURITY-CONSTRAINTS > mac_S_Student	
av_range [UNCLASSIFIED, SECRET]	

If the statement *S_Student.mac_S_Student* = **insert***(Sc_S_Student)* is executed, then the address of *Sc_S_Student* will be added to the set of *mac_S_Student* automatically.

A *content-based constraint* provides a means of classifying data at the object or property value level by using a predicate based on the values of some objects and/or properties. For example, the security-constraint *Sc_Employee* in the following classifies the view *Av_Employee* at either secret or unclassified depending on the value of the property *occupation*. Using the security constraint definition, Example 2 can be redefined as follows.

access-view Av_Employee
select E **from** E: EMPLOYEE
mac-constraint set-of < SECURITY-CONSTRAINTS > mac_Employee

security-constraint Sc_Employee
if (E.occupation =" chancellor" **or** E.occupation =" vice-chancellor")
then Av_Employee.**level** = SECRET
else Av_Employee.**level** = UNCLASSIFIED

If the statement *Av_Employee.mac_Employee* = **insert***(Sc_Employee)* is executed, then the address of *Sc_Employee* will be added to the set of *mac_Employee* automatically. Whenever the view *Av_Employee* is evaluated, the view is classified secret if it contains instances with profession "chancellor" or "vice-chancellor" otherwise is unclassified.

An *aggregate constraint* classifies a collection of property values (say ten or more/less) or relationship among data at a higher security level. For example, if we define a view which retrieves accessories supplied by a specific supplier, the associated security constraint can be declared such that if the number of the accessories is more than ten, the view is classified at a higher level say SECRET.

4 Secure Multi-level View Model

Let us emphasize two points. First, our design supports multi-level objects at the view layer only. The multi-level views will be mapped onto single-level objects. Second, our design for the enforcement of mandatory security properties relies on the underlying mandatory security kernel An OODBSystem is considered secure if (see [16]):

1. No subject is able to obtain information without authorization.
2. No subject is able to modify information without authorization.
3. No mechanism exists whereby a subject authorized to obtain information can communicate that information to a subject not authorized to obtain it.
4. No subject is able to activate a method without authorization.

Requirements (1), (2), and (4) are usually addressed by models that enforce discretionary access control, while properties (1) and (3) are normally addressed by models that enforce mandatory security. Our concern in this paper is to enforce mandatory security using a secure multi-level view object model. Enforcing discretionary access control through view object models is discussed in [2].

We assume that the entities of the security classification are all kinds of objects. The entities include access views, security constraints, classes, objects, methods, and instance variables. Each entity in the database (except atomic object) and each user has an associated *security level*. We also assume that the following methods are available to retrieve security information associated with an entity *e*.

Level(e) - displays the security level of entity *e*.

Lower(e) - displays the lower level range of entity *e*.

Upper(e) - displays the upper level range of entity *e*.

Denote by LUB the least upper bound, and GLB the greatest lower bound.

4.1 View

In general, a view can be constructed in two distinct ways [2, 1]: *top-down* or *bottom-up*. In the *top-down* approach, large classes (or views) are divided into smaller ones via *specialization* (a similar operation in relational systems is to define a view by selecting a subset of tuples from a large table). In the *bottom-up* approach, small classes (or views) are combined to form larger classes via *generalization* (the analogous operation in relational systems is to define a view as the union of several tables). A view constructed in the latter case may contain more information of various level of security and hence, it must be classified at the highest level.

Classification Rule 1 *(View Property). If an access view v is constructed on classes (or views) $v_1, v_2, \ldots, v_n$, the security level of v must satisfy:*
$v.av\_level \geq LUB\{v_1.av\_level, v_2.av\_level, \ldots, v_n.av\_level\};$. *The view range must contain the security level of v,* $\mathbf{Lower}(v) \leq v.av\_level \leq \mathbf{Upper}(v)$.

From now on we refer to properties and methods of an access view as *view facets*. *View facets* can be derived from the base views or classes, or can be defined independently. Moreover, the security level associated with each facet can also be derived or defined.

Classification Rule 2 *(View Facet Property). If x is a facet of a view v and is derived from a base view w, the security level of x must satisfy both* $\text{Level}(v.x) \geq v.\text{av\_level}$, *and* $\text{Level}(v.x) \geq \text{Level}(w.x)$. *If x is defined or redefined in v, the security level of x must dominate the security level of v, i.e.,* $\text{Level}(v.x) \geq v.\text{av\_level}$.

Classification Rule 3 *The security range of a view must contain the security level of all facets contained in the view, i.e., if $x_1, x_2, \ldots, x_n$ are facets of v, then the following must be held:*
$GLB\{\text{Level}(v.x_1), \text{Level}(v.x_2), \ldots, \text{Level}(v.x_n)\} \geq \text{Lower}(v)$, *and*
$LUB\{\text{Level}(v.x_1), \text{Level}(v.x_2), \ldots, \text{Level}(v.x_n)\} \leq \text{Upper}(v)$.

A *view name* is the external representation of an access view. When a user wants to access a view definition, they must first be authorized to access *view name*. Every access view v is defined by a view specification which has a security level, **av_level**. A user is able to access the view if the security level of the view is dominated by the security level of the user.

Classification Rule 4 *(Subject Property). A user u can access an access view v if one of the following holds: a) if the view v is single-level view, then the security level of the user u dominates the security level of the access view v, i.e, ($v.\text{av\_level} \leq \text{Level}(u)$), and b) if the view v is a multi-level view, the lowest security range of the view v is dominated by the security level of the user u (Lower(v) $\leq \text{Level}(u)$).*

Rule 4 is the simple security property specified in the Bell-LaPadula model. Only *read-up* is permitted. Users with security level lower than the security level of a view are not able to access the view definition and consequently the view instances.

Classification Rule 5 *The user u is able to access facet x of the view v if the security level of x is dominated by the security level of the user u ($\text{Level}(v.x) \leq \text{Level}(u)$).*

Classification Rule 6 *(View-instance Property). If a view v consists of objects $o_1, o_2, \ldots, o_n$, then the security level of the view v must satisfy:*
$v.\text{av\_level} \geq LUB\{\text{Level}(o_1), \text{Level}(o_2), \ldots, \text{Level}(o_n)\}$.

The set of database objects contained in the view instance is controlled by the view specification and a set of associated security constraints. Every *security-constraint* is defined by a set of constraints, and a security level (sc_level).

Classification Rule 7 *(Security-Constraint Property). If $x_1, x_2, \ldots, x_n$, and $s_1, s_2, \ldots, s_m$ are facets and their nominated security levels contained in the security-constraint sc, respectively then the following must be held:*
a) The nominated security levels ($s_i 1 \le i \le n$) for the facets contained in constraints of a security-constraint must dominate the security levels of the facets, i.e., $s_i \ge \mathbf{Level}(x_i)$ for $1 \le i \le n$;
b) the security level of the security-constraint is the least upper bound of the security levels of all facets and all nominated security levels contained in the constraints, i.e.,
$sc.sc\_level \ge LUB\{\mathbf{Level}(x_1), \mathbf{Level}(x_2), \ldots, \mathbf{Level}(x_n), s_1, s_2, \ldots, s_m\}$, and $\mathbf{Level}(sc) \ge \mathbf{Level}(v)$ where v is the view associated with sc.

The above rule indicates that the security level of a security constraint must be at least equal to the least upper bound of security levels of information contained by the constraints. Moreover, the security level of the security constraint must dominate the security level of the associated view. For example, by applying the above rule to *Sc_Employee* in Example 2 of Section 2, the security level of the *Sc_Employee* must be at least SECRET. If the **sc_level** is not indicated by the user, the computed security level will be considered for **sc_level**.

Classification Rule 8 *A user u can access a security constraint sc if the security level of sc is dominated by the security level of the user u, $\mathbf{Level(u)} \ge sc.sc\_level$.*

There are four types of inheritance among views: *constraint, strict constraint, proper specialization,* and *specialization* inheritance (for detailed description see [2]). In the *constraint inheritance,* views consist of all object instances of base classes (or views) that satisfy given constraints. We use only selection. In the *strict constraint inheritance,* not only the set of database objects contained by a view is constrained, but the base classes (or views) properties are also projected. We apply selection and projection. In the *specialization inheritance,* not only views inherit all the properties of the base classes (or views), but they also have some new additional properties and methods. In the *proper specialization inheritance,* views filter out the object instances of the base classes and project the properties. They also contain new additional properties or methods.

In the case *constraint* and *strict constraint inheritance,* the amount of the information provided by a sub-view is smaller than the information contained in the base view. So the security level of the sub-view may be the same as the security level of the base view.

Classification Rule 9 *(Hierarchy Property). Suppose a view v is derived from a view w, then their security levels must satisfy one of the following:*
$\mathbf{Level}(v) = \mathbf{Level}(w)$, if the relationship is strict constraint or constraint inheritance,
or $\mathbf{Level}(v) \ge \mathbf{Level}(w)$, if the relationship is proper specialization or specialization inheritance.

The set of the constraints of the view v must contain the set of constraints of the view w.

A view may inherit or derive the facets from one or more super-views or the base-views. In the case of a conflict, the following rule will be used to resolve the conflict.

Classification Rule 10 *(Multiple Inheritance). Assume a view v inherits the facet x from views $v_1, v_2, \ldots, v_n$, then the security level of x in v must dominate the least upper bound of the security levels of x in v_i $(1 \leq i \leq n)$,* **Level**$(v.x) \geq$ $LUB\{$**Level**$(v_1.x),$ **Level**$(v_2.x), \ldots,$ **Level**$(v_n.x)\}$.
If there are more than one such v_j, then a priori rule must be enforced to resolve the conflict.

It is possible to define composite (or aggregation) hierarchy on views. The domain of the instance variable provided with the keyword **additional-properties** for a view property can be a view or a class. A view v may then be a composite of views $v_1, v_2, \ldots, v_n$.

Classification Rule 11 *(Composite Property). Let a view v be a composite of the views (or classes) $v_1, v_2, \ldots, v_n$, then the security level of v must satisfy:* v.av_level $\geq LUB\{v_1$.av_level, v_2.av_level, $\ldots, v_n$.av_level$\}$.

4.2 Derivation Rules

The next two rules are needed to obtain a view from a real single-level database.

Classification Rule 12 *(Single-level View Instantiation). If an object in the database has a security level dominated by the security level of a view, the object may be derived as an instance of the view.*

Note that a view may be evaluated dynamically [4]. Then view instances are created only if a user or process requests it.

Classification Rule 13 *(Multi-level View Instantiation). If the security level of the view property dominates the security level of the corresponding property value of an object in the database, then that object is derived as an instance of the view from the database.*

All derived single-level objects corresponding to the view properties are joined to instantiate a multi-level view. As said in Rule 6, the security level of derived data is at least the least upper bound of the security levels associated with the derived data. Those derived data will be presented to a user whose security levels are dominated by the security level of the user.

4.3 Update and Object Creation

In relational databases, updates of any views are impossible due to the so-called *view update problem* (views only contain the key of their (one) underlying base relation; they can be updated). As shown by Scholl, Laasch and Tresch in [17], the view update problem disappears in object-oriented databases , if the query language preserves objects' identities. This happens because objects have an identity independent of their associated values. Views can almost freely be updated since the objects contained in the result of a query are the base objects. For detailed discussion of the properties of the query language that fulfills the updates of views, the reader is directed to [17].

Classification Rule 14 *(Insertion). The security level of the inserted object is computed according to a set of security constraints associated with views. If the computed security level for the entire object is unique, then the single security level is assigned to the object and the single-level object is stored into the database. But if the computed security levels are different for every properties, the inserted datum is decomposed to single-level objects according the computed security levels, and then stored.*

According to the above rule, if the inserted object is multi-level (note that the underlying database is single-level), the object must be decomposed into several single-level objects. The solution to the object decomposition was proposed in [6].

Classification Rule 15 *(Updating). For every view property value, the value with the computed security level is stored back into the database if either 1) the computed security level dominates the security level associated with the user on whose behalf it was computed, and the computed security level is equal to the security level of the corresponding property in the database; or 2) the downgrade of the property value is authorized and confirmed by the security officer.*

5 Evaluation of the Proposed Model

Gajnak in [10] chooses a view of a multi-level secure database as a set of associated facts, and presents three general principles for a *well formed* multi-level secure database. The principles are: *the granularity principle, the dependency principle,* and *the determinacy principle.* The granularity principle states that the finest level of granularity for the protection purpose should be a structure which correspond to atomic facts. The dependency principle requires that the security level of a fact dominate the security level of any other fact it depends upon. The determinacy principle states that factual dependencies should not be ambiguous.

A *fact* is an encapsulated unit of information. Atomic facts are facts that do not depend on other facts. Six types of atomic facts can be distinguished in a multi-level view model. They are:

1. the fact of the object existence (which is presented by the object identifier),
2. the association between an object and the values of its properties,
3. the association between an object and a view-instance,
4. the association between a view definition and a view name,
5. the association between a view definition and a security constraint, and
6. a hierarchical association (inheritance or composite). The following four factual dependencies also exist among these facts.
1. The fact of association of a property value with an object depends on the existence of the object.
2. The fact of association of a view-instance with object depends on the existence of objects.
3. The fact of association of a view-name with a view-definition depends on the existence of the view definition.
4. The fact of association of a security constraint with a view definition depends on the existence of the view definition.

Now, consider the three principles, *granularity*, *dependency*, and *determinacy*. The granularity principle states that the finest level of labeling granularity must cover all the above mentioned facts. This is exactly the entities that we assumed to be labeling.

The dependency principle states that the security level of an association must dominate the security level of its components. Rules 1-2, 6-7, and 9-11 take care of that.

The determinacy property addresses the problem of interpreting the database in the face of polyinstantiation. As discussed before because we assume that globally unique OIDs are used to identify objects, then the polyinstantiation will not occur in the model.

We believe that the proposed multi-level view model works because we have been able to map its constructs to the set of associated facts and because it directly supports the three principles of multi-level data: granularity, dependency, and determinacy.

6 Conclusions and Remarks

In this paper, our objective was to provide a multi-level view model derived from a single-level secure object-oriented database. The model allows us to use the existing security kernel. It also overwhelms the difficulties of handling object-oriented databases with multi-level objects. One distinct advantage of our approach is that the multi-level view model relies on an underlying security kernel for the enforcement of mandatory security properties.

Our model can be seen as an extension to the view model proposed in [4]. We introduced the notion of *security constraint* which may be associated with each view. We also discussed the usage of the views and security constraints to handle simple, content-dependent, and context-dependent classifications. We have then described the multi-level security properties for a secure multi-level view model based on a secure single-level object-oriented database.

Note that two types of users require access to views: the database *security officers* and *the end users.* It is clear that the security officers require unrestricted access to the views in order to define and maintain the database. End users require some access to the view definitions in order to be able to query the database. Two policy approaches have been proposed for this access: (1) to allow users to browse through all external schema, or (2) to attached discretionary access rights on the views. The first approach violates the least privilege principle. Then the second approach should be used. In other words, a discretionary access control on top of the multi-level view model should be implemented.

References

1. S. Abiteboul and A. Bonner, "Objects and Views," in *Proceedings of the 1991 ACM SIGMOD International Conference on Management of Data* (J. Clifford and R. King, eds.), pp. 238–247, SIGMOD RECORD, ACM Press, 1991.

2. A. Baraani-Dastjerdi, J. Pieprzyk, R. Safavi-Naini, and J. R. Getta, "A Model of Authorization for Object-Oriented Databases based on Object Views," in *Proceedings of The Fourth International Conference on Deductive and Object-Oriented Databases* (T. Ling, A. Mendelzon, and L. Vielle, eds.), vol. 1013 of *Lecture Notes in Computer Science*, (Singapore), pp. 503–520, Springer-Verlag, Dec. 1995.

3. D. Bell and L. LaPadula, "Secure Computer System: Unified Exposition and Multics Interpretation," Technical Report MTR-2997, MITRE Corporation, Bedford, MA, July 1975.

4. E. Bertino, "A View Mechanism for Object-Oriented Databases," in *Proceedings 3rd International Conference on Extending Data Base Technology (EDBT)*, vol. 580 of *Lecture Notes in Computer Science*, (Vienna, Austria), pp. 136–151, Springer-Verlag, Mar. 1992.

5. E. Bertino and S. Jajodia, "Modeling Multilevel Entities Using Single Level Objects," in *Proceedings of the Deductive and Object-Oriented Databases, Third International Conference, DOOD'93*, vol. 760 of *Lecture Notes in Computer Science*, (Phoenix, Arizona, USA), pp. 415–428, Springer-Verlag, Dec. 1993.

6. N. Boulahia-Cuppens, F. Cuppens, A. Gabillon, and K. Yazdanian, "Decomposition of Multilevel Objects in an Object-Oriented Database," in *Computer Security ESORICS 94, Third European Symposium on Research in Computer Security*, vol. 875 of *Lecture Notes in Computer Science*, pp. 375–402, Springer-Verlag, Nov. 1994.

7. U. Dayal, "Queries and views in an Object-Oriented Data Model," *International Workshop on Data Base Programming Languages*, vol. 2, 1989.

8. D. E. Denning and T. F. Lunt, "A Multilevel Relational Data Model," in *Proceedings of Symposium on Computer Security and Privacy*, (Oakland, CA.), pp. 220–234, IEEE Computer Society Press, 1987.

9. P. A. Dwyer, G. D. Jelatis, and M. B. Thuraisingham, "Multilevel Security in Database Management Systems," *Computers & Security*, vol. 6, pp. 252–260, June 1987.

10. G. E. Gajnak, "Some Result from the Entity/Relationship Multilevel Secure DBMS Project," in *Discussions of topics presented at a Workshop held at the Vallombrosa, Conference and Retreat Centre, Menlo Park, CA May 1988, Research Directions in Database Security* (T. Lunt, ed.), pp. 173–190, Springer-Verlag, 1992.

11. S. Heiler and S. Zdonik, "Object Views: Extending the Vision," in *Proceedings 6th Data Engineering Conference*, pp. 86–93, IEEE Compu er Society Press, 1990.

12. S. Jajodia and B. Kogan, "Integrating an Object-Oriented Data Model with Multilevel Security," *IEEE Computer Society Press*, pp. 76–85, 1990.

13. T. F. Keefe and W. T. Tsai, "Prototyping the SODA Security Model," in *Database Security II* (D. L. Spooner and C. E. Landwehr, eds.), pp. 211–235, Elsevier Science Publishers B. V. (North-Holland) IFIP, 1990.

14. T. F. Lunt, "Multilevel Security for Object-Oriented Database Systems," in *Database Security III* (D. L. Spooner and Landwehr, eds.), pp. 199–209, Elsevier Science Publishers B. V. (North-Holland) IFIP, 1990.

15. J. K. Millen and T. F. Lunt, "Security for Object-Oriented Database Systems," in *Proceedings of IEEE computer Society Symposium on Research in Security and Privacy*, (Oakland, CA.), pp. 260–272, IEEE Computer Society Press, May 1992.

16. M. S. Olivier and S. H. V. Solms, "A Taxonomy for Secure Object-Oriented Databases," *ACM Transactions on Database Systems*, vol. 19, pp. 3–46, Mar. 1993.

17. M. H. Scholl, C. Laasch, and M. Tresch, "Updatable Views in Object-Oriented Databases," in *Proceedings of the Deductive and Object-Oriented Databases, Second International Conference, DOOD'91* (C. Delobel, M. Kifer, and Y. Masunga, eds.), vol. 566 of *Lecture Notes in Computer Science*, (München, FRG), pp. 189–207, Springer-Verlag, Dec. 1991.

18. G. W. Smith, "Identifying and Representing the Security Semantics of an Application," in *Proceedings of the Fourth Aerospace Computer Security Applications Conference*, , Dec. 1988.

19. M. B. Thuraisingham, "Mandatory Security in Object-Oriented Database Systems," in *Proceedings International Conference on Object-Oriented Programming Systems, Languages, and Applications (OOPSLA)*, (N∋w Orleans), pp. 203–210, Oct. 1989.

20. J. Wilson, "Views as the Security Objects in a Multilevel Secure Relational Database Management System," in *Proceedings of Symposium on Computer Security and Privacy*, (Oakland, CA.), IEEE Computer Society Press, Apr. 1988.

A Appendix

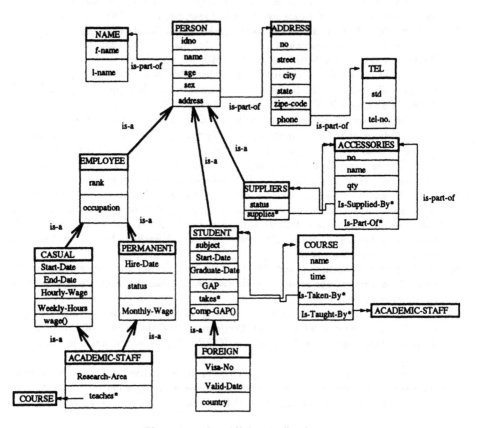

Fig. 1. A portion of University Database.

B   Appendix

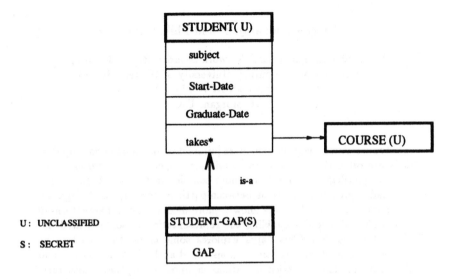

Fig. 2. Simple Constraint.

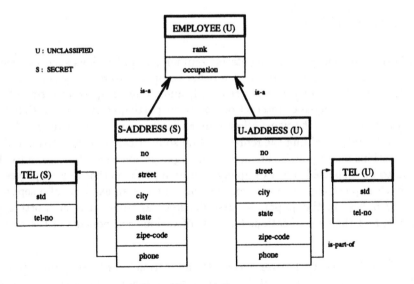

Fig. 3. Content Constraint.

Support for Joint Action Based Security Policies

Vijay Varadharajan[1] and Phillip Allen[2]

[1] Distributed System and Network Security Research Group,
Department of Computing, University of Western Sydney,
Nepean, Kingswood, NSW 2747, Australia
[2] JP Morgan, UK

Abstract. Authorization policy requirements in commercial applications are often richer compared to military applications in terms of the types of privileges required, and more complex in terms of both the nature and degree of interactions between participating objects. Delegation and joint action mechanisms allow a more flexible and dynamic form of access control, thereby enabling the representation of sophisticated authorization policies. This paper explores some issues that need to be addressed when designing joint actions based authorization policies, and their ramifications for trust of various components in the architecture. We consider an example from the medical field, and define attributes relevant to the design of joint action schemes and present three schemes for supporting joint action based authorization policies.

1 Introduction

Authorization policy requirements in commercial applications such as medical, finance and telecommunications, are richer compared to military applications in terms of the types of privileges required, and more complex in terms of both the nature and degree of interactions between participating objects. Delegation and joint action mechanisms allow a more flexible and dynamic form of access control, thereby enabling the representation sophisticated authorization policies. Joint actions based policies are used in an organization where trust in individuals needs to be dispersed. For example, a common policy is that any employee may not both request and authorize expenditure. This is an example of a separation of duty policy. In this paper, we describe some approaches to supporting joint actions, and their ramifications for trust of various components of the architecture.

Section 2 presents the overall context and section 3 compares joint action with delegation. In section 4, attributes used in the design of joint action schemes are introduced. Section 5 presents an example from medical field and outlines joint action requirements. Section 6 describes three joint action schemes in terms of the attributes defined in section 4, and considers issues that arise in each of the three schemes.

2 Joint Actions in Context

Briefly, security deals with the identification of sources of risk and implementation of measures which produce a more acceptable set of assumptions and distribution of trust. Separation of duty policies are a means of spreading trust around individuals, by breaking up what might be considered atomic actions, such as ordering a piece of equipment or admitting a patient to hospital, into packets of work for several people, where those people are able to cross-check each others' contributions.

Such policies are likely to lead to increased safety, in that several people would have to be mistaken for an accident to happen, and increased security, for several people would have to collude to commit fraud.

Generalizing from the DoD TCSEC [1] usage of the term, we will call a policy mandatory on an individual if there is nothing that an individual can do to change it or circumvent it, even though somebody hierarchically superior to that individual might be able to do so. In that sense, separation of duty policies which we have come across are mandatory on the participants. That is, they are required by a level of management hierarchically superior to the participants, and designed to protect the interests of parties other than the participants. In the example which we consider in section 4, the policy is imposed on behalf of the patients.

The disadvantage of separation of duty policies is their co-ordination costs, in describing and understanding the policy in the first place, and in finding a suitable set of participants for any instance of the action. However, co-ordination costs are a concern wherever many people must co-operate to get a task done and where typically the motivation for such a separation is specialization rather than security.

In many organizations, there are several people qualified to perform any one task. The identities of the people able to perform a given task may not be well-known. Moreover, some of those people might be on vacation, ill at home, or following a training course. Given the need for people to co-operate in such organizations, there would seem to be an application for a trader between people as well as between printers, storage media and software components. In distributed systems the concept of a trader is an important one. It is a server well-known to all applicants. Given a description of a service from a client, and perhaps non-functional criteria, it tries to provide a handle for a server which matches the request. For people, skills and permissions denoted by roles held which are traded, rather than simple services. In such a distributed system, it is necessary that the trader will be informed of separation of duty polcies within which its clients have to work. Indeed, a trader may be a pre-requisite for supporting a separation of duty policy based on roles in a large organization.

3   Joint Actions and Delegation

This section is intended to clarify the kinds of policy which joint actions and delegation are intended to support. In particular, it is useful to understand the distinctions between them in the design of secure authorization systems. Both these techniques involve several participants and the flow of authority. However, they are distinct and can be distinguished in terms of the following attributes.

Persistence of Association

Delegations of powers between managers in organizations tend to be long lived; indeed the appointment of managers can be regarded as delegation. Within a system, short-lived delegations take place: for example, processes invoked by a user may effectively adopt the identity of the user, allowing them to access files owned by the user [4]. The delegation can be partial whereby only a subset of privileges are delegated or complete.

There are also applications for corresponding long-lived associations which might be termed "joint delegations". These model such scenarios as the appointment of one member of a committee to act on behalf of the whole committee in some respect. However, such a policy entails exactly opposite effect of the dispersal of powers across several agents. Hence this is a reverse of the joint action policies, and is not considered here.

Participation in an action

Although delegates may be held responsible for the consequences of their actions under the delegation, they have a free hand in performing those actions; the delegator does not participate in them directly. In contrast, joint action implies the active participation of several agents: they each have to agree to some of the details of the action being taken.

Accretion of Rights

Delegation is usually taken to mean the granting of a subset of the authority of the delegator to the delegate. Even if there is complete delegation whereby all the privileges are delegated, no new authority is generated. However in the case of joint action based policies, agents may acquire authority by working in tandem which none in isolation possess. This is how the participation of several agents in performing an action of a given type, that is the separation of duty policy itself, is guaranteed.

Type of Participants

Temporary delegations will usually involve trusted components, except at the start of the chain. There will be little human intervention. In contrast, human intervention is the whole point of separation of duty polices supported by joint actions. Thus, the software agents in the protocol supporting joint actions are likely to be shells or interfaces designed for people doing a specific job within an organization.

Dependent on the role of the participant within an organization, and experience and the frequency of use of IT, it will be necessary to build in varying amounts of flexibility into such an interface.

4   Attributes of Joint Action Schemes

In this section, we introduce attributes in terms of which possible solutions for the joint action problem can be described and compared. In section 5, we discuss three such arrangements via the example given in section 4.

Support

We seek something more than a taxonomy or a descriptive framework: a scheme or schemes which will support the implementation of joint actions. The support offered by a scheme can be characterized in terms of depth and breadth. The depth of support means the work which the scheme saves the developer, and the level of abstraction o f implementation the scheme is able to provide. Breadth means the range of applicability of the scheme, both in terms of the subset of possible problems it will support, and the likely availability of the underlying technology assumed by the scheme.

Intrusiveness

It is important to separate the representation of organizational policy from the behaviour of point application objects as far as possible. Following this principle, the objects participating in a joint action should be able to play a minimal part, and should be protected from protocol details as far as possible by layers of encapsulation and common services.

Ordering

In some formalisms such as Communicating Sequential Processes (CSP), several participants may be involved in the same event. Such a communication primitive would offer the most direct support for joint action policies. However, the vast majority of object models are based on 1-to-1 communication. Also, not all participants may be available at the same moment. Thus, on practical grounds, joint actions must be implemented in terms of individual requests, which must be grouped together and checked for consistency by the mechanism supporting joint actions. This change from the ideal scenario gives rise to two important problems.

- Matching requests which form part of the same joint action.
- Agreement among the participants about the action to be performed.

Matching

We must allow for the spread over a period of time of the requests which correspond to one joint action. This results in problems for any authority responsible for discharging the request. First, corresponding requests from different - as yet undischarged - joint actions may exist simultaneously. Secondly, an unmatchable (e.g. surplus) requests may be generated.

Turning to mandatory policy, what consistency between requests does there need to be for a joint action to be committed ? In general, to answer this question,it is necessary to consider how to describe the matching set of actions (See "Forms of Policy" below).

Agreement

Where participation is not synchronous, there exists a need to inform the participants of which action is being carried out, and to elicit their agreement. In most object models, this means agreeing target of the joint action, the name of the method to be applied to the target object, and the parameters with which the method is called. Indeed, even the set of participants in the action might be something for which agreement cannot be taken as read. This is agreement over and above any mandatory notion of well-formedness which might also need to be enforced.

Note that there must be some asymmetry in the roles of the participants with regard to a particular parameter: just one participant must supply the value for this parameter.

Support for agreement might take two forms: first, the ability to say "yes" or "no" to parameters which have already been specified; secondly, the ability to specify acceptable ranges or other tests for parameters not yet available.

There will be circumstances where the information flow policy intrudes into considerations. In such cases, this would imply partial visibility to a given participant of other participants and the parameters provided.

Forms of Policy

This paper is concerned with the support of joint actions policies rather than some specific ones. The important consideration is that forms of policy should not be excluded by the scheme if possible. Here we are assuming that there is in effect just one joint action service. It is acceptable to provide complementary forms of support, provided that there is no problem with interference. It is also desirable that the consequences of changing between forms of support be restricted locally. On the whole, therefore, the clean provision of several for ms of support might be very difficult to achieve.

Most questions of global policy relate to how participants are chosen. Without judging the possible options, the following have to be supported.

1. Participants may instigate joint actions.
2. Participants may partially or completely specify the parameters of the joint action.
3. Participants may partially or completely specify the identities of some of the other participants.
4. Potential participants may be informed of the need for participation, and subsequently volunteer.
5. Participants may be compelled to participate, once nominated, in that inaction may be taken to mean assent or the provision of default values.

One candidate simplifying assumption, which is independent of protocols, is the treatment of the identity of participants as if they were "ordinary" parameters. In practice, the effect of disagreement with participation which has already taken place ove r-complicates matters, especially for the people involved in trying to retrieve an action stalled in this fashion.

In section 2, we judged that separation of duty policies were likely to be mandatory. Indeed, restrictions on participation in the joint action are likely to be mandatory also, so that the specification of participants by participants may be somewhat restricted, say to selection amongst a small group provided by the trader for e xample. Another simplifying assumption would be that constraints imposed by participants must respect the order of participation.

The global restrictions on participation by people would be based on their skills and permissions, that is, the state of the organization. Other restrictions on participation may not be static with respect to the state of the organization. For in the general case, it may be possible for one agent to take on a set of roles within an orgnization. Then, certain subsets of roles will be possible for an individual. Given an individual who has already participated in a joint action thus binding s ome roles, her participation in other roles for that joint action will depend on checking the allowed subsets against the extended binding which would be created.

Note that the allowed subsets of roles may not be simple upper bounds, such that any subset of an allowed subset is an allowed subset. This brings up the dual notion to separation of duty, namely *continuity of responsibility*. It is the completion of the responsibility associated with a selected role which will determine whether the individual's participation in other roles is achieved. For example, any buyer who purchases components is responsible for checking on the delivery of those components.

Notification and Error Handling

Here mechanisms should be supported for the following design choices:

- How do other intended participants learn of their assumed involvement?
- If participants are only partially specified, so that many agents may satisfy the criteria for completing a joint action, how is the choice among those who come forward regulated?
- How is successful completion or abortion of actions be signalled to the participants? If some timeout is imposed on the various contributions which go to make up a joint action, then all contributors will need to be notified.

5   A Medical Example

In this section, we consider a small example which arises from a medical policy chiefly concerned with the confidentiality of medical records [5].

Two physicians must act together to admit a patient. As far as the information system is concerned, this means that a joint action by two physicians is required before a patient object can be created in the application. The participants considered are two physicians or rather the objects which form their interfaces to the information system, and the factory for patient objects, responsible for creation of patient objects and their registration in the databases. In the second scheme that we discuss below, there is also a joint action authority.

Let us look at policies for the appointment and dismissal of medical staff. For example, the Administrator (or his delegate) plus two physicians must agree before a physician can be hired or fired. The system thus requires agents as a bootstrap. (The kind of policy which applies to the system while it is under development is not discussed.)

Delegation in the Policy

Normal delegation does exist in the system: the administrator may delegate some of her authority to assistants. From a practical point of view, it may also be necessary to allow medical personnel or patients to delegate their rights over the information system to clerical personnel. Techniques for such delegations may centre around reasonably direct changes of centrally maintained ACI (such as database of roles) or around granting capabilities as in [3] and [4].

Assigning Roles to Participants

Of the two admitting physicians, one must take the role of primary physician. Either physician is qualified to do so. This brings in an issue not treated in the general discussion in section 3, namely the persistence of roles of participants beyond the scope of joint action instance. Somehow, they need to agree who is to take this role. In general, there will be many distinct roles to be filled, several of which may be occupied by several participants. The issue is whether the system would be able to support their choice, as the joint action is on the way.

Notification

In this example, we anticipate that the two physicians will work in close proximity, and hence we will not consider complicated notification procedures.

6   Joint Action Schemes

This section discusses some joint action schemes in terms of the attributes described in section 3. We will address the medical example mentioned above.

6.1   Temporary Delegation Based Joint Action

In this scheme, a "one-time" ticket is provided by one physician to another. It is the second physician who requests the registration of the patient, passing the ticket on to the factory, which enforces some policy based on the ticket.

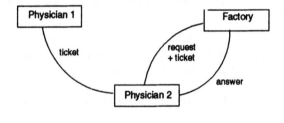

Figure 1

Like any scheme based on ordering participation, this assumes that earlier particiapants do not have to agree to any of the contributions made by the later participants. There are several points to consider here.

- **Validity of the Ticket**

  How far does the factory trust the second physician? If there are no additional trust assumptions, the ticket must be unforgeable by the second physician. If the second physician is in fact a mediating interface object, containing trusted code, then there is no need for an unforgeable ticket, and strictly speaking no need for a ticket at all. However, this does not constitute system support for joint actions.
- **Matching the Ticket against the subsequent request**

  One of the difficulties of matching requests, namely which requests are intended to be grouped together, is taken away in this case. The problem of checking whether the final, aggregated request is well-formed remains.

– **Choice of Participants**

As the originator of the ticket, the first physician has a free choice of the second. Given a longer delegation chain, it might be that the intermediaries also have a free choice. Alternatively, the contents of the tickets might restrict this choice.

– **Choice of Roles of Participants**

Recall that one of the admitting physicians adopts the role of the primary physician. If some fixed scheme, such as "the primary physician is the sender of the request", is chosen, this may result in too little flexibility for the participants. With prior agreement between the physicians, information on the role of primary physician might instead be included in the ticket and the request.

In a general case, the issue is how would one describe how much agreement between participants would be necessary? Where there are several "volunteers" for participation in a particular role, should agreement on participation in the other roles be used as a matching criterion?

Looking at the original question from another direction, one may ask: which roles (here the primary and certifying physicians) are allowed to instigate the joint action?

Given that a certain degree of flexibility is required, an obligation is placed upon the participants, and the implementations of their interfaces to the system, to be flexible in the reception of composite tickets.

– **Participation Notification and Error Handling**

The main advantage of a ticket scheme is that notification of intended involvement is taken care of. In general, however, it may happen that one participant does not want to specify (or does not need to know) precsiely who are the other participants. It seems that this can only be achieved using some explicit broking service.

– **Notification of Successful Completion**

Above in Figure 1, the factory is shown returning an answer to the secondary physician, that is the last participant in the chain. Since the chain of tickets built up contains the identity of all the participants, there is no reason in principle why the factory should not broadcast the notification.

The arrangement described above does not preclude help for the developer (in the form of code) for forming and interpreting these tickets. However, the multiplicity of answers to the questions raised above suggest that a large number of components would have to be produced in order to get reasonable coverage of various organizations' needs.

6.2 Central Authority based Joint Action

In this scheme, trust is moved to some central authority responsible for holding joint action policies. The joint action authority must have some association between sets of actions and participants and the single actions which result. It is trusted with undispersed power, only to act when legitimate groups of agents in the system co-operate. It must be able to match disparate actions into groups, and perform some form of notification.

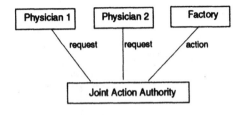

Figure 2

Some of the questions arising from this scheme are as follows.

- How is notification of successful completion given? Here the responsibility of a broadcast notification is lifted from the factory.
- Who has authority over the joint action authority? The explicit inclusion of a joint action authority responsible for the representation and enforcement of policy raises the question of whether this policy is to be modified, and by whom. As such, this is not a problem specific to joint action.
- How is the primary physician selected from the two requestors? The request itself might contain the role information.
- What happens when joint actions are interleaved? This leads to increased problems in matching groups of requests together. One solution is the agreement by the participants on a key which uniquely identifies a particular group. In the medical example, this could be the name of the patient, which must necessarily have been discussed by the two physicians before admission. In other cases, where the instigators of a joint action might wish to take advantage of some broking aspect of the system in finding other participants, such collusion would not be possible.

The advantage of this scheme is that the "many-to-one" nature of the action can be hidden from the implementation of the participants' interfaces to the system, to some extent. (Of course, this is not true of the participants' themselves). No special trust assumptions are needed for the agents involved, apart from the authority itself. The problems of this scheme lie in the general treatment of matching requests and notification, for which a range of solutions is required. Support for the developer in the context of a central authority seems to entail a general mechanism for representing this range of possibilities.

6.3 Moving Responsibility to Factory

In this scheme, we side-step some of the general issues arising in Section 2.3, such as notification and matching, by moving the work of the joint action authority to the factory - that is, to the eventual recipient of the joint action.

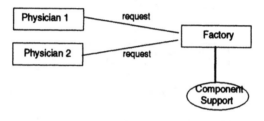

Figure 3

Here, the question arises: in what respect is joint action *supported* by the system? Clearly, "trust" has been moved entirely to the factory. However, this does not preclude support for the application programmer in storing and matching joint messages. This is a component rather than a language based approach, which inevitably leads to more problems in checking the consistency of behaviours chosen at various points by the programmer. However, dealing with such problems on a case-by-case may prove better value for money than some complicated language requiring a great deal of training.

A consequence of this change of provision of support is the loss of an explicit representation of the policy covering joint action, which might have been modified while the system is available, under some form of meta-policy implicit in the construction of the system. Here, the policy is itself implicit, which deprives the system of a degree of freedom.

References

1. Dept. of Defense, Trusted Computer Evaluation Criteria, DoD 5200.28-STD, Dec., 1985
2. ANSAware 3.0 Reference and Implementation manual, Document RM.097.00, APM Ltd. Cambridge, U.K.

3. Gasser, M. and McDermott, E.: An Architecture for a Practical Delegation in a Distributed System. IEEE Symposium on Research in Security and Privacy, Oakland (1990)
4. Varadharajan, V., Allen P., Black, S.: Analysis of Proxy Problem in Distributed Systems. IEEE Symposium on Research in Security and Privacy, Oakland (1991)
5. Ting, T. C.: Application Information Security Semantics : A Case of Mental Health Delivery. Database Security III : Status and Prospects Eds. D.L.Spooner and C.Landwehr, Elsevier.

Access Control: The Neglected Frontier

(Invited Paper)

Ravi Sandhu

George Mason University
ISSE Department, MS 4A4
Fairfax, VA 22033, USA
sandhu@isse.gmu.edu
www.isse.gmu.edu/faculty/sandhu

Abstract. Access control is an indispensable security technology. However, it has been relatively neglected by the research community. Over the past ten years, the doctrine of mandatory and discretionary access controls has slowly become discredited but no dominant doctrine has emerged to replace it. There are promising candidates such as role and task-based access controls but these are still in their formative stages and have not gained wide acceptance. This paper gives my personal perspective on these issues and identifies some of the important access control issues that researchers and practitioners should focus on.

1 Introduction

Information and system security is a multi-faceted discipline. Security presents diverse and conflicting objectives. Availability, confidentiality, integrity and privacy have been explicitly recognized in the security literature for some time. Other objectives such as intellectual property protection, copyright, secure electronic transactions, metering systems and information and resource usage for specific purposes are emerging to the forefront.

Security objectives will continue to be refined, expanded and elaborated over the next decade. Security will have a very different meaning in ten years than it does today. Much of classical security thinking has been oriented towards single organizations. In the future we will increasingly have to consider systems encompassing multiple organizations with different and, often conflicting, security objectives. Even within a single organization, as true enterprise-wide computing emerges we will see such conflicts. All this presents interesting policy and technical challenges.

The principal security technologies today are cryptography, access control, authentication, intrusion detection and recovery, risk analysis and assurance. No single technology can solve real security problems under realistic assumptions. Each addresses a piece of the problem. Fortunately security technologies are mutually supportive and there is no fundamental conflict or incompatibility between them. As technologists we must strive to use the most appropriate mix of technologies to achieve overall security objectives.

These security technologies are all important and all deserving of interest from researchers and funding agencies. However, from my perspective it does appear that access control has been relatively neglected in the last decade compared to other technologies, particularly cryptography. There are many reasons for this. It is not my goal here to analyze these in detail. Rather, I would like to draw the attention of researchers and systems developers to access control as an area with tremendous potential for achieving significant results. It is a frontier that has not yet been heavily mined and offers high payoff in terms of achieving practical security.

This paper gives my personal perspective on the neglected frontier of access control. It begins by reviewing classic access control doctrine which is based on the twin pillars of discretionary and mandatory access control. This is followed by a discussion of what is wrong with this doctrine and what alternatives are being pursued.

2 Discretionary Access Control (DAC)

Discretionary access control (DAC) has its genesis in the academic and research setting from which time-sharing systems emerged in the early 1970's. A classic paper by Lampson [Lam71] introduced the basic ideas. DAC is based on the notion that individual users are "owners" of objects and therefore have complete discretion over who should be authorized to access the object and in which mode (e.g., read or write). Ownership is usually acquired as a consequence of creating the object.

In DAC-think if Alice owns an object it is at her pleasure, whim or fancy that she decides to grant Bob access to it. Later on, should she change her fancy she can revoke Bob's access. There are many subtle issues in DAC. A question that arose almost immediately was whether or not Bob can further grant access to Charlie, so the notion of a "grant option" or "copy flag" was invented [GD72]. In turn this led to problems of cascading revoke [GW76, Fag78]. Furthermore if Alice can grant access to a group of users but at the same time withhold access from Bob even if Bob is a member of that group additional subtleties arise [GSF91, Lun88, RBKW91].

All these subtleties of DAC are still being discussed, debated and refined in the literature. Nevertheless the driving principle of DAC is ownership, so much so we should perhaps be calling it owner-based DAC.

I will leave the readers with a DAC conundrum. Suppose Alice grants a permission X to Bob with the grant option. Bob then grants X to Charlie, followed by a grant X from Alice to Charlie. Now Alice revokes X from Charlie. Should Alice's revoke override Bob's grant or should Bob's grant override Alice's revoke? The exercise is to check what System R [GW76, Fag78] would have done in this situation, and to argue the other alternative is equally, if not more, reasonable.

DAC has an inherent weakness that information can be copied from one object to another, so access to a copy is possible even if the owner of the original does not provide access to the original [SS94]. Moreover, such copies can be propagated by Trojan Horse software without explicit cooperation of users who are allowed access to the original.

3 Mandatory Access Control (MAC)

Mandatory access control (DAC) was invented to enforce lattice-based confidentiality policies [BL75, Den76] in face of Trojan Horse attacks. Subsequently it was shown how to apply MAC for integrity and aggregation objectives (such as Chinese Walls) [Bib77, Lip82, San93]. MAC ensures that even in the presence of Trojan Horses information can only flow in one direction in a lattice of security labels (from low confidentiality to high confidentiality, or equivalently from high integrity to low integrity).

MAC enforces one-directional information flow in a lattice assuming there are no covert channels by which information can flow in prohibited ways. Covert channels are expensive to eliminate even if they could all be identified and analyzed. In the late 1980's it became apparent that many low-level hardware performance improvement technologies, such as cache memory, result in very high speed covert channels [KZB+90]. Information can be leaked through these channels at disk and LAN speeds. Moreover the faster the hardware the faster these covert channels get. The covert channel problem remains a major bottleneck for high assurance MAC. MAC also does not solve the inference problem where high information is deduced by assembling and intelligently combining low information.

4 Beyond MAC and DAC

There has been a persistent criticism of the MAC-DAC doctrine over the past decade. The criticism has not been universally accepted but is has been steady and has come from a number of authors. We can roughly divide the critics into two classes as follows.

4.1 Real MAC is more than classical lattice-based MAC

Traditional lattice-based is a very narrow interpretation of the term "mandatory." Lattice-based MAC cannot enforce integrity policies. A more general notion of MAC is needed for integrity. Various authors have suggested trusted pipelines and type enforcement [BK85], well-formed transactions and constrained data items [CW87] and controls based on static and dynamic properties [San90]. Some of these arguments can be extended to confidentiality applications.

In my view inadequacy of lattice-based MAC stems from its reductionist approach of controlling access in terms of read and write operations. Operations such as credit and debit both require read and write access to the account balance and therefore cannot be distinguished for access control purposes in lattice-based MAC.

Several authors have argued that by appropriate construction of lattices it is possible for lattice-based MAC to accommodate policies that do not appear at first sight to be compatible with MAC. For instance, it had been argued that the Chinese Wall policy cannot be implemented using lattice-based MAC [BN89] but it was subsequently shown how to do this [San93]. Attempts to implement the Clark-Wilson integrity model using lattices were described by [Lee88]. Foley shows how various exceptions to information flows in a lattice can be accommodated by modifying the lattice [Fol92]. The question of how far lattice-based MAC can be pushed to support the information flow component of security policies is still not fully resolved.

4.2 Real DAC is more than classical owner-based DAC

Traditional owner-based DAC is but one form of DAC. A general model for propagation of access rights, commonly called HRU, was proposed by Harrison, Russo and Ullman [HRU76]. Unfortunately this model has very weak safety properties so it is difficult to determine the precise consequences of a propagation policy. Pittelli [Pit87] established the connection between lattice-based MAC and HRU by showing how the former can be simulated in the latter.

A variety of models and policies for propagation of rights were developed [LS77, San88a, MS88]. The SPM model of [San88a] was based on the premise that reducing expressive power may facilitate safety analysis. Of course, if we reduce expressive power too much the resulting model will not be very useful. The take-grant model has efficient safety analysis but very limited expressive power [LS77]. It turns out that SPM has strong safety analysis and has considerable expressive power [San92], including the ability to simulate lattice-based MAC. With a slight extension [AS92] it is formally equivalent to monotonic HRU. For monotonic systems (i.e., systems in which only those permissions that are restorable can be revoked) we can have general safety analysis and expressive power simultaneously.

The typed access matrix (TAM) model introduces types into HRU. Augmented TAM (ATAM) further adds the ability to test for absence of rights [SG93]. A systematic analysis of the relative expressive power of different variations of TAM and ATAM was recently completed [Gan96]. These results indicate that even very simple variations of ATAM retain ATAM's full expressive power. From an implementation viewpoint this is an encouraging result. Simple models should have simple implementations and these can provide complete expressive power.

From a safety perspective the results are disappointing because the premise of the successful SPM work does not extend to non-monotonic systems.

These results need to be interpreted carefully. General algorithms for safety analysis of non-monotonic systems are rather unlikely to exist. However, it is still possible to effectively analyze and develop safety results for individual systems. The analogy is to program verification where general verification algorithms do not exist, but individual programs can be verified. The problem with verification technology is the limited size of program that can be verified. In access control systems the policies should not require millions of lines of ATAM specification, but could perhaps be done in hundreds or thousands of lines. So case-by-case safety (and even liveness) analysis of access control policies might be practical.

5   Role Based Access Control

Role-based access control (RBAC) has recently received considerable attention as a promising alternative to traditional discretionary and mandatory access controls (see, for example, [FK92, SCY96, SCFY96]). In RBAC permissions are associated with roles, and users are made members of appropriate roles thereby acquiring the roles' permissions. This greatly simplifies management of permissions. Roles are created for the various job functions in an organization and users are assigned roles based on their responsibilities and qualifications. Users can be easily reassigned from one role to another. Roles can be granted new permissions as new applications and systems are incorporated, and permissions can be revoked from roles as needed.

An important characteristic of RBAC is that by itself it is policy neutral. RBAC is a means for articulating policy rather than embodying a particular security policy (such as one-directional information flow in a lattice). The policy enforced in a particular system is the net result of the precise configuration and interactions of various RBAC components as directed by the system owner. Moreover, the access control policy can evolve incrementally over the system life cycle, and in large systems it is almost certain to do so. The ability to modify policy to meet the changing needs of an organization is an important benefit of RBAC.

There is similarity between the concept of a security label and a role. In particular, the same user cleared to say Secret can on different occasions login to a system at Secret and Unclassified levels. In a sense the user determines what role (Secret or Unclassified) should be activated in a particular session. In [San96] it is shown how traditional lattice-based MAC can be simulated using RBAC96 model of [SCFY96]. This establishes that traditional MAC is just one instance of RBAC thereby relating two distinct access control models that have been developed with different motivations. It is also practically significant, because it implies that the same Trusted Computing Base can be configured to enforce RBAC in general and MAC in particular. This addresses the long held desire of multi-level security practitioners that technology which meets needs of the larger commercial marketplace be applicable to MAC. The classical approach to fulfilling this desire has been to argue that MAC has applications in the commercial sector. So far this argument has not been terribly productive. RBAC, on the other hand, is specifically motivated by needs of the commercial sector. Its customization to MAC might be a more productive approach to dual-use technology.

In large systems the number of roles can be in the hundreds or thousands. Managing these roles and their interrelationships is a formidable task that often is highly centralized and delegated to a small team of security administrators. Because the main advantage of RBAC is to facilitate administration of permissions, it is natural to ask how RBAC itself can be used to manage RBAC. We believe the use of RBAC for managing RBAC will be an important factor in the long-term success of RBAC. Decentralizing the details of RBAC administration without loosing central control over broad policy is a challenging goal for system designers and architects.

Since RBAC has many components, a comprehensive administrative model would be quite complex and difficult to develop in a single step. Fortunately administration of RBAC can be partitioned into several areas for which administrative models can be separately and independently developed to be later integrated. In particular we can separate the issues of assigning users to roles, assigning permissions to roles and defining the role hierarchy. In many cases, these activities would be best done by different administrators. Assigning permissions to roles is typically the province of application administrators. Thus a banking application can be implemented so credit and debit operations are assigned to a teller role, whereas approval of a loan is assigned to a managerial role. Assignment of actual individuals to the teller and managerial roles is a personnel management function. Design of the role hierarchy relates to design of the organizational structure and is the function of a chief security officer under guidance of a chief information officer.

6 Task Based Access Control

The overriding concern of models we have discussed so far (DAC, MAC, HRU, TAM, ATAM, RBAC) has been the fine-grained protection of individual objects and subjects in the system. This approach has served as a reasonable basis for these model, but it lacks the concepts and expressiveness of an information-oriented model that captures the organizational and distributed aspects of information usage.

Increased automation of organizational functions and workflows, and the subsequent need to computerize information systems that often have distributed processing needs. Increased automation always carries it with the risk of weakened controls, especially when human judgment and paper-based checks and balances are taken out of the loop. The emergence of multi-system applications and information-related services that cross departmental and organizational boundaries, call for modeling constructs and integrity mechanisms beyond those existing for centralized systems.

Modern organizations encompass complex webs of activities (tasks) that often span departmental and organizational boundaries. Tasks are authorized and initiated by users in accordance with their roles, responsibilities, and duties (obligations) in the organization. One can view an organization as a system that is required to maintain a certain state (or standard) of integrity. Organizational procedures and internal controls then have to ensure that the tasks carried out in the organization preserve such a state of integrity. Now when we computerize organizational functions, we are faced with the problem of maintaining the required integrity in our computer-based information systems.

These considerations lead to the notion of task-based authorizations (TBA) and access control (TBAC) [TS94]. TBA is concerned with modeling and management of the authorizations of tasks (activities) in information systems. The central objective is preservation of integrity, but confidentiality applications are also possible. In a paper-based system, authorizations manifest as signatures on documents propagating through the organization. The analog to this in a computerized information system would be digital signatures on electronic documents. As such, we believe that task-based authorizations are central to the successful evolution of the concept of the "paper-less office".

A key element of TBA is the fact that authorization is transient and dependent on organizational circumstances. Consider the ability to issue a check. In RBAC we can associate this permission with a role, say, APM (accounts-payable-manager). This association is long-lived. A user who can exercise this role is capable of issuing many checks. In TBA the authority to issue a check is not directly associated with the role. We can say that role APM is a necessary requirement for issuing checks but it is not sufficient. In addition we require that a suitable authorization should have been obtained for the particular check in question. In the paper world this is achieved by obtaining one or more approval signatures on a voucher prior to issuance of the check. Techniques such as transaction control expressions [San88b] can be used to enforce this one-time permission.

7 Conclusion

In this paper I have given a high-level personal perspective on access control models and their future. I do believe this is a neglected frontier where much interesting and practically useful work remains to be done. I have identified some questions which merit particular attention.

There is considerably more literature than I have cited here. The papers I have cited are the ones that have influenced my own thinking most strongly.

Acknowledgment. My research on access control is partly supported by grants from the National Science Foundation and the National Security Agency.

References

AS92.   P.E. Ammann and Ravi S. Sandhu. The extended schematic protection model. *The Journal Of Computer Security*, 1(3&4):335–384, 1992.

Bib77.   K.J. Biba. Integrity considerations for secure computer systems. Technical Report TR-3153, The Mitre Corporation, Bedford, MA, April 1977.

BK85.   W. Boebert and R. Kain. A practical alternative to hierarchical integrity policies. In *NBS-NCSC National Computer Security Conference*, pages 18–27, 1985.

BL75.   D.E. Bell and L.J. LaPadula. Secure computer systems: Unified exposition and Multics interpretation. Technical Report ESD-TR-75-306, The Mitre Corporation, Bedford, MA, March 1975.

BN89.   D.F.C. Brewer and M.J. Nash. The chinese wall security policy. In *Proceedings IEEE Computer Society Symposium on Security and Privacy*, pages 215–228, Oakland, CA, May 1989.

CW87.   D.D. Clark and D.R. Wilson. A comparison of commercial and military computer security policies. In *Proceedings IEEE Computer Society Symposium on Security and Privacy*, pages 184–194, Oakland, CA, May 1987.

Den76.   D.E. Denning. A lattice model of secure information flow. *Communications of the ACM*, 19(5):236–243, 1976.

Fag78.   R. Fagin. On an authorization mechanism. *ACM Transactions on Database Systems*, 3(3):310–319, 1978.

FK92.   David Ferraiolo and Richard Kuhn. Role-based access controls. In *15th NIST-NCSC National Computer Security Conference*, pages 554–563, Baltimore, MD, October 13-16 1992.

Fol92.   Simon Foley. Aggregation and separation as non-interference properties. *The Journal Of Computer Security*, 1(2):159–188, 1992.

Gan96.   Srinivas Ganta. *Expressive Power of Access Control Models Based on Propagation of Rights*. PhD Thesis, George Mason University, 1996.

GD72.   G.S. Graham and P.J. Denning. Protection – principles and practice. In *AFIPS Spring Joint Computer Conference*, pages 40:417–429, 1972.

GSF91.   Ehud Gudes, Haiyan Song, and Eduardo B. Fernandez. Evaluation of negative, predicate, and instance-based authorization in object-oriented databases. In S. Jajodia and C.E. Landwehr, editors, *Database Security IV: Status and Prospects*, pages 85–98. North-Holland, 1991.

GW76.   P.P. Griffiths and B.W. Wade. An authorization mechanism for a relational database system. *ACM Transactions on Database Systems*, 1(3):242–255, 1976.

HRU76.   M.H. Harrison, W.L. Ruzzo, and J.D. Ullman. Protection in operating systems. *Communications of the ACM*, 19(8):461–471, 1976.

KZB+90.   P.A. Karger, M.E. Zurko, D.W. Bonin, A.H. Mason, and C.E. Kahn. A vmm security kernel for the vax architecture. In *Proceedings IEEE Computer Society Symposium on Security and Privacy*, pages 2–19, Oakland, CA, May 1990.

Lam71.   B.W. Lampson. Protection. In *5th Princeton Symposium on Information Science and Systems*, pages 437–443, 1971. Reprinted in *ACM Operating Systems Review* 8(1):18–24, 1974.

Lee88.   T.M.P. Lee. Using mandatory integrity to enforce "commercial" security. In *Proceedings IEEE Computer Society Symposium on Security and Privacy*, pages 140–146, Oakland, CA, May 1988.

Lip82.   S.B. Lipner. Non–discretionary controls for commercial applications. In *Proceedings IEEE Computer Society Symposium on Security and Privacy*, pages 2–10, Oakland, CA, May 1982.

LS77.   R.J. Lipton and L. Snyder. A linear time algorithm for deciding subject security. *Journal of the ACM*, 24(3):455–464, 1977.

Lun88.   Teresa Lunt. Access control policies: Some unanswered questions. In *IEEE Computer Security Foundations Workshop II*, pages 227–245, Franconia, NH, June 1988.

MS88.   J.D. Moffett and M.S. Sloman. The source of authority for commercial access control. *IEEE Computer*, 21(2):59–69, 1988.

Pit87.   P. Pittelli. The bell-lapadula computer security model represented as a special case of the harrison-ruzzo-ullman model. In *NBS-NCSC National Computer Security Conference*, 1987.

RBKW91.   F. Rabitti, E. Bertino, W. Kim, and D. Woelk. A model of authorization for next-generation database systems. *ACM Transactions on Database Systems*, 16(1), 1991.

San88a.   Ravi S. Sandhu. The schematic protection model: Its definition and analysis for acyclic attenuating schemes. *Journal of the ACM*, 35(2):404–432, April 1988.

San88b.   Ravi S. Sandhu. Transaction control expressions for separation of duties. In *Fourth Annual Computer Security Application Conference*, pages 282–286, Orlando, FL, December 1988.

San90.   Ravi S. Sandhu. Mandatory controls for database integrity. In D.L. Spooner and C.E. Landwehr, editors, *Database Security III: Status and Prospects*, pages 143–150. North-Holland, 1990.

San92.   Ravi S. Sandhu. Expressive power of the schematic protection model. *The Journal Of Computer Security*, 1(1):59–98, 1992.

San93.   Ravi S. Sandhu. Lattice-based access control models. *IEEE Computer*, 26(11):9–19, November 1993.

San96.   Ravi Sandhu. Rationale for the RBAC96 family of access control models. In Ravi Sandhu, Ed Coyne, and Charles Youman, editors, *Proceedings of the 1st ACM Workshop on Role-Based Access Control*. ACM, 1996.

SCFY96.   Ravi S. Sandhu, Edward J. Coyne, Hal L. Feinstein, and Charles E. Youman. Role-based access control models. *IEEE Computer*, 29(2):38–47, February 1996.

SCY96.   Ravi Sandhu, Ed Coyne, and Charles Youman, editors. *Proceedings of the 1st ACM Workshop on Role-Based Access Control*. ACM, 1996.

SG93.   Ravi S. Sandhu and S. Ganta. On testing for absence of rights in access control models. In *IEEE Computer Security Foundations Workshop*, Franconia, NH, June 1993. 109–118.

SS94.   Ravi Sandhu and Pierangela Samarati. Access control: Principles and practice. *IEEE Communications*, 32(9):40–48, 1994.

TS94.   Roshan Thomas and Ravi S. Sandhu. Conceptual foundations for a model of task-based authorizations. In *IEEE Computer Security Foundations Workshop 7*, pages 66–79, Franconia, NH, June 1994.

On the Quantitative Assessment
of Behavioural Security

Erland Jonsson and Mikael Andersson

Department of Computer Engineering and Department of Mathematics
Chalmers University of Technology, S-412 96, Göteborg, Sweden
email: erland.jonsson@ce.chalmers.se, mikael@math.chalmers.se

Abstract. This paper is based on a conceptual framework in which security can be split into two generic types of characteristics, behavioural and preventive. We show that, among the traditional security aspects, availability and confidentiality should be used to denote be havioural security. The third aspect, integrity, is interpreted in terms of fault prevention and is regarded as a preventive characteristic. A practical measure for behavioural characteristics, including reliability and safety, is defined. We show how the measure could be derived using traditional reliability methods, such as Markov modelling. The measure is meant for practical trade-offs within a class of computer systems. It quantifies system performance on user-specified service levels, which may be operational or failed. Certain levels may be related to confidentiality degradations or confidentiality failures. A simple example based on a Reference Monitor is given. Failures resulting from security breaches are normally not exponentially distributed. The calculation method must therefore be extended to handle situations with non-exponential failure rates. This is done by means of phase-type modelling, illustrated by introducing malicious software, such as a Trojan Horse, into the Reference Monitor.

Keywords. Computer Security, Dependability, Confidentiality, Measure, Modelling.

1. Introduction

Security has not traditionally been expressed quantitatively. Instead, security evaluation levels have been used, such as the divisions and classes of the Orange Book [25], which are primarily concerned with the design features and the development process. Lately, some attempts have been made to develop methods for a quantitative assessment of *preventive security* interpreted as the system's ability to prevent intrusions [3], [17]. The assumption is that the *effort* expended to achieve an intrusion could be used as a measure of the preventive security. Practical intrusion experiments were performed in an attempt to obtain data for the purpose of modelling the preventive security [21], [8], [9].

The attacking process exploited in the intrusion experiments is the result of human interaction, which may include e.g. planning and strategic reasoning. It is thus a very complicated process that can not generally be described by a simple stochastic time variable. Still, it seems plausible that the system *behaviour* resulting from the combined processes of intentional attacks, component and other faults as well as fault prevention and error recovery mechanisms could indeed be modelled by a time variable.

This paper attempts to find a measure for those behavioural aspects of security, *behavioural security*, which, as we will show in the paper, can be understood as a subset of dependability, including aspects of reliability and safety. The measure will also include aspects related to *performability*, see e.g. [1], [18], [23] and the comprehensive overview in [19]. It may therefore be applied to systems that exhibit a degradable performance and not only a binary functional characteristic. The measure is represented by a vector derived using traditional Markov modelling. An entry in the vector reflects the expected sojourn time of the system at a certain operational *service level*. The measure of the failed service levels describes the portion of the lifetime of several identical systems that will lead to that failed level, as defined in section 3.

In the following, section 2 describes the conceptual framework for the derivation of the measure, including the system model. Section 3 defines the vectorized measure, and section 4 outlines how non-exponential failure rates can be dealt with. Section 5 summarizes the paper.

2. The Conceptual Framework

2.1 The System Model

This section describes the system model used in the paper. The total system that we consider consists of the *object system* and the *environment*. In general, there are two basic types of interaction between the system and its environment, see figure 1. First, the system interacts with the environment or is *delivering an output* or *service* to the environment. We call this the *system behaviour*. There is also an environmental influence on the system, which means that the system *receives an input* from the environment. The input consists of many different types of interaction. The type of interaction we are interested in here is that which involves a *fault introduction* into the system, in particular intentional, and often malicious faults, i.e., *security breaches*. Since faults are detrimental to the system, we seek to design the system such that the introduction of faults is prevented: fault prevention.

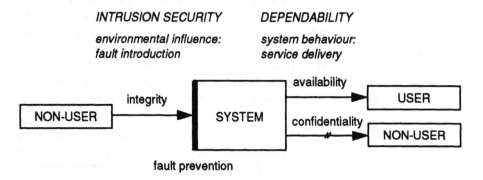

Figure 1: A system model for dependability

There are two different types of receivers of the output delivered by the system: the authorized user and the unauthorized user. The authorized users are the users that are the intended receivers of the service that the system delivers, as specified in the system specification. In the following we shall call the authorized user(s) the **User**. This may be a human or an object: a person, a computer, a program etc. All potential users except the authorized users are unauthorized users. Unauthorized users are called **Non-users**.

2.2 Dependability Attributes

Dependability is traditionally described by four basic attributes which are primarily related to non-degradable systems: *reliability, availability, safety* and *security* [16]. See figure 2.

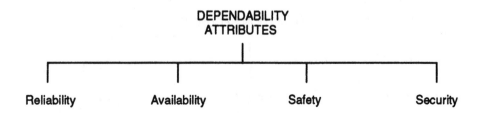

Figure 2: Dependability and its attributes

Furthermore, *performability* is a performance-related measure of dependability used for degradable systems. Except for security, these attributes all refer to the system behaviour, i.e. the service that the system delivers to the environment. Therefore, they form an adequate basis for a behavioural dependability measure. For security, however, the situation is different: The traditional security concept describes not only the system behaviour, i.e. the service that the system delivers to the environment, but also the system's ability to resist external faults, *intrusion security* or *preventive security*. The dependability measure proposed in this paper is intended only to describe the behaviour of the system, and some aspects of security must thus be excluded. As we shall find in the following subsection, there is a certain overlap between some security aspects and dependability attributes that must be clarified. See also [12], [15].

2.3 Interpreting the Security Attribute

Security is normally defined by three different aspects: *confidentiality, integrity* and *availability* [25], [11]. Other definitions exist: in the database literature in particular security is exclusively related to Non-user action, whereas problems arising from User action are called integrity problems [5], [4]. We will adhere in the present paper to the traditional definition. Given the system model for dependable systems in the previous section, we shall now show that the three aspects above are already covered, to a large extent, by existing concepts in the dependability discipline, either as a behavioural or preventive concept. See figure 3, which shows the definitions for information security, which is a subset of security in general.

Availability is primarily defined as the ability of the system to deliver its service to the User, i.e., a behavioural concept. Therefore, availability as a security aspect is clearly a subset of the availability concept in dependability.

Integrity is the prevention of unauthorized modification or the deletion or destruction of system assets. Integrity is violated by means of an attack, which is normally performed by a Non-user, but may also be performed by a User who is abusing his authority. Thus, integrity is a preventive quality of a system and characterizes the system's ability to withstand attacks. In dependability terms, this is called fault prevention.

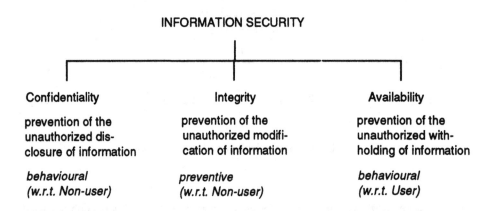

Figure 3: Information security and its aspects

Confidentiality is the ability of the system to deny the Non-user access to confidential information. It is thus a behavioural concept but, unlike other attributes, it defines system behaviour *with respect to a Non-user*. It actually defines to what extent information should be accessible, or rather not accessible, to Non-users. Therefore, confidentiality can be regarded as a new attribute in the dependability discipline, parallel to reliability, availability and safety. Confidentiality can also be understood in a broader sense, i.e., the prevention of the delivery of service to the Non-user, even if this service delivery would not include harm to the User or disclosure of secret information. The term *exclusivity* has been proposed for this broader concept [6].

2.4 Two Types of Security

The conclusion of the discussion above leads to a modified understanding of security as two concepts: preventive security and behavioural security. **Preventive security** is simply regarded as a form of fault prevention, namely fault prevention with respect to intentional faults and attacks. **Behavioural security** is an integrated part of dependability and can not readily be distinguished from it. Thus, measures for behavioural security cannot be separated from measures of dependability. In the following, we will therefore use the term **dependability measure**.

In view of this discussion, we may interpret dependability as being composed of two generic types of behavioural attributes: *reliability/availability* and *confidentiality*. See figure 4. Confidentiality relates to the denial-of-service to Non-users, i.e. unauthorized users shall not be able to obtain information from the system, nor be able to use it in any other way. Reliability and availability have been merged, since they both refer to delivery-of-service to the User. The *safety* attribute characterizes a certain failure mode of the system: it denotes the non-occurrence of catastrophic failures. In analogy, *performability* refers to the degradation of the system.

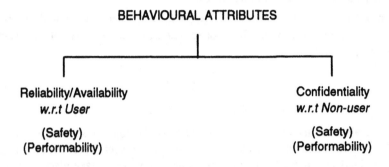

BEHAVIOURAL ATTRIBUTES

Reliability/Availability	Confidentiality
w.r.t User	*w.r.t Non-user*
(Safety)	(Safety)
(Performability)	(Performability)

Figure 4: Behavioural dependability attributes

Note that failures and degradations can be of both a "reliability" type, i.e., related to the User, as well as a "confidentiality" type, i.e., related to the Non-user. Furthermore, any type of degradation or failure can be the result of an accidental ("reliability") fault or an intentional ("security") fault.

3. A Behavioural Dependability Measure

3.1 Definition

This section provides a mathematical definition of the vectorized dependability measure for the set behavioural dependability and security attributes as defined previously. It also presents the method and equations to be used for the calculations. The method is based on a user-defined set of service levels and a set of corresponding failure rates which quantify the rate of transitions between levels.

We shall assume that the state of the system can be modelled as a continuous time Markov process $\{X_t\}_{t \geq 0}$ with a finite state space E, in which each service level, SLn, can be identified with a subset of states in E. Thus, E is the disjoint union $SL0 + \ldots + SLk$, where l is the number of service levels. Further, we imagine that service levels $0, \ldots, k$ correspond to operational states O, i.e. the states in which the system functions, in the sense that it delivers a full or degraded service to the user. Service levels $k+1, \ldots, l$ correspond to the *failed* states F, i.e., states in which the system is not functioning, meaning that it is not delivering any service of interest to the user. That is [1],

$$E = O + F \tag{1}$$
$$O = SL0 + \ldots + SLk \tag{2}$$
$$F = SL(k+1) + \ldots + SLl \tag{3}$$

---

[1]   here as usual "+" means union of disjoint sets

In the simplest case, corresponding to the traditional operational-failed model, O consists of just one single state o and F consists of just one single state f. In more complex situations, the different states in O represent different full or degraded service levels and F represents different types of failed states.

Transitions $i \rightarrow j$ have intensity λ_{ij}, $(i, j \in E, i \neq j)$, and the initial probability $P(X_0 = i)$ is denoted by π_i. In most situations, the system will always start in a fixed state i_0 so that

$$\pi_j = \begin{cases} 1, j = i_0 \\ 0, j \neq i_0 \end{cases} \tag{4}$$

We shall also assume that the system starts at the highest service level, so that $i_o \in SL0$. Transitions between operational states represent degradations, and transitions to a failed state represent failures. No transition will ever take place from a failed state, i.e., after entering a failed state, the system stops evolving. Therefore, failed states are absorbing so that $\lambda_{fj} = 0$ for $f \in F$ and all $j \in E$.

For mathematical convenience we shall use the notation that the intensity for leaving state i is $\lambda_i = \sum \lambda_{ij}$, for $i \neq j$, and we write $\lambda_{ii} = -\lambda_i$.

Assuming that O has n states and F m, we suggest the $n+m$ vector

$$w = ((u_i)_{i \in O}, (v_i)_{i \in F}), \tag{5}$$

as a measure of dependability of the system (the "dependability" vector). Here u_i is the mean time in state i or **Mean Time To Degradation** (= MTTD) and v_i is the Mean Time To Failure (=MTTF), i.e. the sum of the MTTDs of the operational states, divided by the probability p_i of ending up in the failed state i. The measures v_i that we allocate to the failed states represent a splitting of the mean *operational* lifetime of a sequence f identical systems, so that a more probable failed state receives a smaller allocation than a less probable state. This can be used to model failures of different severities, where the most severe one, the *catastrophic* failure, is normally represented by the lowest state [13]. Obviously we want the value for a catastrophic state to be as large as possible. The measures v_i can also be used to model failures of different types, e.g., confidentiality failures and reliability failures, which may represent different costs to the User. In formal mathematical terms we have

$$u_i = E \int_0^\infty I(X_t = i)dt, \quad i \in O \tag{6}$$

$$v_i = \frac{1}{p_i} \sum_{j \in 0} u_j, \quad i \in F \tag{7}$$

where I is the indicator function (i.e., $I(X_t = i) = 1$ in equation (6) if $X_t = i$ and $I(X_t = i) = 0$ if $X_t \neq i$).

For computational purposes, we note that u_i, and the probability p_i that the system ever enters i can be denoted

$$u_i = \frac{p_i}{\lambda_i}, \quad p_i = P(\tau_i < \infty), \quad i \in E = O + F \tag{8}$$

since we have assumed that there is no feedback. Here, $\tau_i = \inf\{t \geq 0 : X_t = i\}$, ($\tau_i = \infty$ if no t with $X_t = i$ exists) is the hitting time of i, i.e. the time of first entry. In order to illustrate the use of the dependability measure (5) a simple example is given below. Here the computations can be made by hand. For computational procedures in more complicated (and thus more realistic) examples, see [14].

3.2 Example: Reference Monitor

Example 1. Consider a Reference Monitor, RM, in a computer system with enhanced security characteristics. See figure 5. The Reference Monitor acts as a special type of filter between a user and the system that checks accesses to the system. The function of the Reference Monitor is to ensure that a particular user U_1 can only access such information that (s)he is authorized to access. Suppose that U_1 has no special privileges. If (s)he, despite that, attempts to access secret information, the request will be turned down by the Reference Monitor.

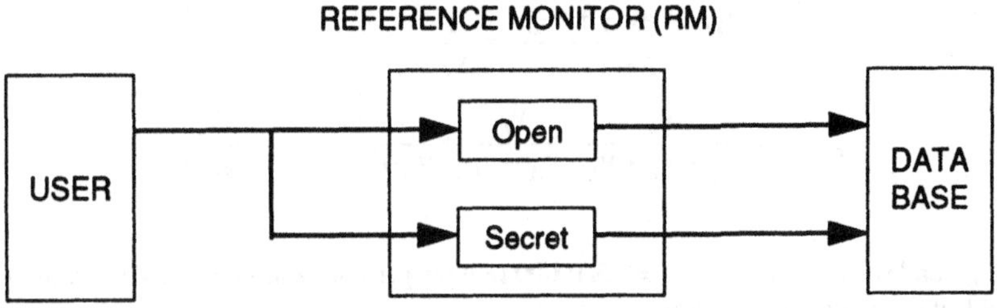

REFERENCE MONITOR (RM)

Figure 5: A Reference Monitor unit for access control

Suppose that the system contains information of two classes: classified (secret) and unclassified (open), and that the RM has two units, RM_S, which checks accesses to secret information, and RM_O, which connects the user to the information bank with open information.

The RM_S unit has two failure modes: mode B, in which it stops all accesses to the secret information, and mode C, in which no "secret" accesses are stopped. The RM_O unit has only one failure mode A, in which all accesses to the open information are stopped. In view of the discussion above, the service levels may be defined as

$SL0 = ABC$	(full service)
$SL1 = AbC$	(degraded service - no secret information available)
$SL2 = aBC$	(degraded service - no open information available)
$SL3 = abC$	(system failure - no information at all available)
$SL4 = **c$	(confidentiality failure - secret information availabe to all users.)

We can identify E with $\{SL0,SL1,SL2,SL3,SL4\}$, where $O = \{SL0,SL1,SL2\}$, $F = \{SL3,SL4\}$. A state diagram for the system is given in figure 6.

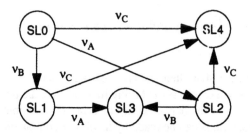

Figure 6: State diagram for the Reference Monitor unit

Straightforward calculations lead to expressions of the following types:

$$u_{SL0} = \frac{1}{v_A + v_B + v_C}$$

$$p_{SL1} = \frac{v_B}{v_A + v_B + v_C}, \qquad u_{SL1} = p_{SL1} \cdot \frac{1}{v_A + v_C}$$

$$v_{SL4} = \frac{u_{SL0} + u_{SL1} + u_{SL2}}{p_{SL4}} = \frac{1}{v_C}$$

$$p_{SL3} = \frac{v_B}{v_A + v_B + v_C} \cdot \frac{v_A}{v_A + v_C} + \frac{v_A}{v_A + v_B + v_C} \cdot \frac{v_B}{v_B + v_C}$$

Inserting numerical values $v_A = v_B = 9.5/10000$ failures per hour and $v_C = 1/10000$ failures per hour yields the dependability vector

$$w = (500\ 452\ 452\ 1634\ 10000)$$

hours, where the higher figure on the lowest level represents the fact that confidentiality failures are much less probable than other failures.

3.3 Introducing Rewards

The above set-up is also convenient for computing performance measures defined in terms of Markov reward processes, cf. [10], [24]. That is, we assume that a reward is earned at rate r_i whenever the system is in state $i \in O$. The total reward gained is then

$$R = E\int_0^\infty r_{X_t} dt = \sum_{i \in O} r_i u_i,$$

as is seen by decomposing the integral into the contributions from the sojourns in the individual states $i \in O$.

We could also extend the concept of reward by introducing for $i \in F$ a penalty or cost c_i (typically ≥ 0) imposed if the system ends up in some failed state i. Here, c_i is a non-recurring or time-independent quantity since, according to the model, the entered state is absorbing and the system will remain there forever. The total reward gained is then

$$R = \sum_{i \in O} r_i u_i - \sum_{i \in F} c_i p_i,$$

The reward figure thus calculated may serve as a single figure-of-merit for systems with a well-defined design or as a means of comparison and trade-off between different design alternatives.

Example 2. Consider again the Reference Monitor in example 1. A reward rate, in terms of e.g. USD, could be mapped onto each operational service level, normally with a decreasing amount for "lower" levels, i.e., higher "SL-values". Furthermore, a cost could be introduced for the failed levels, typically a zero value for the normal failure level and a considerable cost for the catastrophically failed level. Thus, a single figure of merit could readily be calculated.

4. Modelling of Non-Exponential Lifetimes

The analysis above is based on the assumption that the lifetimes of the components in the object system are exponentially distributed. This may often be quite unrealistic, especially when dealing with degradations or failures related to security breaches. Using phase-type distributions instead of the exponential distribution allows us to dispense with this assumption at the expense of a higher complexity of the involved calculations. Still, phase-type assumptions allow the possibility of remaining within the universe of Markovian modelling by introducing some additional states to the system, as originally suggested by [20]. Thus, the process that describes the behaviour of a system of components with phase-type distributed lifetimes can be regarded as a special case of a semi-Markov process. Despite this restriction, there is no essential loss of generality, since any distribution can be approximated arbitrarily closely by a phase-type distribution.

4.1 Definition

A random variable Y is said to be *phase-type distributed* if it can be described as the time to absorption of a terminating Markov process $\{J_t\}_{t \geq 0}$ with fixed transition rates and n states, where one state is absorbing and all others are transient (figure 7). If we let $\{n\}$ be the absorbing state, this means that $\lambda_{ni} = 0$ and $\lambda_i > 0$ for $1 \leq i \leq n-1$. Hence, we can never leave state n but we must always leave the others sooner or later. By introducing the initial distribution π, where $\pi_i = P(J_0 = i)$, we can formally express Y as $Y = \min\{t: J_t = n\}$.

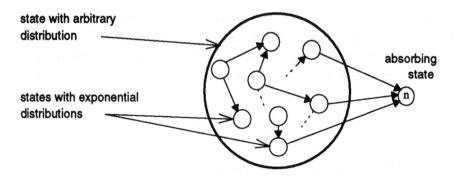

Figure 7: Principle of a phase-type distribution

What this means in practice is that we can extend any Markov representation of a physical system by replacing any fixed state with a number of phase-type states and internal transition rates. The sojourn times of the physical states then become phase-type distributed rather than exponentially distributed. For a more complete exposition, see [20].

Note that the *hyperexponential distribution* and the *Erlang distribution* can both be regarded as special cases of phase-type distributions.

4.2 Example: Trojan Horse

Example 3. Consider again the Reference Monitor of Example 1 in section 3.2. Suppose there is a certain probability that the RM_S unit that checks accesses to secret information has been tampered with. A software module may have been replaced with a modified version of the same module, and the modified module has a hidden function that, once initiated, will force the unit to grant all access requests to secret information, which is a typical confidentiality failure. The triggering condition for the hidden function is unknown. Also, it is not clear whether a certain Reference Monitor really has been tampered with at all, but let us suppose that the probability for this can be estimated.

This situation can be modelled using a phase-type distribution, and in the simplest case with a *hyperexponential distribution* H_r. Such a distribution with r parallel channels is a mixture of r exponential distributions with rates $v_1, \dots, v_r$ so that the density is

$$\sum_{i=1}^{r} \pi_i v_i e^{v_i x},$$

where $\pi_1 + \dots + \pi_r = 1$.

In general, this could model the lifetime of an item which may be of one of r types, the i:th type having exponential lifetime with parameter v_i. In our example, we may assign $r = 2$, where type 1 represents a unit with a normal software and type 2 a unit with a Trojan horse, which is prone to early failure due to the hidden function.

We may take $O = \{1,2\}$, where state 2 represents a unit with a Trojan Horse, state 1 a unit with an error-free program, and F consisting of a single state f. Thus the state diagram is as given in figure 8, with initial probabilities $\pi_1 = 1 - p$, $\pi_2 = p$ for states 1 and 2, respectively.

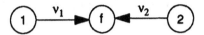

Figure 8: State diagram for a simple phase-type distribution, the hyperexponential

The assumption that state 2 is exponentially distributed is in reality not realistic. The Trojan Horse would normally be activated by some triggering condition, such as a specified time or a certain event. The triggering condition could also be stochastic with some other distribution than the exponential. This situation could be handled by modelling state 2 with a phase-type distribution. This is possible since phase-type distributions are dense: distributions of arbitrary complexity can be modelled by choosing the number of phases large enough. However, for the simplicity of this example, we will adhere to the exponential assumption.

If we thus apply the phase-type reasoning above to the secret access function C of the RM_S unit, we get the following state diagram:

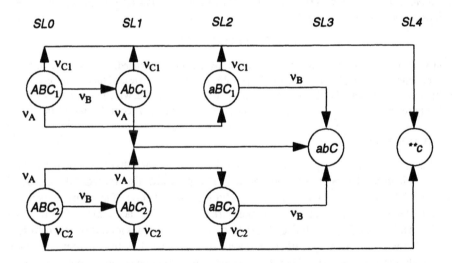

Figure 9: State diagram for a Reference Monitor unit with a possible Trojan Horse

Here state abC is attained by pooling states abC_1 and abC_2. For this case, the operational states $SL0$, $SL1$ and $SL2$ have each been split into two different states according to the assumption of hyperexponential distribution. This implies that the calculations become similar to the ones in Example 1, except that we have twice as many terms.

We obtain expressions of the following types

$$u_{SL0} = (1-p)\frac{1}{v_A + v_B + v_{C_1}} + p\frac{1}{v_A + v_B + v_{C_2}}$$

$$u_{SL1} = (1-p)\frac{v_B}{v_A + v_B + v_{C_1}} \cdot \frac{1}{v_A + v_{C_1}} + p\frac{v_B}{v_A + v_B + v_{C_2}} \cdot \frac{1}{v_A + v_{C_2}}$$

$$u_{SL2} = (1-p)\frac{v_A}{v_A + v_B + v_{C_1}} \cdot \frac{1}{v_B + v_{C_1}} + p\frac{v_A}{v_A + v_B + v_{C_2}} \cdot \frac{1}{v_B + v_{C_2}}$$

$$v_{SL3} = \frac{u_{SL0} + u_{SL1} + u_{SL2}}{p_{SL3}}$$

$$v_{SL4} = \frac{u_{SL0} + u_{SL1} + u_{SL2}}{p_{SL4}}$$

Assume that the untampered program still has a failure rate of $v_{C1} = 1/10000$ failures per hour and that there is a probability of $p = 0.05$ that there is a Trojan Horse, in which case the failure rate increases to $v_{C2} = 500/10000$. If we let the other values be unchanged at $v_A = v_B = 9.5/10000$ failures per hour, we get the dependability vector

$$\mathbf{w} = (476 \ 430 \ 430 \ 1636 \ 7281)$$

hours. We note that the three operational entries are virtually unchanged, as well as the entry for the "reliability" failure. However, due to the possible existence of a Trojan Horse in the confidentiality function, the entry for the "confidentiality" failure is decreased, which means that a system will only function for 7281 hours on an average, before exhibiting a "confidentiality" failure.

5. Summary

In this study we have proposed a way to model security, such that a practical measure for its behavioural aspects can be defined. The measure also includes behavioural aspects of dependability and is therefore valid for a broad class of systems, which can be modelled in behavioural terms. The measure also takes into account the fact that this service is normally degradable, i.e., it can be delivered to various amounts or at different levels. However, reliability, rather than availability, is used as the attribute denoting service delivery to the user. Therefore, aspects of maintenance and repair are not in themselves covered by the measure. We have also outlined how components with ncn-exponential failure rates, such as Trojan Horses, can be modelled using phase-type distributions.

Acknowledgement

We are most grateful to Professor Søren Assmusen, Department of Mathematical Statistics, Lund Institute of Technology, Sweden, for our fruitful cooperation in earlier phases of this work. This paper is partly based on some of his ideas and suggestions.

References

[1] M. D. Beaudry, "Performance-Related Reliability Measures for Computing Systems". IEEE Transactions on Computers, Vol. C-27, No. 6, June 1978.

[2] S. Brocklehurst and B. Littlewood, "New Ways to Get Accurate Reliability Measures", IEEE Software, vol. 9, No. 4, pp. 34-42, 1992.

[3] S. Brocklehurst, B. Littlewood, T. Olovsson, E. Jonsson: "On Measurement of Operational Security", in Proceedings of the Ninth Annual IEEE Conference on Computer Assurance, COMPASS'94, Gaithersburg, Maryland, USA, June 29-July 1, pp. 257-266. 1994.

[4] S. Castano, M. G. Fugini, G. Martella, P. Samarati, "Database Security", Addison-Wesley, 1995. ISBN 0-201-59375-0.

[5] C. J. Date, "An Introduction to Database Systems", Vol. 1, 5th edition, pp. 429ff, Addison-Wesley 1990, ISBN 0-201-51381-1.

[6] D. E. Denning, "A New Paradigm for Trusted Systems", Proceedings of the IEEE New Paradigms Workshop, pp. 36-41. 1993.

[7] G. Grimmet, D. R. Stirzaker, "Probability and Random Processes". ISBN 0-19-853666-6. Clarendon Press. p. 396ff. 1992.

[8] U. Gustafson, E. Jonsson, T. Olovsson: "Security Evaluation of a PC Network based on Intrusion Experiments". Proceedings of the 14th International Congress on Computer and Communications Security, SECURICOM '96, Paris, France, pp. 187-203, June 4-6, 1996.

[9] U. Gustafson, E. Jonsson, T. Olovsson: "On the Modelling of Preventive Security Based on a PC Network Intrusion Experiment". Proceedings of the Australasian Conference on Information Security and Privacy, ACISP'96, Wollongong, Australia, June 24-26, 1996.

[10] R.A. Howard, "Dynamic Probabilistic Systems", New York Wiley 1971, ISBN 99-0002431-1. 1971.

[11] *Information Technology Security Evaluation Criteria (ITSEC)*, Provisional Harmonized Criteria, December 1993. ISBN 92-826-7024-4.

[12] E. Jonsson, T. Olovsson, "On the Integration of Security and Dependability in Computer Systems", IASTED International Conference on Reliability, Quality Control and Risk Assessment, Washington, Nov. 4-6, 1992. ISBN 0-88986-171-4, pp. 93-97.

[13] E. Jonsson, S. Asmussen, "A Practical Dependability Measure for Embedded Computer Systems", Proceeedings of the IFAC 12th World Congress, Sydney, Vol. 2, July 18-23, 1993. pp. 647-652.

[14] E. Jonsson, M. Andersson, S. Asmussen, "A Practical Dependability Measure for Degradable Computer Systems with Non-exponential Degradation", Proceedings of the IFAC Symposium on Fault Detection, Supervision and Safety for Technical Processes, SAFEPROCESS'94, Espoo, Finland, vol. 2, June 13-15, 1994. pp. 227-233.

[15] E. Jonsson, T. Olovsson, "Security in a Dependability Perspective", Nordic Seminar on Dependable Computing Systems 1994 (NSDCS'94), Lyngby, Aug. 24-26, 1994. pp. 175-186.

[16] J. C. Laprie et al.: *Dependability: Basic Concepts and Terminology,* Springer-Verlag, ISBN 3-211-82296-8, 1992.

[17] B. Littlewood, S. Brocklehurst, N.E. Fenton, P. Mellor, S. Page, D. Wright, J.E. Dobson, J.A. McDermid and D. Gollmann, "Towards Operational Measures of Computer Security", Journal of Computer Security, vol. 2, no. 3. 1994.

[18] J.F. Meyer, "On Evaluating the Performability of Degradable Computing Systems", IEEE Transaction on Computers, Vol. C-29, pp. 720-731. 1980.

[19] J.F. Meyer, "Performability: a Retrospective and Some Pointers to the Future" in Performance Evaluation 14, North-Holland, 1992. pp.139-156.

[20] M. F. Neuts, "Matrix-Geometric Solutions in Stochastic Models", Johns Hopkins University Press, Baltimore. 1981.

[21] T. Olovsson, E. Jonsson, S. Brocklehurst, B. Littlewood, "Data Collection for Security Fault Forecasting: Pilot Experiment", Technical Report No 167, Department of Computer Engineering, Chalmers University of Technology, 1992 and ESPRIT/BRA Project No 6362 (PDCS2) First Year Report, Toulouse Sept. 1993, pp. 515-540.

[22] T. Olovsson, E. Jonsson, S. Brocklehurst, B. Littlewood: "Towards Operational Measures of Computer Security: Experimentation and Modelling", in B. Randell et al. (editors.): *Predictably Dependable Computing Systems*, ESPRIT Basic Research Series, Springer Verlag, 1995, ISBN 3-540-59334-9, pp 555-572.

[23] R.M. Smith, K.S. Trivedi, "A Performability Analysis of Two Multi-Processor Systems", Proc. 17th IEEE Int. Symp. on Fault Tolerant Computing, FTCS-17, Pittsburg, Pennsylvania, 1987. pp. 224-229.

[24] E. de Souza e Silva, H.R. Gail, "Calculating Availability and Performability Measures of Repairable Computer Systems Using Randomization", Journal of the ACM, vol. 36, no. 1. 1989.

[25] *Trusted Computer System Evaluation Criteria* ("orange book"), National Computer Security Center, Department of Defense, No DOD 5200.28.STD, 1985.pd

On the Modelling of Preventive Security Based on a PC Network Intrusion Experiment

Ulf Gustafson, Erland Jonsson and Tomas Olovsson
Department of Computer Engineering
Chalmers University of Technology, S-412 96, Göteborg, Sweden
email: ulfg, jonsson, olovsson@ce.chalmers.se

Abstract. This paper describes a realistic intrusion experiment intended to investigate whether such experiments can yield data suitable for use in quantitative modelling of *preventive security*, which denotes the system's ability to protect itself from external intrusions. The target system was a network of Personal Computer clients connected to a server. A number of undergraduate students served as attackers and continuously reported relevant data with respect to their intrusion activities. This paper briefly describes the experiment and presents a compilation of all the types of data recorded. A first interpretation and classification of the data are made, and its possible use for modelling purposes is discussed. Summaries of breach parameters and a number of informtive diagrams and tables reflecting the intrusion process are presented.

Keywords. Security, Vulnerability, Intrusion, Experimentation, Modelling.

1. Introduction

The present position in security evaluation seems to be based on the classes, or rankings, of various Security Evaluation Criteria, such as the European ITSEC [7] and the American "Orange Book" [13]. These classes primarily reflect static design properties and the development process of the system, but do not incorporate the uncertainty and dependence of the operational environment in a probabilistic way, similar to the way in which reliability is commonly expressed. It is clear that those static factors will influence operational security, and they may very well be beneficial in producing secure systems, but they do not facilitate a quantitative evaluation of the actually achieved operational security. Another approach is a security evaluation using so-called "Tiger Teams", a method used for quite some time [1], [6], [4]. A Tiger Team is a group of people that is very skilled and knowledgeable in the security domain and has deep knowledge of the system and an awareness of potential vulnerabilities. However, this type of evaluation reflects the same static factors as do the security evaluation criteria, with the difference that tiger teams focus on the absence of certain factors rather than their presence. The assumption used in this paper is that the operational security of the system, understood as 'the ability of the system to resist attack' may be reflected by a measure *effort* [11]. The advantage of this operational approach is that it allows us to obtain measures of the actual interaction between a system and its environment [8], rather than merely of some static properties. For the purpose of modelling the effort parameter, we have gathered data by means of carrying out intrusion experiments.

The experiment presented here, which was the third in a series, was conducted over a fourweek period in November/December 1994. The target system in the two previous experiments was a Unix-based network [2], [12]. This time, in an attempt at diversification, a PC network was selected in the hope that the change of target system would lead to new findings with respect to security modelling and, in particular, give us a better knowledge of the various aspects of the effort parameter.

In the following, section 2 gives a brief conceptual background and discusses the idea of practical intrusion experiments and section 3 presents the experimental set-up. The parameters recorded are classified and summarized in sections 4 and 5 respectively. An analysis of the data collected and a discussion of its potential use in the modelling of the effort parameter is given in section 6. Section 7 gives an overall conclusion.

2. The Security Model

The security of a computer system is normally understood to be its ability to withstand illegal intentional interaction or attacks against *system assets* such as data, hardware or software. This notion of security normally assumes a hostile action from a person, the *attacker*, who often tries to gain some kind of personal benefit from his actions. Security is traditionally defined by three different aspects: *confidentiality, integrity* and *availability* [7], [13]. The conceptual framework of this paper is a security model in which security is split into two different parts: *preventive* and *behavioural* [8]. Behavioural security deals with those system aspects that reflect the system behaviour with respect to the user of the system, informally the "output" of the system, mainly encompassing availability and confidentiality aspects [9]. Preventive security refers to the system's ability to avoid detrimental influence from the environment, in particular influence originating from security breaches into the system, informally the "input" to the system. Our approach is to try modelling preventive security on the basis of empirical data by means of letting normal users attack a real system under controlled conditions.

3. The Intrusion Experiment

3.1 The System

The target system was a "standard" Novell NetWare 3.12 system with eight Intel 486 PCs connected to a file server. All PC clients had DOS 6.2 and Windows 3.1 installed. The server was secured in a well-protected room, to which only the system administrator had access. Intruder detection and lockout features in NetWare were enabled, and the number of login attempts was restricted to 10. Network (NCP) packet signatures were not used, nor did the hardware support hardware-based passwords. The system was not yet in production use, meaning that there were no outside users during the experiment. No special security-enhancing modifications were made to the system, but "standard" security features were active. For details on the experimental set-up, see [5].

3.2 The Actors

Three different actors are involved in this kind of experimentation: the *attackers*, the *system administrator* and the *coordinator*. The *attackers* were last-year university students who conducted this experiment as part of an undergraduate course in Applied Computer Security. There were 14 attackers divided into eight groups (six groups with two persons, two groups with one person). Each group was given an account with ordinary user privileges. We expected the attackers to spend about 40 hours each of effective working time during a four-week calendar period, although there was no absolute requirement to spend exactly this amount of time. The attackers were to judge by themselves when a breach was achieved, but were subjected to certain restrictions: They were not allowed to physically damage the equipment or disturb other users and they were forbidden to cooperate with members of other attacking groups. The *coordinator's* role was to monitor and coordinate all activities during the experiment. In particular, he was to make sure that the attackers and the system administrator complied with the experimental rules. The *system administrator* was to behave as realistically as possible. He interfered only when the system did not behave normally.

3.3 Reporting

There were three main reports of the type "fill-in forms": the *background report*, the *activity report* and the *evaluation report*. Finally, there was a *written final report* that each group had to submit at the end of the experiment. The *background report* was submitted before the experiment started. Its aim was to document the attackers' background, knowledge in computer security, prior experience with NetWare and DOS etc. In the *activity report*, the attackers were asked to document and describe every specific activity, such as how long they had worked on an activity and what resources they had used, e.g., literature, personal help, Internet or BBSs. They were also asked to record their subjective reward for every activity they carried out. After the experiment, the attackers filled in the *evaluation report*. Here, the attackers described how the work had been performed. This report gave an estimate of the validity of the recorded data. In the *written final report*, the attackers summed up all activities and described the outcomes of the experiment rather freely. The intention was for them to describe their main leads and explain what vulnerabilities they had found during the experiment. They were also required to describe their intrusions and all intrusion attempts in detail.

3.4 Breach Summary

The definition of a breach was rather informal: *A security breach occurred whenever the attackers succeed in doing something they were not normally allowed to do*, for example reading or modifying a protected file, using another user's account or disturbing the normal system function. Table 1 shows a summary of breaches and unsuccessful attempts. Of a total of ten breaches, five were not actually carried through completely, the *discontinued* breaches. They are still classified as breaches however, since they would have resulted in a full breach if there had been real outside users on the system. Three attacks that would have succeeded if the file server had been physically accessible are counted separately.

The vulnerabilities exploited are grouped as follows: *Use of System Administration Programs:* -Programs that are used by the system administrator to administer the system. *Network Attacks:* -Reading or modifying packets transmitted over the network. *Attacks against the Login Process:* -Password guessing or replacing the NetWare login command on the client with a Trojan Horse. *Keyboard Snooping:* -Listening to the keyboard through TSR (Terminate and Stay Resident) programs. And last *File server Attacks:* -Attacks against the file server, provided that the server is physically available to the attacker.

Most vulnerabilities arise from the fact that local clients are insecure. On the clients, it is possible to directly modify any file, including system files. By this, it is possible to install trojan horses, terminate and stay resident programs, remote control programs etc. The second most vulnerable part of the system is the network, since all network traffic except passwords are transmitted in cleartext..

Table 1: Number of Breaches per Category

Category	Unsuccessful Attempts	Breaches		Total no. of Attempts
		discontinued	accomplished	
Use of System Administration Programs (AP)	4	-	-	4
Keyboard Snooping (KS)	4	3	2	9
Network Attacks (NA)	5	-	1	6
Attacks against the Login Process (AL)	3	2	2	7
File server Attacks (FS)	3	-	-	3
Total	16+3	5	5	29

4. Classification of the Recorded Parameters

4.1 Skill Level

The more experienced the attacker is and the more specific knowledge he possesses, the easier it will be for him to break into a system. We have chosen to call this characteristic "skill". To obtain an estimate of this characteristic, we required the attackers in the experiment to state their *skill level* in their initial background report on a scale between 1 and 10.

Table 2: Skill Levels

Group No n	Attacker skill level		Group skill level mean $\bar{S}_n$
	S_{nA}	S_{nB}	
1	6	4	5.0
2	5	9	7.0
3	6	6	6.0
4 (alone)	6	-	6.0
5	7	5	6.0
6	4	8	6.0
7	4	8	6.0
8 (alone)	7	-	7.0

The *skill level*, S_{nX}, $X \in A, B$ is understood as their level of attainment in computer security in relation to an "average" undergraduate student at the same university level (Table 2). Because most of the attackers were working in groups of two we must also derive a group skill level, which we have chosen as the arithmetic mean value of the group members. Input Parameters

Input parameters are the attack-related parameters that were recorded during the performance of the experiment. These are summarized in Table 3. The table also gives the quality of reporting for those parameters. Our classification suggests two types of data related to human working time, these being the *time parameters*: specifically, *working time*(1) for the attackers and *on-line time*(2) on the target workstation. We also have several types of resource-related data. These types could be further subdivided into *informative* resources and *active* resources.

Table 3: Input Parameters

Class	Sub-Class		Parameter	Quality of reporting none/some/fair/good
Time Data	time parameters	1	working time	good
		2	on-line time	fair
Resource Data	informative resources	3	network resources	good
		4	other written media	some
		5	human resources	some
	active resources	6	programs developed by the attacker	good
		7	existing programs	good
		8	external computers	fair

Among the informative resources is found: *network resources* (3), i.e. resources available via the network, which in practice means the Internet. Examples are World Wide Web (WWW) News articles and Bulletin Board Systems (BBS's). *Other written media*(4) are for example manuals, books and journals. *Human resources* (5) consisted primarily of friends (co-operation with other attack groups was not permitted).

Active resources are those that are used actively in the intrusion process and not primarily for information purposes. The main categories are *programs developed by the attacker*(6) and *existing programs*(7), whether freeware, shareware or other. The last active resource is *external computers*(8), i.e. computers available to the attackers by other means than the experimental setup.

5. Collected Data

5.1 Background Data Summary

The background report gave us information about the knowledge in computer security of the attackers, the skill level discussed previously. The total number of attackers were 14 and, according to the background report most had worked with PCs before the experiment started. Four of the attackers reported their experience with DOS and Windows to be "some" and eight considered their experience to be "good". The remaining two reported their experience with those operating systems to be "very good". Most of the attackers claimed that they had no previous knowledge at all of security vulnerabilities in NetWare. However, six of the attackers had prepared themselves before the experiment by trying to figure out potential NetWare vulnerabilities. Some came up with attacking ideas such as utilizing unprotected clients, bad passwords and the possibility of attacking a physically accessible file server. Among those six attackers the pre-experiment preparation time was distributed between 0.5 and 6 hours.

5.2 Activity Report Summary

During the experiment, we received 80 activity reports describing a total of ten security breaches of various kinds. In total, the groups reported 318 hours of group working time, i.e. time during which at least one group member was active. A total of 424 man-hours was spent, which is an average of 30 man-hours per individual attacker. It is important not to forget that these numbers include time learning about the system. A total of 89 hours of on-line time was reported.

Table 4 presents a summary of effort parameters per breach. It is organized with one row per breach giving the group skill level, type of activity, various time data and type of breach. The time parameters are discussed in section 6.2. It is noteworthy that many

Table 4: Working Time per Successful Breach

Group No	Group skill level	Breach No.	Activity[1]	Preparation time[2] t_p	Attack time[3] t_a	Accumulated working time[4] t_{gw}	Breach[5]
1	5	1	fetch/install KS	6.5	2.5	9	KS
		2	improving KS	0	2.5	11.5	KS
2	7	3	passwd guessing	9	0.5	9.5	AL
3	6	4	fetch/install/ rewrite hack.exe	18	12	30	NA
4 alone	6	5	fetch/install KS	25	3	28	KS
5	6	6	coding/install TH	6	7.5	13.5	AL
		7	coding/install KS	7	11.5	32	KS
6	6	8	fetch/install KS	12.5	14	26.5	KS
		9	coding/install TH	17.5	1	44	AL
7	6	10	coding/install TH	27	17	44	AL
8 alone	7	-	-	22	16.5	38.5	-

1. Activity: *KS* Keyboard snooper, *TH* Trojan Horse.
2. Preparation time is the time spent learning before the actual attack started.
3. Attack time is equal to working time per breach: An estimate made from the activity reports of how many working hours the team spent developing the particular attack.
4. Accumulated working time is also an estimate from the activity reports. This is the total working time from the start of the experiment to the actual breach.
5. *KS* keyboard snooper, *AL* attacks against the login process, *NA* network attack.

attackers, after having accomplished a successful breach, started to improve the intrusion method instead of searching for new ways to break into the system. This was done in spite of the fact that the objective was to make as many breaches as possible. Also, many groups tried to write their own software, even though they knew that similar software was available elsewhere. Note that group 8 did not accomplish any breach, so the time data listed for this group is the total time up to the end of the experiment. Figure 1 shows the group working time and on-line time for each group.

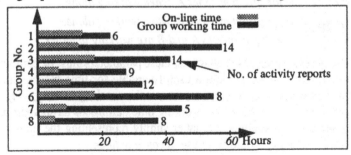

Figure 1: Accumulated working time per group

The distribution of time between group members is shown in figure 2.

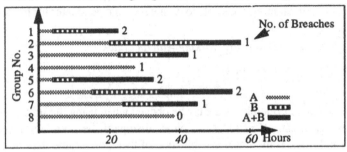

Figure 2: Accumulated working time per group member

It is interesting to note that the attackers cooperated during only 84 hours, which is about 26% of the working time (or, if we consider that 1 hour of cooperative work is 2 man-hours, cooperation accounts for 40% of the working time). The members of the most cooperative group worked together during 67% of the overall working time. We see that there is a tendency towards a greater number of breaches among groups with much cooperation as opposed to groups with little cooperation or single-attacker groups. This indicates that cooperation should be favourable for achieving many breaches.

6. Evaluation of the Recorded Parameters

6.1 Introduction

There are no evident rules or policies that can serve as a guideline for selecting the parameters that would form the measure of *effort*. In this section, we discuss how the recorded data may be interpreted and used. We also evaluate possible ways of finding an appropriate method for compounding the effort parameter. Clearly, the time parameters are the most tangible ones. They are easily measured and recorded, even if the accuracy may vary. In contrast, the resource-related data are more difficult to quantify. Furthermore, it is not clear how the value of different resources should be combined. Given the inherent uncertainty and inaccuracy of much of the input data, the discussion must be heuristic rather than mathematically stringent.

6.2 Time Parameters

The recorded time parameters are:

t_A = *working time for group member A, when working alone.*

t_B = *working time for group member B, when working alone.*

t_{A+B} = *time when group members A and B are working together.*

As a result of the grouping of the attackers, these parameters must be combined. Thus, there are two obvious ways to yield a useful variable for time measurement: either *group working time*, $t_{gw} = t_A + t_B + t_{A+B}$, or *attacker working time*, $t_{aw} = t_A + t_B + 2 \times t_{A+B}$. It is possible to use either of the working time definitions as an appropriate variable in the process of modelling, the latter simply representing the number of man-hours. Another possibility would be to use the *on-line time*:

$t_o = t_{gw}$ | *the group is logged in to the target workstation.*

On the other hand, the on-line time could also be regarded as an (active) resource parameter. Another parameter possible to define is a time parameter for the active intrusion work, an *attack time*:

$t_a = t_{gw}$ | *the group is actively working with an attack.*

The *preparation time* is defined as:

$t_p = t_{gw} - t_a.$

There are two main principles that may be obeyed when using the time parameters in the composition of an effort measure, either disregard the correlation between the parameters and make some linear combination of them, or assume that the correlation is high, in which case (any) one of the types could serve as a suitable time parameter. We support the latter alternative, and some evidence for this standpoint from a Unix experiment is found in [10]. We thus suggest using the *group working time* as the main parameter in the effort measure. This standpoint is further supported by the fact that on-line time could, as mentioned, be regarded as an active resource parameter.

6.3 Active Resources

The availability of a certain active resource represents a value that should be accounted for in the modelling of the attacking process. For example, the use of a certain program, such as *hack.exe*, should in principle represent a value corresponding to the time it would take for the attacker to develop such a program. However, for a number of reasons, it is quite hard to make this estimation. Not only is it difficult to make the estimation in itself, but the development time may also differ between different attackers, and it not obvious which time would be the "correct" time. A solution to this would be to use different values for different attackers, but this approach seems both complicated and unattractive. Another complication is that different attackers would probably develop different versions of the program, versions that would not be comparable. To sum up, the use of an active resource represents a contribution to the effort measure and should thus be quantified and incorporated. However, it has not been possible to do so at this stage, primarily owing insufficient reporting, but also because no good suggestions exists for how to quantify and integrate differing effects of different resources. We therefore believe that it is more consistent to disregard all types of such data, rather than to arbitrarily incorporate them only in those cases when we are aware of the use of the actual resource. This omission will increase the inaccuracy of the measure we wish to establish.

6.4 Informative Resources

Informative resources could probably be traded off against the skill factor, since the use of these resources will, in general, gradually increase the skill of the attacker. However, this "skill growth" is most certainly not linear, and it is expected that after some time a "saturation level" is asymptotically approached. The saturation level results from the fact that the total available knowledge in intrusion-related matters is limited. In theory, if this level should be reached, further progress must be made by means of the innovative skills of the attacker himself. A further complication is that the knowledge and skill level is not a constant, but varies with respect to different breaches. For example, knowledge of cryptography and certain types of mathematics is valuable in an attempt to break a crypto, but is of little use in trying to create a dummy root account by means of spoofing the operating system. It could be argued that the skill levels of the attackers are so comparatively high that the skill growth resulting from the use of informative resources is negligible. This is most probably not true however, mainly because the information that the attackers really do find is in many cases very specific to the intrusion situation and has a very high value for the attackers. Table 5 shows the use of informative resources as compared with a number of other variables, i.e. skill level, number of attempts and number of breaches, discontinued as well as accomplished. The right-hand side of the table shows which informative resources were actually used and for how long. It also shows the percentage of group working time spent on informative resources versus the total amount of group working time. (The total amount of group working time is found in Figure 1 and Figure 2.)

Table 5: Informative Resources

Group No	Group skill level	No of attempts	No of breaches	Informative resources[1]	Use of Informative Resources (hours)	Inf.res/ Working time (%)
1	5	4	2	pers, WWW, lit	9.5	41
2	7	4	1	lit, News, WWW	21	37
3	6	2	1	WWW, News	20	48
4 alone	6	4	1	WWW, books, lit, pers, News, BBS	21.5	77
5	6	2	2	News, WWW, Magazines, lit	8	26
6	6	8	2	Books, Articles, News, WWW, BBS	36	65
7	6	2	1	Lit, WWW, News, BBS	21	48
8 alone	7	3	0	News, Books	17	45

1. Resource: *WWW*, World Wide Web. *lit*, supplied literature on Novel NetWare 3.12. *pers*, external personal help. *News*, USENET News. *BBS*, Bulletin Board System.

It is difficult to distinguish any correlation between the use of informative resources (or the skill level) and the number of breach attempts or successful breaches. Thus, there is no basis for considering informative resources separately, at least not for the time being. Furthermore, the use of informative resources seems to be rather "evenly" distributed among the groups, both the information sources as such and the use of these in percentage of the working time. Therefore, our preference is to let these resources be represented by an approximately constant "offset" to the measure, an offset that could be disregarded, at least comparatively.

6.5 Discussion on Quantitative Modelling

Clearly, the recorded data from this only experiment is vastly insufficient for any decisive judgement with respect to the modelling of the preventive measure or as an input to the problem of how the effort parameter should actually be composed. The only data that can readily be handled are the time data. Disregarding that the fact that these data are far from statistically significant, we can still derive a value for the Mean Time To Breach (MTTB) under the assumption that the pre-conditions for such a derivation are fulfilled. Thus, we obtain an MTTB of 30 hours (298 hours divided by 10 breaches) if we use the total group working time, t_{gw}, expended by all groups up to the second breach, or the total time if none or one breach was made. Considering active attack time, t_a, only yields a significantly reduced MTTB figure of seven hours. But how do we incorporate the skill level? By weighting the time data? And how do we consider resource parameters? It is evident that we need data from a number of different experiments, performed on different target systems, to be able to even arrive at a plausible suggestion. During the process of finding the measure, it is likely that several methods and combinations of methods must be tested before we can find a solution that satisfies our requests. It is also possible that there may not be only one single solution to this problem, but rather many different solutions that would all lead to some acceptable measure. Despite these major obstacles, we believe that, given sufficient data, we can arrive at some useful ideas, whose correctness we then hope to be able to show in retrospect.

7. Conclusion

This paper has brought us one step further towards the goal of finding an appropriate way to model preventive security. A classification of the recorded data was made, and the possible use of the data for modelling purposes was discussed. Still, it is evident that substantial problems remain, problems that we hope to be able to counter in the future by means of combining data from several experiments. A qualitative result of the experiment is that, despite the fact that the system owner claimed to have adequate security mechanisms, the system was easily and successfully attacked and breached by novice users who exploited the insecurity of the clients and the network. The experiment also shows the importance of Internet access for information gathering. Although all information is available through alternative sources, the use of Internet makes an enormous amount of, often dedicated, information available very easily and rapidly. It is often this fact that reduces the effort required for a successful intrusion.

8. References

[1] C. R. Attanasio. P. Markstein and R. J. Phillips: *Penetrating an Operating System: A Study of VM/370 Integrity*, IBM Systems J., 15 (1), pp. 102-16, 1976.

[2] S. Brocklehurst, B. Littlewood, T. Olovsson and E. Jonsson: *On Measurement of Operational Security*, in COMPASS 94 (9th Annual IEEE Conference on Computer Assurance), (Gaithersburg), pp.257-66, IEEE Computer Society, 1994.

[3] D. E. Denning: *An Intrusion-Detection model*, IEEE Trans. Software Engineering, 12 (2), pp.222-32, 1987.

[4] P. D. Goldis: *Questions and Answers about Tiger Teams*, EDPACS, The EDP Audit, Control and Security Newsletter, October 1989, Vol XVII, No. 4.

[5] U. Gustafson, E. Jonsson, T. Olovsson: *Security Evaluation of a PC Network based on Intrusion Experiments.* In the Proceedings of the 14th International Congress on Computer and Communications Security, SECURICOM '96 , 4-6 June 1996, Paris, France.

[6] I. S. Herschberg: *Make the Tigers Hunt for You*, Computers and Security, 7, pp. 197-203, 1988.

[7] *Information Technology Security Evaluation Criteria (ITSEC)*, Provisional Harmonized Criteria, December 1993. ISBN 92-826-7024-4.

[8] E. Jonsson, T. Olovsson: *On the Integration of Security and Dependability in Computer Systems*, IASTED International Conference on Reliability, Quality Control and Risk Assessment, Washington, Nov. 4-6, 1992. ISBN 0-88986-171-4, pp. 93-97.

[9] E. Jonsson, M. Andersson: *On the Quantitative Assessment of Behavioural Security.* Presented at the Australasian Conference on Information Security and Privacy, 24-26 june 1996, Wollongong, Australia.

[10] E. Jonsson, T. Olovsson: *An Empirical Model of the Security Intrusion Process.* In the Proceedings of the 11th Annual IEEE Conference on Computer Assurance, COMPASS '96, 17-21 June 1996, Gaithersburg, Maryland, USA.

[11] B. Littlewood, S. Brocklehurst, N.E. Fenton, P. Mellor, S. Page, D. Wright, J.E. Dobson, J.A. McDermid and D. Gollmann: *Towards operational measures of computer security*, Journal of Computer Security, vol. 2, no. 3. 1994.

[12] T. Olovsson, E. Jonsson, S. Brocklehurst, B. Littlewood: *Towards Operational Measures of Computer Security: Experimentation and Modelling*, in Predictably Dependable Computing Systems (editor B. Randell et al.), Springer Verlag, ISBN 3-540-59334-9, 1995.

[13] *Trusted Computer System Evaluation Criteria* ("Orange Book"), National Computer Security Center, Department of Defense, No DOD 5200.28.STD, 1985.

Evidential Reasoning in Network Intrusion Detection Systems

Mansour Esmaili   Reihaneh Safavi-Naini   Josef Pieprzyk

Center for Computer Security Research
University of Wollongong
Wollongong, NSW 2522
Australia.
{mansour,rei,josef}@cs.uow.edu.au

Abstract. Intrusion Detection Systems (IDS) have previously been built by hand. These systems have difficulty successfully classifying intruders, and require a significant amount of computational overhead making it difficult to create robust real-time IDS systems. Artificial Intelligence techniques can reduce the human effort required to build these systems and can improve their performance. AI has recently been used in Intrusion Detection (ID) for anomaly detection, data reduction and induction, or discovery, of rules explaining audit data [1]. This paper proposes the application of evidential reasoning for dealing with uncertainty in Intrusion Detection Systems. We show how dealing with uncertainty can allow the system to detect the abnormality in the user behavior more efficiently.

1   Introduction

The importance of securing the data and information maintained by a corporation has become a driving force in the development of numerous systems that perform computer security audit trail analysis [2, 3]. These systems are generally classified as "*intrusion detection*" systems. The primary purpose of an intrusion detection system is to expose computer security threats in a timely manner.

Intrusion detection and *network security* are becoming increasingly more important in today's computer-dominated society. As more and more sensitive information is being stored on computer systems and transferred over computer networks, more and more *crackers* are attempting to attack these systems to steal, destroy or corrupt that information. While most computer systems attempt to prevent unauthorized use by some kind of access control mechanism such as passwords, encryption, and digital signatures, there are several factors that make it very difficult to keep these crackers from eventually gaining entry into a system [4, 5, 6, 7, 8]. Most computer systems have some kind of security flaw that may allow outsiders (or legitimate users) to gain unauthorized access to sensitive information. In most cases, it is not practical to replace such a flawed system with a new, more secure system. It is also the case that it is very difficult, if not impossible, to develop a completely-secure system. Even a supposedly secure system can still be vulnerable to insiders misusing their privileges, or it can be compromised by improper operating practices. While many existing systems may be designed to prevent specific types of attacks, other methods to gain unauthorized access may still be possible.

Intrusion detection systems (IDS) require that basic security mechanisms are in place which enforce authorization controls over system, data and other resource access on computer or network, and that an audit trail be available to record a variety of computer usage activity. ID systems attempt to identify security threats through the analysis of these computer security audit trail. Intrusion detection systems are based on the principle that an attack on a computer system (or network) will be noticeably different from normal system (or network) activity [9]. An intruder to a system (possibly masquerading as a legitimate user) is very likely to exhibit a pattern of behavior different from the normal behavior of a legitimate user. The job of IDS is to detect these abnormal patterns by analyzing numerous sources of information that are provided by the existing systems. The two major methods used to detect intrusions are *statistical analysis* and rule-based *expert systems analysis* [2, 3]. The statistical method attempts to define *normal* (expected) behavior, while the expert system defines *proper* behavior. The expert system also searches for breaches of policy. If the IDS notices a possible attack using either of these methods, then the *System Security Officer* (SSO) is notified. The SSO may then take some action against the aggressor.

In this paper, our interest is to extend the IDS paradigm to include specific models of proscribed activities. These models would imply certain activities with certain observables which could then be monitored. This would allow to actively search for intruders by looking for activities which would be consistent with a hypothesized intrusion scenario. But not always the evidence can be matched perfectly to a hypothesized intrusion. Therefore, a determination of the *likelihood* of a hypothesized intrusion would be made based on the combination of evidence for and against it. The security of such an explicit model should be easier to validate. However, the system must be able to deal with information that can be uncertain.

Various numerical calculi have been proposed as methods to represent and propagate uncertainty in a system. Among the more prominent calculi are *probabilistic* (in particular *Bayesian*) methods, the *evidence theory of Dempster-Shafer*, *fuzzy set theory*, and the MYCIN and EMYCIN *calculi* [10]. In this paper we look at the application of evidential reasoning in computer intrusion detection.

2 The Dempster-Shafer Theory

In the 1960s, A. Dempster laid the foundation for a new mathematical theory of uncertainty. In the 1970s, this theory was extended by G. Shafer to what is now known as *Dempster-Shafer Theory*. This theory may be viewed as a generalization of probability theory. Contrary to the subjective Bayesian method and the certainty factor model [10], Dempster-Shafer theory has not been specially developed for reasoning with uncertainty in expert systems. Only at the beginning of 1980s it became apparent that the theory might be suitable for such a purpose. However the theory cannot be applied in an expert system without modification. Moreover, the theory in its original form has an exponential computational complexity. For rendering it useful in the context of expert systems Lucas and Van Der Gaag in [11] proposes several modifications of the theory.

2.1 The Probability Assignment

As it was mentioned before, the Dempster-Shafer theory may be viewed as a generalization of probability theory. The development of the theory has been motivated by the observation that probability theory is not able to distinguish between *uncertainty* and *ignorance* owing to incompleteness of information. In probability theory probabilities have to be associated with individual atomic hypotheses. Only if these probabilities are known, the computation of other probabilities of interest are possible. In the Dempster-Shafer theory, however, it is possible to associate measures of uncertainty with *sets of hypotheses*, interpreted as disjoints, instead of with the individual hypotheses only. This nevertheless makes it possible to make statements concerning the uncertainty of other sets of hypotheses. Note that, in this way, the theory is able to distinguish between uncertainty and ignorance.

The strategy followed in the Dempster-Shafer theory for dealing with uncertainty roughly amounts to starting with an initial set of hypotheses. Then for each piece of evidence associating a measure of uncertainty with certain subsets of the original set of hypotheses. This continues until measures of uncertainty may be associated with all possible subsets on account of the combined evidence. The initial set of all hypotheses in the problem domain is called the *frame of discernment*. In such a frame of discernment the individual hypotheses are assumed to be disjoint. The distribution of a unit of belief over a frame of discernment is called a *mass distribution* [12]. A mass distribution, m_Θ, is a mapping from subsets of a frame of discernment, Θ, into the unit interval. The impact of a piece of evidence (*body of evidence*) on the confidence or belief in certain subsets of a given frame of discernment is described by means of a function which is defined in the following definition [11].

Definition 1. Let Θ be a frame of discernment. If with each subset $x \subseteq \Theta$ a number $m_\Theta(x)$ is associated such that:

(1) $m_\Theta(x) \geq 0$
(2) $m_\Theta(\emptyset) = 0$
(3) $\sum_{x \subseteq \Theta} m_\Theta(x) = 1$

then m_Θ is called a *basic probability assignment* (or mass distribution) on Θ. For each subset $x \subseteq \Theta$, the number $m_\Theta(x)$ is called the *basic probability number* of x. □

There are two other notions which should be defined.

Definition 2. Let Θ be a frame of discernment and let m_Θ be a mass distribution on Θ. A set $x \subseteq \Theta$ is called a *focal element* in m_Θ if $m_\Theta(x) > 0$. The *core* of m_Θ, denoted by $\kappa(m)$, is the set of all focal elements of m_Θ.   □

Notice the similarity between a basic probability assignment (mass distribution) and a probability function. A probability function associates each element in Θ with a number from the interval $[0, 1]$ such that the sum of these numbers equal 1. Figure 1 shows the lattice of all possible subsets for a typical set Θ. A mass distribution (basic probability) associates a number in the interval $[0, 1]$ with each element in 2^Θ a such that once more the sum of the numbers equal 1.

$$m_\Theta : \quad 2^\Theta \quad \longmapsto \quad [0, 1]$$

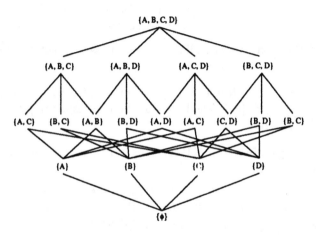

Fig. 1. Lattice of all possible subsets of the universe $\Theta = \{A, B, C, D\}$.

A probability number $m_\Theta(x)$ expresses the confidence or belief assigned to precisely the set x. It does not express any belief in subset of x. It will be evident, however, that the total confidence in x is not dependent on the confident assigned to subsets of x. For a given basic probability assignment, [11] defines a function describing the cumulative belief in a set of hypotheses.

Definition 3. Let Θ be a frame of discernment, and let m_Θ be a mass distribution on Θ. Then the *belief function* (or *credibility function*) corresponding with m_Θ is the function $\mathrm{Bel}:2^\Theta \longmapsto [0, 1]$ defined by

$$\mathrm{Bel}(x) = \sum_{y \subseteq x} m_\Theta(y)$$

for each $x \subseteq \Theta$.   □

Several properties of this belief function can easily be proved:

(1) $\mathrm{Bel}(\Theta) = 1$ since $\displaystyle\sum_{y \subseteq \Theta} m_\Theta(y) = 1$.

(2) For each $x \subseteq \Theta$ containing exactly one element, $\mathrm{Bel}(x) = m_\Theta(x)$.

(3) For each $x \subseteq \Theta$, we have $\mathrm{Bel}(x) + \mathrm{Bel}(\bar{x}) \leq 1$, since

$$\mathrm{Bel}(\Theta) = \mathrm{Bel}(x \cup \bar{x}) = \mathrm{Bel}(x) + \mathrm{Bel}(\bar{x}) + \sum_{\substack{x \cap y \neq \emptyset \\ \bar{x} \cap y \neq \emptyset}} m_\Theta(y) = 1.$$

Furthermore, the inequality $\mathrm{Bel}(x) + \mathrm{Bel}(y) \leq \mathrm{Bel}(x \cup y)$ holds for each $x, y \in \Theta$.

Some special belief functions follows. Recall that a basic probability assignment (mass distribution) describing lack of evidence had the following form:

$$m_\Theta(x) = \begin{cases} 1 \text{ if } x = \Theta \\ 0 \text{ otherwise.} \end{cases}$$

The belief function corresponding to such an assignment has been given a special name [11].

Definition 4. Let Θ be a frame of discernment, and let m_Θ be a mass distribution such that $\kappa(m_\Theta) = \{\Theta\}$. The belief function corresponding to m_Θ is called a *vacuous belief function* □

The following definition from [11] concerns functions corresponding with mass distribution of the form:

$$m_\Theta(x) = \begin{cases} 1 - C_1 \text{ if } x = \Theta \\ C_1 \quad\;\; \text{if } x = A \\ 0 \quad\quad \text{otherwise} \end{cases}$$

where $A \subset \Theta$, and $0 < C_1 < 1$ is a constant.

Definition 5. Let Θ be a frame of discernment, and let m_Θ be a mass distribution such that $\kappa(m_\Theta) = \{A, \Theta\}$ for a certain $A \subset \Theta$. The belief function corresponding to m_Θ is called a *simple support function*. □

A belief function provides a lower bound for each set x to the 'actual' belief in x. It is also possible that belief has been assigned to a set w such that $x \subseteq w$. Therefore, in addition to the belief function the Dempster-Shafer theory defines another function corresponding with a basic probability assignment (mass distribution).

Definition 6. Let Θ be a frame of discernment, and let m_Θ be a mass distribution on Θ. Then the *plausibility function* corresponding to m_Θ is the function $\mathrm{Pl}: 2^\Theta \longmapsto [0,1]$ defined by

$$\mathrm{Pl}(x) = \sum_{x \cap y \neq \emptyset} m_\Theta(w)$$

for each $x \subseteq \Theta$. □

A function value $\text{Pl}(x)$ indicates the total confidence *not* assigned to $\bar{x}$, so $\text{Pl}(x)$ provides an upperbound to the 'real' confidence in x. It can be shown that, for a given basic probability assignment m_Θ, the property

$$\text{Pl}(x) = 1 - \text{Bel}(\bar{x})$$

for each value $x \subseteq \Theta$, holds for the belief function Bel and the plausibility function Pl corresponding to m_Θ. The difference $\text{Pl}(x) - \text{Bel}(x)$ indicates the confidence in the sets w for which $x \subseteq w$ and therefore expresses the uncertainty with respect to x.

Definition 7. Let Θ be a frame of discernment and let m_Θ be a mass distribution on Θ. Let Bel be the belief function corresponding to m_Θ, and let Pl be the plausibility function corresponding to m_Θ. For each $x \subseteq \Theta$, the closed interval $[\text{Bel}(x), \text{Pl}(x)]$ is called the *belief interval* of x. □

The lower bound of a belief interval indicates the degree to which the evidence supports the hypothesis, while the upper bound indicates the degree to which the evidence fails to refute the hypothesis, i.e., the degree to which it remains plausible.

Example 1. Let Θ be a frame of discernment and let $x \subseteq \Theta$. Now, consider a basic probability m_Θ on Θ and its corresponding functions Bel and Pl.

- If $[\text{Bel}(x), \text{Pl}(x)] = [0,1]$, then no information concerning x is available.
- If $[\text{Bel}(x), \text{Pl}(x)] = [0,0]$, then x has been completely denied by m_Θ.
- If $[\text{Bel}(x), \text{Pl}(x)] = [0,0.8]$, then there is some evidence against x.
- If $[\text{Bel}(x), \text{Pl}(x)] = [1,1]$, then x has been completely confirmed by m_Θ.
- If $[\text{Bel}(x), \text{Pl}(x)] = [0.3,1]$, then there is some evidence in favor of the hypothesis x.
- If $[\text{Bel}(x), \text{Pl}(x)] = [0.15, 0.75]$, then there is some evidence in favor as well as against x.

□

If $\text{Pl}(x) - \text{Bel}(x) = 0$ for each $x \subseteq \Theta$, then we are back to conventional probability theory. In such a case, the belief function is called a Bayesian belief function. This notion is defined more formally in the following definition from [11].

Definition 8. Let Θ be a frame of discernment and let m_Θ be a mass distribution such that the core of m_Θ consists only of singleton sets. The belief function corresponding to m_Θ is then called a *Bayesian belief function*. □

Dempster's rule of combination The Dempster-Shafer theory provides a function for computing from two pieces of evidence and their associated basic probability assignment a new basic probability assignment describing the combined influence of these pieces of evidence. This function is known as *Dempster's rule of combination*. The more formally definition is as follows [11].

Definition 9. Let Θ be a frame of discernment, and let m_Θ^1 and m_Θ^2 be basic probability assignments on Θ. Then $m_\Theta^1 \oplus m_\Theta^2$ is a function $m_\Theta^1 \oplus m_\Theta^2 : 2^\Theta \longmapsto [0,1]$ such that

(1) $m_\Theta^1 \oplus m_\Theta^2(\emptyset) = 0$, and

(2) $m_\Theta^1 \oplus m_\Theta^2(x) = \dfrac{\displaystyle\sum_{y \cap z = x} m_\Theta^1(y) \cdot m_\Theta^2(z)}{\displaystyle\sum_{y \cap z \neq \emptyset} m_\Theta^1(y) \cdot m_\Theta^2(z)}$ for all $x \neq \emptyset$.

$\text{Bel}_1 \oplus \text{Bel}_2$ is the function $\text{Bel}_1 \oplus \text{Bel}_2 : 2^\Theta \longmapsto [0,1]$ defined by

$$\text{Bel}_1 \oplus \text{Bel}_2(x) = \sum_{y \subseteq x} m_\Theta^1 \oplus m_\Theta^2(y). \tag{1}$$

$\square$

2.2 Evidential Reasoning

The goal of evidential reasoning is to assess the effect of all available pieces of evidence upon a hypothesis, by making use of domain-specific knowledge. *The first step in applying evidential reasoning to a given problem is to delimit a propositional space of possible _situations_.* Within the theory of belief functions, this propositional space is called the *frame of discernment*. A frame of discernment delimits a set of possible situations, exactly one of which is true at any one time. Once a frame of discernment has been established, propositional statements can be represented by subsets of elements from the frame corresponding to those situations for which the statements are true. Bodies of evidence are expressed as probabilistic opinions about the partial truth or falsity of propositional statements whose granularity is appropriate to the variable evidence.

Evidential reasoning provides a number of formal operations for assigning evidence [12], including:

1. **Fusion** — to determine a consensus from several bodies of evidence obtained from independent sources. Fusion is accomplished through Dempster's rule of combination (Eq. 1):

$$m_\Theta^3(A_h) = \frac{1}{1-k} \sum_{A_i \cap A_j = A_h} m_\Theta^1(A_i) \imath n_\Theta^2(A_j), \tag{2}$$

$$k = \sum_{A_i \cap A_j = \phi} m_\Theta^1(A_i) m_\Theta^2(A_j).$$

Dempster's Rule is both commutative and associative (meaning evidence can be fused in any order) and has the effect of focusing belief on those propositions that are held in common.

2. **Translation** — to determine the impact of a body of evidence upon elements of a related frame of discernment. The *translation* of a BOE from frame Θ_A to frame Θ_B using the compatibility relation $\Theta_{A,B}$ is defined by:

$$m_{\Theta_B}(B_j) = \sum_{\substack{C_{A \mapsto B}(A_k) = B_j \\ A_k \subseteq \Theta_A, \, B_j \subseteq \Theta_B}} m_{\Theta_A}(A_k), \qquad (3)$$

where $C_{A \mapsto B}(A_k) = \{b_j \mid (a_i, b_j) \in \Theta_{A,B}, a_i \in A_k\}$.

3. **Projection** — to determine the impact of a body of evidence at some future (or past) point in time. The *projection* operation is defined exactly as translation, where the frames are taken to be one time-unit apart.

4. **Discounting** — to adjust a body of evidence to account for the credibility of its source. Discounting is defined as

$$m_{\Theta}^{discounted}(A_j) = \begin{cases} \alpha \cdot m_{\Theta}(A_j), & A_j \neq \Theta \\ 1 - \alpha + \alpha \cdot m_{\Theta}(\Theta), & \text{otherwise} \end{cases} \qquad (4)$$

where α is the assessed credibility of the original BOE ($0 \leq \alpha \leq 1$).

Independent opinions are expressed by multiple bodies of evidence. Dependent opinions can be represented either as a single body of evidence, or as a network structure that shows the inter-relationships of several BOEs. The evidential reasoning approach focuses on a body of evidence, which describes a meaningful collection of interrelated beliefs, as the primitive representation. In contrast, all other such technologies focus on individual propositions.

2.3 Analysis Using an Example

To illustrate the evidential reasoning method described above in an intrusion detection system, we use the following example.

> A user successfully logs in from a remote host after trying several bad passwords and usernames. The user enters several wrong command names and arguments and tries to look at some directories and files for which permissions for him is denied. The user also several times uses commands such as 'finger' to find out about other system users and activities. The user also copies the /bin/csh file into /usr/spool/mail/root where the root's mail directory resides, and makes it a setuid file by chmod 4755 /usr/spool/mail/root command. After a few minutes, the user leaves. Who was this? Could it be an intruder or just an inexperienced user who was experimenting with the system?

In evidential reasoning the first step is to construct the sets of possibilities (the frame of discernment) for each unknown. For example, the user could either be an intruder or not; these possibilities can be represented in the **Intruder?** frame:

$$\{\textit{Yes, No}\}$$

Other frames could also be constructed; `Location` will be included for user's location containing the possibilities:

$$\{\textit{Local, Remote}\}.$$

Two types of location for a user is distinguished — local (i.e., physically at the keyboard) and remote. Because the majority of intruders do not have direct physical access to the locally connected terminals, a local keyboard is considered to indicate normal use and not an intruder. Most intrusions originate from remote internet sites. However, because an intruder can jump from host to host, intrusive behavior is also likely to appear originating from local hosts. Thus, activity originating from any location other than the local keyboard is considered equally indicative of intrusive behavior, so only the single category 'remote' will be used for this. For remote user, it can not be distinguished whether the user is an intruder based on this dimension of behavior alone.

An intruder is expected to be somewhat paranoid, therefore a frame, **Fear**, is included to capture paranoia level

$$\{\textit{Paranoid, Calm}\}.$$

A paranoid intruder (one who is afraid of being caught) will probably have very short sessions (eg., lasting under two minutes), because the longer the session the greater the risk of discovery. A paranoid intruder will also commonly check to see who is logged in and what they are doing. Thus, for example, in Unix an ordinate number of 'who', 'ps', and 'finger' commands can be expected to indicate a paranoid intruder. User sessions can be characterized as having a high degree of this sort of activity if two or more such commands are used. Therefore, short sessions and two or more "surveillance" commands are considered to be strong indicators of fear.

An intruder may also be unfamiliar with the system, so another frame, **familiarity**, will be defined to contain:

$$\{\textit{Familiar, Unfamiliar}\}.$$

A person who is unfamiliar with the computer system under attack is likely to have a relatively large number of invalid commands, resulting from attempts to execute commands that are not recognized by the system. Such a person is also likely to have a relatively large number of errors resulting from invalid command usage, for example, too few arguments or invalid parameters. But this alone can not be a good measure to condemn a user to be an intruder, since the user might be inexperienced. This frame should be looked at in conjunction with other frames. A relatively large number of file permission errors, resulting from attempting to read, write, or execute files or directories when permission is denied, is also indicative of a person unfamiliar with the computer system under attack. Therefore, relatively large numbers of errors of these several types are considered to be strong indicators of unfamiliarity with the system. Conversely, low error rates for all of these categories of error strongly suggest a normal, nonintrusive user.

Another frame can be constructed for the actions which raise the suspicion level, such as copying a file from /bin directory or trying to access somebody else's mail file, or etc. These actions can be represented in the Actions frame.

{*Malicious, Normal*}

Authentication errors result from the use of an invalid username or password during login. A high rate of authentication errors (greater than three failed login attempts for a given username in a certain time period) is considered to be strongly suggestive of an intrusion attempt.

Once the frames are defined, the next step is to construct the compatibility relations that define the domain-specific relationships between the frames. A connection between two propositions A_1 and B_1 indicates that may co-occur (in other words, $(A_1, B_1) \in \Theta_{A,B}$).

Figure 2 shows the frames and compatibility relations used in determining whether the user is an intruder.

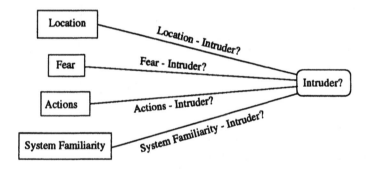

Fig. 2. Frames and compatibility relations.

Once the frames and compatibility relations have been established, the evidence can be analyzed. The goal of the analysis is to establish a line of reasoning from the evidence to determine belief in a hypothesis, in this case that the user is an intruder.

The first step is to assess each piece of evidence relative to an appropriate frame of discernment. Each piece of evidence is represented as a mass distribution, which distributes a unit of belief over subsets of the frame. For example, the fact that user logged in from a remote host is pertinent to the Location frame, and 1.0 is attributed to *Remote* to indicate the complete certainty on this point.

The fact that the user had a high number of authentication errors leads to the belief that the user may be an intruder. Based on this, a likelihood of 0.75 is assigned to the possibility that the user is an intruder.

The high number of command usage and file permission errors gives information about **Familiarity**. Based on the number and types of errors, a belief of 0.7 is assigned to the possibility, *Unfamiliar*, the remaining 0.3 is assigned to *Familiar*.

The user tried some commands that at least two of them can be interpreted as his malicious intentions that might give information about **Actions**. Based on this belief, a likelihood of 0.8 is given to the possibility that user's actions have been malicious.

The last piece of evidence, that the user used several "surveillance" commands and had a short session, give information about **Fear**. It might be assessed as giving 0.75 support that the user is paranoid and 0.25 that the user is *Calm* and that is usual behavior for that user (perhaps the user is a system administrator).

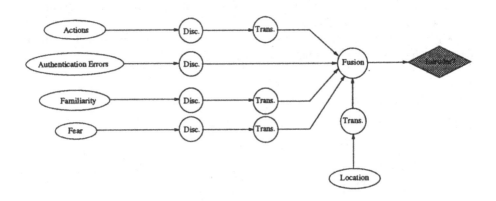

Fig. 3. Frames and compatibility relations.

In this example, beliefs about paranoia levels, system familiarity, actions and, in the case of authentication errors, directly from interpretations over various types of audit data are drawn. These processes can also be represented directly in evidential reasoning, at the cost of some additional complexity. In practice, reasoning processes will be required to include more extensive analysis of this sort.

Evidence from these sources will provide the inputs to the analysis. Many of these determinations are judgments that may not be of equal validity. In order to be able to weight them differently, a means for discounting the impact of the evidence through the discounting operation will be provided. This will allow the change in their relative weights.

The final step is to construct the actual analysis of the evidence as shown in Figure 3 to determine its impact upon the question at hand. In this case the question of whether the user is an intruder can be answered by an assessment of belief over elements in the **Intruder?** frame. Evidential operations can be used to derive a body of evidence providing beliefs about whether the user is an intruder.

In the analysis in Figure 3, all sources except Location source are discounted. The *Authorization Errors* source provides information directly about the likelihood of an intruder, but the others must all be translated to the Intruder? frame. After translation, these independent BOEs are represented relative to a common frame and can be combined using the *fusion* operation (i.e., Dempster's Rule). Fusing the mass distributions yields a mass distribution relative to the Intruder? frame, from which conclusions as to whether the user is an intruder can be drawn.

3 Conclusion

In this paper we attempted to demonstrate the applications of AI techniques specifically Expert Systems in Intrusion Detection Systems. We also showed how using evidential reasoning can allow the system to detect abnormality in the user behavior more efficiently. The use of Expert System technology allows certain intrusion scenarios to be specified much more easily and naturally than is the case using other technologies. However, expert system technology provides no support for developing models of intrusive behavior and encourages the development of *ad hoc* rules.

4 Acknowledgement

Authors wish to thank Dr. Muthukumar Balachandran and Prof. Svein Knapskog for their invaluable discussions and feedbacks.

References

1. J. Frank, "Artificial intelligence and intrusion detection: Current and future directions," in *Proceedings of 17th National Computer Security Conference*, vol. 1, (Baltimore, Meryland), pp. 22–33, 11 - 14 oct 1994.
2. M. Esmaili, R. Safavi-Naini, and J. Pieprzyk, "Computer intrusion detection: A comparative survey," Tech. Rep. TR-95-07, Department of Computer Science, University of Wollongong, Australia, 1995.
3. M. Esmaili, R. Safavi-Naini, and J. Pieprzyk, "Intrusion detection: A survey," in *Proceedings of Twelfth International Conference on Computer Communication ICCC'95*, vol. 1, (Seoul, Korea), pp. 409– 414, 21-24 August 1995. Sponsored by International Council for Computer Communication.
4. D. S. Bauer and M. E. Koblentz, "NIDX - an expert system for real-time network intrusion detection," in *Proceedings of the IEEE Computer Networking Symposium*, pp. 98–106, 1988.
5. A. Brignone, "Fuzzy Sets: An answer to the evaluation of security systems?," in *Proceedings of Fourth IFIP TCII International Conference on Comp. Sec. (IFIP/Sec'86)*, (Monte Carlo, Monaco), pp. 143–151, 2-4 Dec. 1986.

6. H. Debar and B. Dorizzi, "An application of a recurrent network to an intrusion detection system," in *Proceedings of International Joint Conference on Neural Networks*, pp. II478–II483, 7-11 June 1992.

7. H. Debar, M. Becker, and D. Siboni, "A neural network component for an intrusion detection system," in *proceddings of the 1992 IEEE Computer Society Symposium on Research in Security and Privacy*, pp. 240–250, 4 - 6 May 1992.

8. T. F. Lunt, "IDES: An intelligent system for detecting intruders," in *Proceedings of the Symposium: Computer Security, Threat and Countermeasures*, (Rome, Italy), November 1990.

9. D. Anderson and et. al, "Next generation intrusion detection expert system (NIDES): User manual for security officer user interface (SOUI)," technical report, SRI International, 26 March 1993.

10. S. J. Henkind and M. C. Harrison, "An analysis of four uncertainty calculi," *IEEE Transactions on Systems, Man, and Cybernetics*, vol. 18, pp. 700 – 714, Sept./Oct. 1988.

11. P. Lucas and L. Van Der Gaag, *Principles of Expert Systems*. Addison-Wesley Publishing Company, 1991.

12. J. D. Lowrance and T. D. Garvey, "A framework for evidential-reasoning systems," in *Readings in Uncertain Reasoning* (G. Shafer and J. Pearl, eds.), pp. 611 – 618, San Mateo, California: Morgan Kaufmann Publishers, Inc., 1990.

A Group-Oriented *(t, n)* Undeniable Signature Scheme Without Trusted Center[+]

Chu-Hsing Lin* Ching-Te Wang** Chin-Chen Chang**

*Department of Computer and Information Sciences, Tunghai University,
Taichung, Taiwan 407, R.O.C.

**Institute of Computer Science and Information Engineering,
National Chung Cheng University
Chiayi, Taiwan 62107, R.O.C.

ABSTRACT

We shall propose a new group-oriented undeniable signature scheme without the assistance of any trusted center in this paper. A single group public key, instead of all public keys of individual members, is used to represent the group. A transformation is applied to convert the group public key and the message to be signed into a public parameter and thus been transmitted out. With this public parameter, a paying customer will be able to authenticate the validity of the signature with the consent of all signers. It is worth mentioning that the new group-oriented undeniable signature possesses all of the characteristics that listed in Harn and Yang's. Besides, Harn and Yang's scheme is generalized in this research.

[+]This research is partially supported by National Science Council, Taiwan, Republic of China, under grant NSC-84-2213-E-029-001.

Keywords: Group-Oriented Cryptography, Undeniable Signature, Chaum's Scheme, Group-Oriented Undeniable Signature

1. Introduction.

A group-oriented undeniable signature always incorporates two concepts, the group-oriented cryptography and the undeniable signature scheme. The concept of group-oriented cryptography was first proposed by Y. Desmedt [3] in 1987. The group-oriented cryptography problem refers to the study of ciphering scheme for secure communications among groups. In the system, each group, instead of all internal members in the group, publishes a single group public key. An outsider can use this public key to send a confidential message to the group, but only a specified subset of the group members, in a cooperative manner, can reveal this message. There are several schemes to solve the problem of group-oriented cryptography. Essentially, the solution is to determine the group public key and generate the corresponding private key for internal members such that the above requirements are satisfied. For instance, the group-oriented cryptosystems in [3,4,5,6,10] require a mutually trusted center and the schemes in [2,8,9] do not need any trusted center within the group. Recently, the latter case becomes more and more popular in some commercial applications when there do not exist any third party which can be trusted by all members in a group.

The concept of undeniable signature scheme was introduced by D. Chaum and H. van Antwerpen[1] in 1989. An undeniable signature like a digital signature, is a public parameter which is generated from a message and the secret key of the signer. However, the undeniable signature can only be authenticated by a paying customer in cooperation with the signer. An undeniable signature scheme consists of two phases: the commitment phase and the verification phase. In the commitment phase, the signer uses his secret key to produce a commitment signature. In the verification phase, an outsider uses the public key to authenticate the validity of the signature with the cooperation of the signer.

In 1993, Harn and Yang[8] proposed a group-oriented undeniable signature scheme which does not need the assistance of a mutually trusted party. They tried to solve the problem of the group-oriented (t, n) undeniable signature scheme, where n is the total number of group members and t is the threshold value. The group, instead of each member within the group, publishes a single group public key. Basically, for their undeniable signature scheme, in the commitment phase, at least t internal members in the group will be able to sign a message by working together; and in the verification phase, an outsider can authenticate the validity of the signature by the help of all signers working cooperatively. They proposed two schemes for two different threshold values n and 1, respectively. However, the general (t, n) case is still an open problem. In this paper, we shall propose a general group- oriented (t, n) undeniable signature without the assistance of a mutually trusted party. By the use of our scheme, we can handle the case that any t members in the group have the ability to help the verifier to authenticate the validity of the undeniable signature. Besides, breaking the system based on exhaustive searching will be very difficult. Our basic idea is simple and easy to be implemented.

In the next section, we will review Chaum's scheme[1] and Harn and Yang's method[8]. In Section 3, we propose a general group-oriented (t, n) undeniable signature scheme. Some features and security of the (t, n) scheme are discussed and analyzed in the Section 4. Finally, we have some conclusions in the last section.

2. Background of the Scheme

Firstly, we discuss some properties of Chaum's undeniable signature scheme in this section. Consider a large prime number P and a primitive element g in $GF(P)$, which are publicly known. User A selects a random exponent x as his secret key, where $x \in [1, P - 1]$ with $gcd[x, P - 1] = 1$, and computes his public key y, $y = g^x (mod P)$. Assume that user A wants to sign a message

$M, (0 < M \leq P - 1)$ to a verifier B, they have the following two processing phases:

Commitment Phase:

User A calculates as his undeniable signature of message M and sends Z and M to the verifier.

Verification Phase:

A challenge and response protocol is applied between the signer A and the verifier to establish the signature validity.

Step 1. The verifier randomly selects two integers a, b in $GF(P)$, and calculates $W = Z^a y^b \bmod P$. W is sent to A.

Step 2. A calculates $R = W^{x^{-1}} \bmod P$, where x^{-1} is the multiplicative inverse of x. R is sent back to the verifier.

Step 3. The verifier calculates $R' = M^a g^b \bmod P$. If $R' = R$, the signature has been verified.

Harn and Yang[8] used Chaum's undeniable signature scheme to propose the group-oriented undeniable signature scheme without the assistance of a mutually trusted party. The scheme consists of three phases: the group public key generation phase, the group undeniable signature generation phase, and the group undeniable signature verification phase. In the scheme, we assume there are n members in the group, each member U_i selects a random exponent x_i in $GF(P)$ as his own private key, where $x_i \in [1, P-1]$ with $gcd(x_i, P-1) = 1$. In the group public key generation phase, we assume the members are connected in the order $U_1, U_2, \ldots, U_n$. The group public key y will be generated as following:

$$y = y_n = (y_{n-1})^{x_n} \bmod P = ((y_{n-2})^{x_{n-1}})^{x_n} \bmod P = \cdots = (y_1)^{x_2 x_3 \cdots x_n} \bmod P$$
$$= g^{x_1 x_2 \cdots x_n} \bmod P.$$

In the group undeniable signature generation phase, it requires the group undeniable signature to be mutually generated by all members in the group. The scheme is similar to the group public key generation phase. Assuming the members want to sign a message M, the group undeniable signature can be obtained as following:

$$Z = Z_n = (Z_{n-1})^{x_n} \bmod P = ((Z_{n-2})^{x_{n-1}})^{x_n} \bmod P = \cdots = (Z_1)^{x_2 x_3 \cdots x_n} \bmod P$$
$$= M^{x_1 x_2 \cdots x_n} \bmod P.$$

Harn and Yang[8] have proposed two protocols to generate the group undeniable signature. The first one cannot verify the content of a message before they sign it while the second one enables all signers to validate the group undeniable signature before releasing it to the public.

In the group undeniable signature verification phase, using Chaum's scheme, all group members and a verifier execute the verification procedure. Firstly, the verifier employs the single group public key y and the group undeniable signature Z to compute the value $W = Z^a y^b \bmod P$, where a and b are two random values in $GF(P)$. Then, all group members are connected in any order and compute the value $R = W^{x_1^{-1} x_2^{-1} \cdots x_n^{-1}} mod P$. Finally, the verifier computes $R' = M^a g^b mod P$, if $R' = R$, the group undeniable signature has been verified.

3. Group-Oriented (t, n) Undeniable Signature Scheme.

In this section, we will present a solution to solve the general group-oriented (t, n) undeniable signature problem. In our scheme, without the assistance of any mutually trusted center, all members generate a group public key, instead of all individual public keys, to represent all internal members in the group. Based on the individual secret keys of all members, the t members can generate the group undeniable signature for a message. On the other hand, with the consent of t group members, a verifier can authenticate the validity of the message. For related research on (t, n) case, one can consult that presented in [12]. Using the part of the scheme, we will introduce the concept of the shadow secret keys [7], which can be used to construct the general group-oriented (t, n) undeniable signature scheme. Now, we will describe the scheme in the following.

In the group-oriented (t, n) undeniable signature scheme, we assume that there are n members in the group $A(|A| = n)$, and t is the threshold value. Firstly, without the assistance of mutually trusted center, the system selects public parameters P, P', q, g, α, where P, P' are two large prime numbers, g is a generator with order P' in $GF(P')$, q is a large prime factor of $P' - 1$, α is a generator with order q in $GF(P')$. Then, each member $U_i, i = 1, 2, \ldots, n$ selects
1. a public integer number x_i from $GF(q)$;
2. a polynomial function $f_i(x)$ with degree $t - 1$,
$$f_i(x) = C_{i,0} + C_{i,1}x + C_{i,2}x^2 + \cdots C_{i,t-1}x^{t-1} mod q,$$
where $0 < C_{i,j} < q, j = 0, 1, \ldots t - 1$, as an individual secret function.
For the lack of a trusted center, each member U_i should execute the following steps to obtain their individual secret key, respectively.
1. Computes $V_{ij} = f_i(x_j), j = 1, 2, \ldots n$ and keeps it secretly.
2. Computes $S_{ij} = \alpha^{V_{ij}} mod P', j = 1, 2, \ldots n$, where S_{ij} is a secret value for the member U_i to the member U_j.
3. Sends S_{ij} to U_j through a secure channel.
4. After receiving the value S_{ji} from the member $U_j, j = 1, 2, \ldots n, j \neq i$, U_i computes the individual secret key S_i,

$$S_i = \Pi_{j=1}^n S_{ji} mod P' = \Pi_{j=1}^n \alpha^{f_j(x_i)} mod P' = \alpha^{\Sigma_{j=1}^n f_j(x_i)} mod P'.$$

We may consider the secret parameters S_{ji} as the shadow keys [7] of the secret key S_i.

(1) Constructing the Group Public Key
For the purpose of convenience, all group members are connected in a ring structure and mutually generate the group public key. Assume that the members are connected in the order of $U_1, U_2, \ldots, U_n$. Firstly, each member U_i computes $y_i = \alpha^{f_i(0)} mod P'$, keeps it secretly and executes the following procedure:

Procedure 3.1

Step 1. Member U_i computes $Y_{i,1} = g^{y_i} \bmod P$. $Y_{i,1}$ is transmitted to his successor U_{i+1}.

Step k. $(2 \leq k \leq n-1)$. U_i receives $Y_{i-1,k-1}$ from his predecessor U_{i-1}, and computes:

$$Y_{i,k} = (Y_{i-1,k-1})^{y_i} \bmod P.$$

Then, $Y_{i,k}$ is transmitted to his successor U_{i+1}.

Step n. The group public key Y can be obtained by U_i as

$$Y = (Y_{i-1,n-1})^{y_i} \bmod P.$$

After executing n steps of the procedure, each member can obtain the same group public key Y individually. Then, each member is required to release the group public key Y, and authenticates the correctness of his Y value. Here we note that the group public key will be of the form $Y = g^{\alpha^{\sum_{i=1}^{n} f_i(0)}} \bmod P$.

(2) Generating the Undeniable Group Signature

In generation phase, let there be t group members want to mutually generate the group undeniable signature. Suppose that the subset $B(|B| = t)$ of group A contains t members $U_{B1}, U_{B2}, \cdots, U_{Bt}$. By executing the following procedure, they can generate the undeniable signature Z for a sent message M :

Step 1. Each member U_{Bi} computes the modified individual secret key

$$S'_{Bi} = (S_{Bi})^{\prod_{j=1, j \neq i}^{t} \frac{-x_{Bj}}{x_{Bi} - x_{Bj}}} \bmod P'.$$

Step 2. Each member U_{Bi} executes the Procedure 3.1, the parameters y_i are replaced with S'_{Bi}, g is replaced with M, $Y_{i,k}$ is replaced with $Z_{i,k}$, and obtains the undeniable group signature Z :

$$Z = M^{\prod_{i=1}^{t} S'_{Bi}} \bmod P = M^{\prod_{i=1}^{t} \alpha^{(\sum_{k=1}^{n} f_k(x_{Bi}))\prod_{j=1, j \neq i}^{t} \frac{-x_{Bj}}{x_{Bi} - x_{Bj}}}} \bmod P.$$

Similarly, after executing the t steps of Procedure 3.1, each member can obtain the same group undeniable signature Z individually. Then, the t members are required to release the group signature Z, which will be of the form:

$$Z = M^{\alpha^{\sum_{i=1}^{t} (\sum_{k=1}^{n} f_k(x_{Bi}))\prod_{j=1, j \neq i}^{t} \frac{-x_{Bj}}{x_{Bi} - x_{Bj}}}} \bmod P. \tag{3.1}$$

We observe that the scheme enables the t members to validate the group undeniable signature before releasing it to the public.

(3) Authenticating the Undeniable Group Signature

In the verification phase, we will authenticate the validity of an undeniable signature for a message. With the agreement of the group, a verifier together with any t members, $U_{B'1}, U_{B'2}, \ldots, U_{B't}$, of a subset B' will be able to execute the authentication procedure by using the group public key Y, and the undeniable signature Z as the following:

Step 1. The verifier V randomly selects two integers a, b in $GF(P)$, and computes $W = Z^a \cdot Y^b mod P$, where Y is the group public key, and Z is the group undeniable signature. W is then transmitted to the group.

Step 2. Each member $U_{B'i}$ computes the modified individual secret key $U_{B'i}, i = 1, 2, \ldots t$, and executes Procedure 3.1, where the parameter y_i is replaced with $S'_{B'i}$, g is replaced with W, $Y_{i,k}$ is replaced with $R_{i,k}$, and obtains the value R

$$R = W^{\Pi_{i=1}^t (S'_{B'i})^{-1}} mod P. \tag{3.2}$$

R is transmitted to the verifier V.

Step 3. The verifier computes $R' = M^a \cdot g^b mod P$. If $R' = R$, the group undeniable signature has been verified.

To verify correctness of the signature, by Lagrange interpolation, given any t ordered pairs $(x_i, f(x_i)), i = 1, 2, \ldots, t$, we can find a $t - 1$ polynomial function:

$$f(x) = \Sigma_{i=1}^n f(x_i) \Pi_{j=1, j \neq i}^t \frac{x - x_j}{x_i - x_j} mod q.$$

Lemma 3.1 Let $f(x) = \Sigma_{k=1}^n f_k(x) mod q$, then the exponent in (3.1) is $f(0)$.

proof: Since

$$f(x) = \Sigma_{k=1}^n f_k(x) mod q = f_1(x) + f_2(x) + \cdots + f_n(x) mod q,$$

we have $f(0) = f_1(0) + f_2(0) + \cdots + f_n(0) mod q$, and

$$f(x_i) = f_1(x_i) + f_2(x_i) + \cdots + f_n(x_i) mod q.$$

By Lagrange interpolation, the exponent in (3.1)

$$\Sigma_{i=1}^t (\Sigma_{k=1}^n f_k(x_{Bi})) \Pi_{j=1, j \neq i}^t \frac{-x_{Bj}}{x_{Bi} - x_{Bj}} mod q = f(0). \quad \text{(Q.E.D.)}$$

Similarly, the exponent in (3.2)

$$\Pi_{i=1}^t (S'_{B'i})^{-1} = (\Pi_{i=1}^t S'_{B'i})^{-1} = (\alpha^{f(0)})^{-1}.$$

Theroem 3.1 For Step 3 of the verification phase, if $R = R'$, then the verifier can authenticate the validity of the group undeniable signature Z.

Proof: In the first phase, all group members generate the group public key as follows:

$$Y = g^{\alpha^{\Sigma_{i=1}^n f_i(0)}} mod P.$$

In the second phase, by Lemma 3.1, the t group members generate the group signature:

$$Z = M^{\alpha^{\Sigma_{i=1}^t (\Sigma_{k=1}^n f_k(x_{Bi})) \Pi_{j=1, j \neq i}^t \frac{-x_{Bj}}{x_{Bi} - x_{Bj}}}} mod P = M^{\alpha^{f(0)}} mod P.$$

In the third phase, the verifier randomly selects two integers a, b, and uses the signature Z and the group public key Y to compute the value W:

$$W = Z^a Y^b \bmod P = (M^{\alpha^{f(0)}})^a \cdot (g^{\alpha^{f(0)}})^b \bmod P = M^{a\alpha^{f(0)}} \cdot g^{b\alpha^{f(0)}} \bmod P.$$

In Step 2,

$$R = W^{\Pi_{i=1}^{t}(S'_{B'i})^{-1}} \bmod P = W^{(\alpha^{f(0)})^{-1}} \bmod P = M^a g^b \bmod P. \tag{3.3}$$

In Step 3,

$$R' = M^a g^b \bmod P. \tag{3.4}$$

Comparing (3.3) with (3.4), the values R and R' are identical. Because $R = R'$, the value S'_{Bi} is the secret key of the member U_{Bi}, the value $S'_{B'i}$ is the secret key of $U_{B'i}$, and a, b are two random values only known to the verifer. Therefore, the signature Z is authenticated. (Q.E.D.)

4. Discussion

Based on Chaum's verification scheme, we have established a general (t, n) undeniable group signature. Now, we will depict some features of our proposed scheme in the following:

1. The group public key is constructed by all of the group members. Any t members in a group can represent the group to sign a message. Further, with the consent of any t members in the original group, a verifier can authenticate the validity of the undeniable signature.

2. The general (t, n) undeniable signature scheme is constructed without the assistance of any mutually trusted center. However, each member need to compute the individual secret key and generate the group public key in a cooperative manner. This is suitable for a practical utilization.

3. Any less than t members in the original group can't generate the legitimate undeniable group signature or authenticate the validity of the signature.

Moreover, the security of the presented scheme should also be considered here. Both in the signature generating phase and the message authenticating phase, each member U_i uses his secret key S_i as an exponent and releases the computed modulus exponentiation result as a partial group signature. The difficulty of revealing the secret key is the same as solving the discrete logarithm problem. If an outsider wants to compute the secret key S_i by knowing the associated public key Y and the signature Z, he should try to solve the discrete logarithm problem first. In fact, our authenticating method is based on Chaum's undeniable signature scheme. Therefore, the security of our method is also rest on Chaum's scheme. Further, suppose that there are less than t members want to attack the system with conspiracy, by the property of Lagrange interpolation, they can't derive the correct group secret key $y = \alpha^{\Pi_{i=1}^{n} f_i(0)}$. Therefore, they can't generate the undeniable group signature. Thus, they can't authenticate the validity of the signature.

5. Conclusions

We have proposed a new group-oriented (t, n) undeniable signature scheme without the assistance of mutually trusted center. By the use of our scheme, an undeniable signature for a sent message can be obtained. Further, the undeniable signature scheme can be applied in a group-oriented system. The basic idea is simple and the method is easy to be implemented. The features and security of the presented scheme are also described. Besides, we also use the concept of the shadow parameters of a secret key to sign the group undeniable signature. Therefore, we do believe that the presented scheme is feasible in practice and secure in theoretical consideration.

References

[1] D. Chaum and H. van Antwerpen, "Undeniable Signatures," In Advances in Cryptology, Proc. of Crypto'89, Santa Barbara, August 20-24, 1989, pp. 212-216.

[2] C. C. Chang, and H. C. Lee, "A New Generalized Group-Oriented Cryptoscheme without Trusted Centers," IEEE Proceedings on Selected Areas in Communications, Vol. 11, No 5, June 1993, pp.725-729.

[3] Y. Desmedt, "Society and Group Oriented Cryptography: A New Concept," In Advances in Cryptology, Proc. of Crypto'87, Santa Barbara, August 1988, pp. 120-127.

[4] Y. Desmedt, and Y. Frankel, "Threshold Cryptosystem," In Advances in Cryptology, Proc. of Crypto '89, Santa Barbara, August 1989, pp. 307-315.

[5] Y. Desmedt, and Y. Frankel, "Shared Generation of authenticators and Signatures," In Advances in Cryptology, Proc. of Crypto'91, Santa Barbara, August 1991, pp. 457-469.

[6] Y. Frankel. "A Practical Protocol for Large Group Oriented Networks," in Advances in Cryptology, Proc. of Eurocrypt'89, Houthalen, Belgium, April 1989, pp56-61.

[7] L. Harn, "Group-oriented (t, n) Threshold Signature and Digital Multisignature", IEE Proceedings -Computers and Digital Techniques, Vol. 141, No. 5, Sept. 1994, pp.307-313.

[8] L. Harn and S. Yang, "Group Oriented Undeniable Signature Schemes Without the Assistence of a Mutually Trusted Party," In Advances in Cryptology, Proc. of Auscrypt'92, Gold Coast, Australia, Dec. 1992, pp. 3.22-27.

[9] I.Ingemarsson, and G. J. Simmons, "A Protocol to Set up Shared Secret Schemes without the Assistance of a Mutually Trusted Party," In Advances in Cryptology, Proc. of Eurocrypt '90, May 1990, pp. 245-254.

[10] C. H. Lin, "On the Group Oriented Secure Communication Schemes," The Third Conference on Information Security, Chiayi, Taiwan, May 1993.

[11] C. H. Lin, and C. C. Chang, "A Method for Constructing a Group-Oriented Cryptosystems," Computer Communications, Vol. 17, No.11, 1994, pp.805-808.

[12] C. C. Tsai, "Group-Oriented Cryptosystems and Signature Scheme", Master Thesis, Institute of Information Engineering, National Chung-Kung University, Tainan, Taiwan, 1995.

Cryptosystems for Hierarchical Groups *

Hossein Ghodosi
Josef Pieprzyk
Chris Charnes
Rei Safavi-Naini

Center for Computer Security Research
Department of Computer Science
University of Wollongong
NSW 2522
Australia
e-mail: hossein/josef/charnes/rei@cs.uow.edu.au

Abstract. This paper addresses the problem of information protection in hierarchical groups. Higher level groups of participants can control the information flow (the decryption ability) to lower level groups. If a higher level group decides to allow a lower level group to read the message, it passes a go ahead ticket so the lower level group can decrypt the cryptogram and read the message. The formal model of top-down hierarchical cryptosystems is given.

Two practical and efficient schemes are described. The first is based on the ElGamal system. The second applies the RSA system. In proposed schemes the dealer publishes a public key such that an individual can use it to send an encrypted message to the (hierarchical) group. Publication of both the group public key and the encryption method does not reveal the decision of the group. The proposed cryptosystems are immune against conspiracy attack.

The lack of verifiability of retrieved messages in threshold ElGamal cryptosystems is also discussed.

1 Introduction

The concept of society or group-oriented threshold cryptography was originally formulated by Desmedt [2] for "democratic" groups where every participant has equal rights in performing cryptographic operations. There were numerous attempts to generalize the group-oriented cryptography for an arbitrary access structure. To illustrate the problem consider the following quotation from Desmedt [2].

---

* Support for this project was provided in part by the Australian Research Council under the reference number A49530480

"Suppose that the group decided that the messages first to be accessed by the supervisor and then afterwards by all other members at the same moment. The group publishes the corresponding public key. The question now is: *does the publication of this key and of the encryption method, reveal what the decision of the group is?* If that would be the case, then everybody knows which hierarchy the group has, or more generally knows which kind of society that corresponds with the group."

Cryptographic operations in hierarchical groups can be done in two different ways: from the top to the bottom or from the bottom to the top. "Top-down" cryptographic operations have to be performed after "the go ahead" is given by the higher level. The permission (or denial) for the lower level may depend upon the result of the operation performed on the higher one. "Bottom-up" cryptographic operations permit the delegation of a specific right from lower levels to the top. The top level successfully executes the cryptographic operation when all lower levels have prepared their partial results.

This paper investigates the problem of a secure top-down information flow. The proposed cryptosystems are designed in such way that the information flow can be stopped by higher level participants. The proposed "top-down" hierarchical cryptosystems are based on the ElGamal and the RSA cryptosystems.

The paper is organized as follows. Section 2 briefly describes both the ElGamal and the RSA cryptosystems. The next section examines the way the top-down information flow can be controlled. Sections 4 and 5 discuss possible implementations. In Section 6, the lack of verifiability of retrieved messages in threshold ElGamal cryptosystems is discussed. Section 7 considers the security of the proposed systems. Finally, conclusions are given in Section 8.

2 Background

The ElGamal [6] and the RSA [7] cryptosystems are based on the discrete logarithm problem. Computations in the ElGamal system are done in a Galois field – the modulus is a prime number. This cryptosystem can be seen as an extension of the Diffie-Hellman public key distribution system [5]. On the other hand, computations in the RSA public key cryptosystem are done in a ring – the modulus is a composite number. The two different definitions of the moduli in these cryptosystems have a profound influence into their structure and security. The security of ElGamal system is equivalent to the difficulty of computing a corresponding an instance of the discrete logarithm. While the security of the RSA system is related to the difficulty of factoring the modulus.

2.1 The RSA cryptosystem

The original RSA system uses the modulus $N = pq$ where p, q are strong primes, that is $p = 2p' + 1$ and $q = 2q' + 1$ (p' and q' are large and distinct primes).

The creator of the system R who knows p and q, selects a random integer e such that e and $\lambda(N)$ are coprime ($\lambda(N)$ is the least common multiplier of two integers $p-1$ and $q-1$ so $\lambda(N) = 2p'q'$). Next R publishes e and N as the public parameters of the system, while keeps secret the values of p, q, and d (d has the property that $ed = 1 \bmod (\lambda(N))$). It is clear that computing of d is easy for R, who knows $\lambda(N)$, but is difficult (equivalent to the factoring of N) for someone who does not know the $\lambda(N)$.

Suppose an individual wants to send R a message M, where $0 \leq M < N$. The sender computes the cryptogram $C = M^e \bmod N$ and communicates it to R. Since for every integer a ($gcd(a,p) = 1$ and $gcd(a,q) = 1$) the equation $a^{\lambda(N)} = 1 \bmod N$ holds, R can decrypt the cryptogram using:

$$C^d = (M^e)^d = M^{ed} = M^{1(\bmod \lambda(N))} = M \bmod N$$

Desmedt and Frankel [3] proposed a (t, n) threshold RSA cryptosystem which incorporates the Shamir [8] secret sharing scheme. In their proposed scheme, any subset B ($|B| = t$) of participants can retrieve the secret

$$s = \sum_{i \in B} s_i \bmod \lambda(N)$$

from their shares s_i. Note that $s = d - 1$. Now, to decrypt the cryptogram $C = M^e \bmod N$, all participants of an authorized subset, B, compute their own partial decryption, C^{s_i}, and send to the combiner. The combiner can decrypt the cryptogram using:

$$\prod_{i \in B} C^{s_i} \times C = C^{(\sum_{i \in B} s_i + 1)} = C^d = (M^e)^d = M \bmod N.$$

2.2   The ElGamal cryptosystem

Computations in the ElGamal cryptosystem are performed in $GF(p)$, where p is a sufficiently large prime number. The system is set up by the receiver R who selects a primitive element g – a generator of $GF(p)$ – and chooses an integer $s \in GF(p)$ at random ($s \neq 0$). Next R computes $y = g^s \bmod p$ and publishes g, p, and y as the public parameters of the system. The secret component s is known to R only.

Suppose that a user P wants to send a message M to R, where $1 \leq M < p$. First P chooses a random integer $k \in GF(p)$ and sends the pair $C = (g^k, My^k)$ as the cryptogram of M. In order to decrypt the cryptogram C and retrieve the message, R needs only raise g^k to the power s (to get y^k). Then the multiplicative inverse, y^{-k}, can be used to decrypt the cryptogram as

$$My^k \times y^{-k} = M \bmod p.$$

Desmedt and Frankel [4] proposed a threshold cryptosystem which uses the ElGamal system. In their construction the dealer uses the Shamir (t, n) threshold scheme to distribute the secret s among the shareholders, such that for any subset B of t participants $s = \sum_{i \in B} s_i$ (mod $p - 1$), where $p - 1$ is a *Mersenne prime*. All computations in their threshold ElGamal system are performed in $GF(p)$, where $p = 2^l$. To decrypt the cryptogram $C = (g^k, M y^k)$, each $P_i \in B$ calculates $(g^k)^{s_i}$ mod p and sends the result to the combiner. The combiner computes $y^k = g^{ks} = \prod_{i \in B} g^{ks_i}$ and decrypts the message.

3 Top-down cryptography in hierarchical groups

In (t, n) threshold cryptography, any participant contributes equally to the cryptographic process. This means that threshold cryptography is well suited for democratic organizations with a "flat" structure. Unfortunately, most organizations exhibit a hierarchical structure. In hierarchical groups cryptography can be used to control the flow of information between two consecutive levels. If cryptography is used to protect information flowing from top to bottom, we talk about "top-down" cryptography. If the flow of information is from bottom to top, we deal with "bottom-up" cryptography. In this paper we consider top-down cryptography.

In top-down cryptography participants on higher levels of the hierarchy are required to perform cryptographic operations before the lower level participants. Lower level participants can only proceed with the cryptographic processing after receiving a "go ahead" from the higher levels. Let us consider the case when a cryptogram is broadcast to all participants of a hierarchical organization. A top-down cryptosystem should allow the participant on the top of the hierarchy to retrieve the plain-text message from the cryptogram. Once the higher level participants have decrypted the cryptogram, they may allow the lower level participants to decrypt the cryptogram by sending a suitable permission – a *go ahead ticket*.

3.1 The model

We assume that the hierarchy of participants is described by a tree structure, where $\mathcal{P}(0)$ is the set of the highest level participants. Every participant $P_j \in \mathcal{P}(0)$ is the head of the group $\mathcal{P}(1; j)$. In general each participant of the group $\mathcal{P}(i; j_1, \ldots, j_i)$ is the head of the group $\mathcal{P}(i + 1; j_1, \ldots, j_i, j_{i+1})$. Any participant can be a member and the head of a single group. It is easy to see that each path in the hierarchy is uniquely identified by a suitable sequence of heads (supervisors). We also assume that all participants of the group defined on the i-th level have the same power, i.e. we are dealing with threshold access structures.

The following notation will be used. The set of all participants is $\mathcal{P}$. The set $\mathcal{E}$ is the collection of all cryptographic keys (either for encryption or decryption).

The set of cryptograms is C, and the set of plain-text messages is M. T denotes the set of all go ahead tickets. The share set is S.

Definition 1. A top-down cryptographic system consists of a single trusted dealer and a collection of combiners. Each combiner on the i-th level acts on behalf of a specific group.

- The trusted dealer (D) sets up the system. D chooses both the encryption and decryption keys. The encryption key $e \in \mathcal{E}$ is public. The decryption key $d \in \mathcal{E}$ is secret and is used to calculate shares. Next D determines shares for all participants (of all groups of the hierarchy) according to the following function:

$$f_D^{(i)} : \mathcal{E} \times \mathcal{P}(i; j_1, \ldots, j_i) \to \mathcal{S} \times \mathcal{T}, \quad (i = 0, 1, \cdots, \ell - 1)$$

for the groups on the i-th levels, and

$$f_D^{(\ell)} : \mathcal{E} \times \mathcal{P}(\ell; j_1, \ldots, j_\ell) \to \mathcal{S}$$

for the lowest level. All secret sharing schemes used for the groups are designed for threshold access structures. The value of the threshold may be different for each group. Note that the dealer sets up all static secret elements which do not change throughout the life time of the system.
- The combiner of the top level group (on 0-th level) collects partial results from participants of the group $\mathcal{P}(0)$ and applies the function

$$f_C^{(0)} : f_0(C, S_1) \times \cdots \times f_0(C, S_t) \to M \times f_{tic}^0(C, T)$$

where $f_{tic}^0(\bullet, \bullet)$ is a function which assigns a go ahead ticket for a cryptogram and a static ticket. If the number of participants who have provided their shares to combiner equals or exceeds the threshold, $f_C^{(0)}$ generates the plain message and a valid go ahead ticket which may be given to the lower level groups.

A combiner on the i-th level collects partial results of its group $\mathcal{P}(i; j_1, \ldots, j_i)$, a go ahead ticket from the higher level and uses the function

$$f_C^{(i)} : f_i(C, S_1) \times \cdots \times f_i(C, S_t) \times f_{tic}^{i-1}(C, T) \to M \times f_{tic}^{i+1}(C, T) \quad (i = 1, 2, \cdots, \ell-1)$$

to retrieve the message and compute the ticket for the lower level groups. The message and the ticket are valid only when the number of contributing participants equals or exceeds the threshold of the group. The combiner on the lowest level, ℓ, retrieves the message (using partial results of the participants and the go ahead ticket) as:

$$f_C^{(\ell)} : f_\ell(C, S_1) \times \cdots \times f_\ell(C, S_t) \times f_{tic}^{\ell-1}(C, T) \to M.$$

The structure of a top-down cryptographic system reflects the hierarchy of the whole group. To avoid unnecessary details and to simplify our presentation, we will examine the simplest case of hierarchical cryptosystems where there is only one participant on every level of the hierarchy (for general case see the final version of this paper). Consequently, there is only one path in the hierarchy and the go ahead decision is taken by a single participant who acts as the combiner.

Go-ahead tickets may be computed by their immediate predecessors – groups one level above the current ones. More specifically, for the group $\mathcal{P}(i; j_1, \ldots, j_i)$, only the group $\mathcal{P}(i-1; j_1, \ldots, j_{i-1})$ can produce the valid ticket. This system is called a *hierarchical cryptosystems with sequential go ahead*.

The alternative to the above are hierarchical systems in which any higher level group belonging to a single path can authorize the lower level groups by passing go ahead tickets to them. So the group $\mathcal{P}(i; j_1, \ldots, j_i)$ can obtain a valid ticket from a one of the following groups: $\mathcal{P}(i-1; j_1, \ldots, j_{i-1})$, ..., $\mathcal{P}(1; j_1)$, and $\mathcal{P}(0)$. These systems are called *hierarchical cryptosystems with jumping go ahead*.

Next we will discuss two implementations of hierarchical cryptosystems. The first is based on the ElGamal cryptosystem. The second uses the RSA system.

4 ElGamal hierarchical cryptosystem

Assume that we have a group of participants $\mathcal{P} = \{P_0, P_1, \ldots, P_\ell\}$. P_0 is on the top of the hierarchy. P_1 belongs to the second level, P_2 to the third, etc. P_ℓ is at the bottom of the hierarchy. Every level is occupied by a single participant.

Algorithm for the dealer.

Let p be a large prime and g be a primitive element of $GF(p)$. The dealer selects $\ell + 1$ random integers $s_0, s_1, \cdots, s_\ell$, where $s_i \in GF(p)$ $(i = 0, 1, \cdots, \ell)$. Next the dealer computes $s = \sum_{i=0}^{\ell} s_i \bmod \phi(p)$, $y = g^s \bmod p$ and publishes p, g and y as the public parameters of the system. The secrets, (degenerate shares) $s_j, s_{j+1}, \cdots, s_\ell$, are distributed via a secure channel to the participant P_j $(j = 0, \cdots, \ell)$.

Algorithm for the sender.

The sender follows the steps of the ElGamal system. First he selects a random integer $k \in GF(p)$. Next he computes the cryptogram $C = (C_1, C_2) = (g^k, My^k)$ for the message $M \in \mathcal{M} = GF(p)$. Finally the sender broadcasts the cryptogram C.

Algorithm for the combiner

All participants P_i can be considered as degenerate combiners $(i = 0, \ldots, \ell)$. The combiner P_0 knows the parameter $s = \sum_{i=0}^{\ell} s_i$ so he can compute $(g^k)^s$ and retrieve the message M. After reading the message, P_0 may send a go ahead ticket $(g^k)^{s_0}$ (mod p) to participant P_1 (enabling P_1 to retrieve the message).

In general the participant P_i $(i = 1, \ldots, \ell)$ knows $s_i, s_{i+1}, \cdots, s_\ell$ and after receiving the ticket $(g^k)^{s_0 + \cdots + s_{i-1}}$ from P_{i-1}, can compute $(g^k)^s$ and decrypt the cryptogram. P_i decides whether to send the go ahead ticket $(g^k)^{s_0 + \cdots + s_i}$ to P_{i+1}.

Note that this system is with jumping go ahead. After receiving a valid go ahead ticket, any higher level participant can issue a correct go ahead ticket to any lower level participant.

4.1 Variant of the hierarchical ElGamal cryptosystem

Assume that the dealer distributes only two secret elements to each participant. So P_0 knows the pair $(s_0, S_0 = \sum_{j=0}^{\ell} s_j)$. In general each $P_i (i = 1, 2, \cdots, \ell)$ knows the pair $(s_i, S_i = \sum_{j=i}^{\ell} s_j)$. Note that s_i is used to create a valid go ahead ticket for one level below and S_i along with the ticket from one level above allows P_i to decrypt cryptogram.

After receiving a valid ticket from the higher level, each participant can decrypt the cryptogram. The participant P_j $(j = 1, \ldots, \ell)$ who has received a correct ticket from the higher level, may issue a valid ticket (using the received ticket and s_j) for the $(j+1)$-th level. Hence, in this variant, each level can only issue a go ahead ticket to its immediate lower level. The obvious way to skip to lower levels, is by an interaction of two levels. Suppose that the j-th level wants to pass a ticket to level i $(j < i, j \neq i - 1)$. First the participant on the level i passes $(g^k)^{S_i}$ to level j. Now the j-th level can compute the ticket:

$$\frac{(g^k)^{S_0}}{(g^k)^{S_i}}$$

and passes it to the level i.

Hierarchical ElGamal cryptosystems (both with jumping and sequential go ahead) have a property which allows higher level participants to modify their tickets so that lower level participants will retrieve false messages. This property is discussed in the following theorem.

Theorem 2. *In hierarchical ElGamal cryptosystems, the higher levels can modify their go ahead tickets in such a way that the lower levels retrieve a false message M_f from the cryptogram C.*

Proof. Assume that P_i wants P_j to accept $M_f \neq M$. Although P_i cannot change the cryptogram $C_2 = My^k$, they can compute the following (invalid) ticket

$$\alpha = \frac{C_2}{M_f(g^k)^{S_j}}$$

provided P_i knows $(g^k)^{S_j}$. It is easy to see that P_j retrieves M_f using the ticket α.

Note that if P_j retrieves a false message M_f (instead of the original M), all lower levels will also retrieve M_f (for valid tickets). These schemes can be useful for hierarchical groups where lower levels always expect to be able to read the cryptogram. Higher levels can always choose whether to send the original or false message.

5 RSA hierarchical cryptosystem

Assume as before that we have a group of participants $\mathcal{P} = \{P_0, P_1, \ldots, P_\ell\}$. P_0 is on the top of the hierarchy. P_1 creates the second level, P_2 the third, etc. P_ℓ is at the bottom of the hierarchy. Every level is occupied by a single participant.
Algorithm for the dealer.

Let $N = pq$ where p, q are two safe primes, that is $p = 2p' + 1$ and $q = 2q' + 1$ (p' and q' are large and distinct primes). The dealer (D) who sets up the system, selects p, q and a random integer e such that e and $\lambda(N)$ are co-prime ($\lambda(N)$ is the least common multiple of $(p-1)$ and $(q-1)$, and therefore, is equal to $2p'q'$). Next D publishes e and N as public parameters of the system while the values of p, q, and d (d has the property that $ed \equiv 1 \pmod{\lambda(N)}$) are kept secret. Also the dealer selects ℓ random integers $s_1, s_2, \cdots, s_\ell$ and solves the equation $s_0 + s_1 + \cdots + s_\ell = d \bmod \lambda(N)$ for unknown integer s_0 (note that in this scheme $\sum_{i=0}^{\ell} s_i$ is equal to d, instead of $d-1$). Finally D distributes the secret elements (degenerated shares) via secure channels to all participants. P_0 gets the sequence $(s_0, \ldots, s_\ell)$. The participant P_i receives $(s_i, \ldots, s_\ell)$ where $i = 1, \ldots, \ell$.
Algorithm for the sender.

As in the original RSA system, the sender computes cryptogram C as $M^e \bmod N$. The cryptogram is broadcast to all participants of the hierarchy.
Algorithm for the combiner.

Participants can be considered as degenerate combiners. Combiner P_0 knows that $s_0 + s_1 + \cdots + s_\ell = d \bmod \lambda(N)$ and can decrypt the cryptogram C as $C^d = (M^e)^d = M \bmod N$. After reading the message, he can control P_i's ability to read the message by sending a correct ticket $C^{s_0 + \cdots + s_{i-1}}$. After receiving the ticket $C^{s_0 + \cdots + s_{i-1}}$, the combiner P_i can retrieve the message as he also knows the sequence $(s_i, \ldots, s_\ell)$ therefore

$$C^{s_0 + \cdots + s_{i-1}} \times C^{(s_i + \cdots + s_\ell)} = C^d = M \bmod N.$$

5.1 Variant of the hierarchical RSA cryptosystem

Assume that the dealer distributes only two secret elements to each participant. So P_0 knows the pair $(s_0, S_0 = \sum_{j=0}^{\ell} s_j)$ and in general each $P_i (i = 1, 2, \cdots, \ell)$ knows the pair $(s_i, S_i = \sum_{j=i}^{\ell} s_j)$. Note that in case of the ElGamal hierarchical cryptosystem, the dealer may compute the sum modulo $\phi(p)$, but in this case the addition must be done in Z (otherwise, some levels may collaborate to find the $\lambda(N)$, however, due to exponentiation the decryption is actually performed in Z_N). The number s_i is used to create a valid go ahead ticket for one level below and S_i along with the go ahead ticket from one level above allows P_i to decrypt the cryptogram.

After receiving a valid ticket from the higher level, each participant can decrypt the cryptogram. A participant P_j, who has received a correct ticket from the higher level, can issue a valid ticket (using the received ticket and s_j) for the $(j+1)$-th level. Hence, in this variant, each level can only issue a go ahead ticket for the levels immediately below it. The obvious way to skip lower levels, is by the collaboration of two levels. Suppose the j-th level wants to pass a ticket to the i-th level where $j < i$ and $j \neq i - 1$. First the level i passes $(C)^{S_i}$ up to the level j. Now the j-th level can compute the ticket (assuming that $(C)^{S_i}$ and N are coprime):

$$\frac{M}{(C)^{S_i}} \bmod N$$

and passes it to the level i. Again, this type of skipping requires interaction between the levels.

The RSA implementation of hierarchical cryptosystems allows every participant in the hierarchy to verify if the retrieved message is genuine. This is explained by the following theorem.

Theorem 3. *In hierarchical RSA cryptosystems, the recipients of tickets can always distinguish valid tickets from corrupt ones.*

Proof. Assume that P_i wants P_j to accept $M_f \neq M$. Although P_i cannot change the cryptogram $C = M^e$ (it has been broadcast to every participants), they can compute a corrupt ticket

$$\alpha = \frac{M_f}{(C)^{S_j}} = M_f \times C^{-S_j} \bmod N$$

provided P_i knows C^{S_j}. Having got the ticket, P_j retrieves the message $M_f = \alpha \times C^{S_j}$. Now P_j can verify the validity of both retrieved message M_f and the ticket α by computing $M_f{}^e \bmod N$ which is not equal to the cryptogram C (as long as $M_f \neq M$ and $M_f < N$).

Hierarchical RSA cryptosystems can be used where all participants want to verify the validity of retrieved messages and received tickets.

6 Verifiability in ElGamal threshold cryptosystem

Consider some repercussions of Theorem 2 in the context of security of ElGamal threshold cryptosystems. The ElGamal threshold cryptosystem was proposed by

Desmedt and Frankel [4]. They request the shares and modified shares (shadows) to be secret. Partial results (generated from modified shares) are assumed to be protected by exponentiation – an attacker faces an intractable instance of the discrete logarithm problem. The partial results are communicated to a trusted combiner who retrieves the message from a public cryptogram. Note that in the light of Theorem 2, the retrieved meaningful message cannot be verified – it has to be accepted as genuine.

This becomes crucial if the message source has no redundancy. A dishonest participant can send a random (invalid) partial result and "force" the combiner to generate a random (but meaningful) false message. There is also another possible attack on the system if one of the participants is able to see other participants partial results. For an arbitrary false message M_f, this participant can modify his or her partial result in such a way that the combiner retrieves the false message M_f. This is the case when one of participants agrees to be the combiner.

Although this attack does not break the system, it compromises the goal of threshold cryptography. On the other hand, in hierarchical cryptosystems this feature is a useful property which enables higher level participants to keep lover level participants informed and in the "dark" at the same time.

7 Security of the proposed systems

In this section we briefly discuss the security of the proposed hierarchical cryptosystems. Since the highest level knows the secret, any attacks which requires the collaboration of the highest level makes no sense. Moreover, as mentioned earlier, in the ElGamal based hierarchical cryptosystems higher levels can force lower levels to decrypt a false message. The same as existing threshold cryptosystems, we assume that all participants keep their shares secret. Moreover, we assume that partial information (tickets) are sent to the combiner (or to other participant) privately, otherwise, an eavesdropper can learn the message (this is the common problem of existing RSA and ElGamal threshold cryptosystems in which knowing partial decryption can help an eavesdropper to learn the message).

Any attacks, trying to obtain the secret of the system (using the cryptogram) or to obtain a particular share (using partial decryption values), is equivalent of solving an instance of an intractable (factorization or discrete logarithm) problem.

7.1 Immunizing the systems against conspiracy attacks

Corresponding threshold cryptosystems (the RSA and the ElGamal threshold cryptosystems) suffer from the conspiracy attack (for more detail on this attack over RSA threshold cryptosystem see [1]). Immunizing the proposed hierarchical cryptosystems against this type of attack requires the following small modification. Let the partial shares s_i in both hierarchical cryptosystems have selected such that their integer sum is exactly equal to the secret. Thus, combination of all partial shares (without considering the highest level) is still insufficient to retrieve the secret, and therefore, collaboration of any subset of participants is insufficient to break the system.

8   Conclusions

We have proposed two cryptosystems as solutions to open problems related to hierarchical groups. The dealer publishes a public key such that an individual can use that public key to send an encrypted message to the whole (hierarchical) group. Publication of both the group public key and the encryption method does not reveal the decision of the group. We have assumed that higher level groups (or participants) can allow lower level groups (or participants) to retrieve the message by passing go ahead tickets. We have proposed two solutions: one based on the ElGamal system and the other based on the RSA system. The cryptosystems are efficient, practical and non-interactive. The security of the proposed schemes are discussed and it is shown that the proposed systems are immune against the conspiracy attack.

Note that in both solutions there are two classes of secret elements (or shares): static (used for decryption), and dynamic (used for ticket generation). Static shares are more or less used in traditional way to compute partial results which later are sent to the combiner who can compute the final result for an authorized subset of participants. On the other hand, dynamic shares are used to control the information flow from one level to another.

There is a crucial difference between threshold and hierarchical cryptosystems. In threshold cryptosystems the dealer distributes all shares in one go and does not allow participants to interfere in the distribution process. On the contrary, the dealer in hierarchical cryptosystems sets up the skeleton of the system by the distribution of static shares and deposits dynamic shares with higher level participants so they can make proper decisions about the distribution of dynamic shares later.

In the final version of the paper, we will generalize the concept of dynamic shares and we will show how this generalization can be used to design a cryptosystem for an arbitrary hierarchy.

This paper has also revealed a weakness in existing ElGamal based threshold cryptosystems.

References

1. T. Hwang C.M. Li and N.Y. Lee. Remark on the Threshold RSA Signature Scheme. In *Advances in Cryptology - Proceedings of CRYPTO '93, Ed. D. Stinson, Lecture Notes in Computer Science, Vol. 773*, pages 413–419. Springer-Verlag, 1993.

2. Y. Desmedt. Society and group oriented cryptography: A new concept. In *Advances in Cryptology - Proceedings of CRYPTO '87, Ed. C. Pomerance, Lecture Notes in Computer Science, Vol. 293*, pages 120–127. Springer-Verlag, 1988.

3. Y. Desmedt. Threshold Cryptosystems. In *Advances in Cryptology - Proceedings of AUSCRYPT '92, Eds. J. Seberry and Y. Zheng, Lecture Notes in Computer Science, Vol. 718*, pages 3–14. Springer-Verlag, 1993.

4. Y. Desmedt and Y. Frankel. Threshold cryptosystems. In *Advances in Cryptology - Proceedings of CRYPTO '89, Ed. G. Brassard, Lecture Notes in Computer Science, Vol. 435*, pages 307–315. Springer-Verlag, 1990.

5. W. Diffie and M.E. Hellman. New Directions in Cryptography. *IEEE Trans. on Inform. Theory*, IT-22(6):644–654, November 1976.

6. T. ElGamal. A Public Key Cryptosystem and a Signature Scheme Based on Discrete Logarithms. *IEEE Trans. on Inform. Theory*, IT-31:469–472, 1985.

7. A. Shamir R.L. Rivest and L. Adleman. A Method for Obtaining Digital Signatures and Public-Key Cryptosystems. *Communications of the ACM*, 21(2):120–126, 1978.

8. A. Shamir. How to Share a Secret. *Communications of the ACM*, 22(11):612–613, November 1979.

On Selectable Collisionful Hash Functions

S. Bakhtiari, R. Safavi-Naini, J. Pieprzyk

Centre for Computer Security Research
Department of Computer Science
University of Wollongong, Wollongong
NSW 2522, Australia

Abstract. This paper presents an attack on Gong's proposed collisionful hash function. The weaknesses of his method are studied and possible solutions are given. Some secure methods that require additional assumptions are also suggested.

1 Introduction

Hash functions have been used for producing secure checksums since 1950's. A hash function maps an arbitrary length message into a fixed length message digest, and can be used for message integrity [1, 5, 8]. For this purpose, a sender calculates the message digest of the transmitting message and sends it appended to the message. The receiver verifies the checksum by recalculating it from the received message and comparing it with the received checksum.

Another application is for protection against spoofing, where the checksum is protected by a key to thwart any modification by an opponent. This application has recently motivated the new term *Keyed Hash Functions* [3]. A keyed hash function uses a symmetric key and the checksum can only be calculated and verified by the insiders — people who know the key.

Berson, Gong, and Lomas [3] have introduced *Collisionful Hash Functions*. In a collisionful hash function many keys result in the same checksum of a given message, and hence, the probability of determining the correct key, used by the communicants, is reduced.

Gong [6] has given a construction of collisionful hash functions to be used for software protection. This paper analyzes this construction and shows how an enemy can modify a message and its corresponding checksum. The weaknesses of other variations of his method are also studied and experimental results that support our approach are included.

Gong's method is described in Section 2. Section 3 examines the problems of Gong's hashing scheme and demonstrates how to attack the system. Practical experiments which support our claims (attacks) are presented in Section 4. Two secure methods that require additional assumptions are given in Section 5. Finally, we conclude the paper in Section 6.

2 Gong's Collisionful Hash Function

Gong uses polynomial interpolation to construct a collisionful hash function. The nice idea behind this construction is that the user can select key collisions to better reduce the probability of guessing attack. However, we show that application of this method for protection against modification is not secure.

2.1 Background

A keyed hash function is a class of one-way and collision resistant hash functions, indexed by a key. The hash value depends on the key, and the computation of the key should be infeasible when pairs of [message, digest] are available.

Collisionful hash functions provide an additional property which is the possibility of having the same hash value of a given message under several keys. This property prevents the opponent to uniquely determine the correct key.

Gong [6] has given a construction of collisionful hash functions with the collision accessibility property. This property allows a user to choose a set of keys that satisfy a given [message, digest] pair, and is desirable when the key belongs to a distinguishable subset of key space (eg. meaningful words).

A similar approach is used by Zheng et al. [11] in construction of *Sibling Intractable Function Family (SIFF)*. It is used for providing secure and efficient access in hierarchical systems and has proven security properties. We show that Gong's construction can be turned into a SIFF which results in a proof of the security of that construction when a large password space is used.

2.2 Notations and Assumptions

- A and E are the user (Alice) and the intruder (Eve), respectively.
- M is a message (a system binary code) to be authenticated by A.
- k_1 is A's password — from a **small** space $\mathcal{K}$ (subset of all possible passwords $\mathcal{P}$) so that an attacker can perform an exhaustive search [6, Sections 1 and 4].
- $k_2, \ldots, k_n \in \mathcal{P}$, $n \in \mathbb{N}$, are selected password collisions. Let $\Delta = \{k_1, k_2, \ldots, k_n\}$.
- $GF(p)$ is the Galois Field of p elements, where p is a prime.
- K is a random key chosen by Alice from a large space (eg. $GF(p)$, for a large prime p).
- $g()$ is a secure keyed hash function, where $g(k, x)$ denotes the hash value of a message x under a key k.
- $|\mathcal{X}|$ denotes the size of a set $\mathcal{X}$.
- '$\|$' denotes string concatenation.

It is assumed that $g()$ produces integer hash values and $g(k_i, M) > n$, $\forall k_i \in \Delta$. Furthermore assume $g(k_i, M) \neq g(k_j, M)$, $\forall k_i \neq k_j$, where $k_i, k_j \in \Delta$. Note that, $\mathcal{K} \subset \mathcal{P}$ is the set of passwords that are commonly used by the users. In general $|\mathcal{P}|$ may not be small, but $\mathcal{K}$, the set of passwords that are often used by the users (called *poorly chosen passwords*), is usually small, and therefore, weak against dictionary attack. It should be emphasized that Gong's construction assumes a small password space that can be exhaustively searched:

> " ... *collisionful hash functions are useful in generating integrity check-sum from user passwords, which tend to be chosen from relatively small space that can be exhaustively searched.*" [6, Section 4]

2.3 Computing the Checksum

Alice (A) chooses a random key K and defines $w(x) = K + a_1 \cdot x + \cdots + a_n \cdot x^n$ (mod p), where p is a suitable large prime number, and the n coefficients $a_1, \ldots, a_n$ are calculated by solving the following n equations. (Equation 'i' is $w(g(k_i, M)) = k_i$, and all calculations are performed in $GF(p)$.)

$$\begin{cases} K + a_1 \cdot g(k_1, M) + \cdots + a_n \cdot g(k_1, M)^n = k_1 \\ K + a_1 \cdot g(k_2, M) + \cdots + a_n \cdot g(k_2, M)^n = k_2 \\ \quad\vdots \qquad\qquad\qquad\qquad\qquad\quad \vdots \\ K + a_1 \cdot g(k_n, M) + \cdots + a_n \cdot g(k_n, M)^n = k_n \end{cases} \tag{1}$$

Using K and $w(x)$, the checksum will be:

$$w_1 \| w_2 \| \cdots \| w_n \| g(K, M), \tag{2}$$

where $w_i = w(i)$, $i = 1, \ldots, n$. Alice does not need K and $w(x)$, and may forget them after producing the above checksum.

2.4 Verifying the Checksum

Alice can verify the checksum as follows. She solves the following $(n+1)$ equations in $GF(p)$ and finds the $(n + 1)$ variables $b_0, b_1, \ldots, b_n$.

$$\begin{cases} b_0 + b_1 \cdot g(k_1, M) + \cdots + b_n \cdot g(k_1, M)^n = k_1 \\ b_0 + \quad b_1 \cdot 1 \quad + \cdots + \quad b_n \cdot 1^n \quad = w_1 \\ \vdots \qquad\qquad \vdots \qquad\qquad\qquad \vdots \\ b_0 + \quad b_1 \cdot n \quad + \cdots + \quad b_n \cdot n^n \quad = w_n \end{cases} \tag{3}$$

Then, she calculates $g(b_0, M)$ and compares it with $g(K, M)$ in the checksum. In the case of a match, she will accept the checksum as valid.

One restriction to this method is that whenever k_1 (A's password) is used to calculate checksums for other messages, the same password collisions should be used. Otherwise, the intruder can guess k_1 from the intersection of different password collision sets, with a high probability (cf. Section 3.4).

Also, Alice should be careful when k_1 is used for other purposes, since some information about k_1 is always leaked from Gong's method. For instance, if k_1 is also used for logging into a system, the enemy can use our attack (see below), guess all possible password collisions, and try them one by one until she logs into the system.

3   Attacking Gong's Method

Since passwords are from a small space $\mathcal{K}$ (Gong's assumption), E can exhaustively search $\mathcal{K}$ (cf. Section 2.2). For each candidate password $k \in \mathcal{K}$, she solves Equation 3 in $GF(p)$, by replacing k_1 with k, and finds $b_0, b_1, \ldots, b_n$. If $g(b_0, M)$ is the same as that in the checksum, she keeps k as an *applicable* password. After exhaustively testing $\mathcal{K}$, Alice will find m applicable passwords $\Gamma = \{k_{r_1}, \ldots, k_{r_m}\}$.

Theorem 1. $(\Delta \cap \mathcal{K}) \subseteq \Gamma$, *and therefore, m is greater than or equal to the number of passwords chosen from $\mathcal{K}$.*

Proof: For any $k_i \in (\Delta \cap \mathcal{K})$, Equation 3 becomes,

$$\begin{cases} b_0 + b_1 \cdot g_i + \cdots + b_n \cdot g_i{}^n = k_i \\ b_0 + b_1 \cdot 1 + \cdots + b_n \cdot 1^n = w_1 \\ \vdots \qquad\qquad \vdots \qquad\qquad \vdots \\ b_0 + b_1 \cdot n + \cdots + b_n \cdot n^n = w_n \end{cases}$$

where $g_i = g(k_i, M)$, $w_i = w(i)$, for $i = 1, \ldots, n$. Similarly, the Equations 1 and 2 can be summarized as,

$$\begin{cases} a_0 + a_1 \cdot g_i + \cdots + a_n \cdot g_i{}^n = k_i \\ a_0 + a_1 \cdot 1 + \cdots + a_n \cdot 1^n = w_1 \\ \vdots \qquad\qquad \vdots \qquad\qquad \vdots \\ a_0 + a_1 \cdot n + \cdots + a_n \cdot n^n = w_n \end{cases}$$

where $a_0 = K$ and $g_i = g(k_i, M)$, $w_i = w(i)$, for $i = 1, \ldots, n$. From the above two equations, we have:

$$\begin{cases} (b_0 - a_0) + (b_1 - a_1) \cdot g_i + \cdots + (b_n - a_n) \cdot g_i{}^n = 0 \\ (b_0 - a_0) + (b_1 - a_1) \cdot 1 + \cdots + (b_n - a_n) \cdot 1^n = 0 \\ \vdots \qquad\qquad \vdots \qquad\qquad \vdots \\ (b_0 - a_0) + (b_1 - a_1) \cdot n + \cdots + (b_n - a_n) \cdot n^n = 0 \end{cases} \tag{4}$$

This results in:

$$\begin{vmatrix} 1 & g_i & \cdots & g_i{}^n \\ 1 & 1 & \cdots & 1^n \\ \vdots & \vdots & \ddots & \vdots \\ 1 & n & \cdots & n^n \end{vmatrix} \neq 0 \quad \Longleftrightarrow \quad a_j = b_j, \; j = 0, 1, \ldots, n$$

In other words, $a_j = b_j$, $j = 0, 1, \ldots, n$, if and only if the determinant of Equation 4 is non-singular. Since we have $g_i > n$, $i = 1, \ldots, n$, and because p (the modulo reduction) is prime, the above determinant is non-singular, and therefore, $a_j = b_j$, $j = 0, 1, \ldots, n$. This proves that $b_0 = K$ is the real key which was chosen by Alice. Hence, k_i is an applicable password, and so, $(\Delta \cap \mathcal{K}) \subseteq \Gamma$.

This implies that m $(= |\Gamma|)$ is greater than or equal to the number of passwords chosen from $\mathcal{K}$ $(= |\Delta \cap \mathcal{K}|)$. □

In the following, we consider Gong's basic and extended constructions and give our attack in each case. Alternative methods with higher security are also suggested. It is important to notice that our attack is aimed at forging a valid checksum without requiring the specific value of k_1.

3.1 Attacking the basic scheme ($m \leq n$)

It is not unexpected to have $m \leq n$. When $K, a_1, \ldots, a_n$, and M are given, it is improbable to find a password $k \notin \Delta$ such that $K + a_1 \cdot g(k, M) + \cdots + a_n \cdot g(k, M)^n = k$, because $|\mathcal{K}|$ is usually much smaller than $|GF(p)|$ and there is no guarantee to find a K' $(\neq K)$ such that $g(K', M) = g(K, M)$. (Note that, $g()$ is not a collisionful hash function.) Furthermore, $k_2, \ldots, k_n$ are selected from $\mathcal{P}$ $(\supseteq \mathcal{K})$, and an exhaustive search on $\mathcal{K}$ might not give all passwords k_2 to k_n. This decreases m, the number of applicable passwords. (Note that, $k_1 \in \mathcal{K}$.)

If $m < n$, the attacker E randomly selects $(n - m)$ passwords $(\notin \Gamma)$ and adds them to Γ. Using the resulting n $(= m)$ passwords (which include k_1), the opponent uses the procedure given in Section 2.3 to calculate the checksum for an arbitrary message M' and a randomly chosen key K'. Contrary to Gong's claim that the chance of a successful guess is as most $\frac{1}{n}$ [6, Page 169], the probability of a successful attack is 1, when $m \leq n$.

As mentioned before, choosing k_i's, $2 \leq i \leq n$, from $\mathcal{P}$ will reduce the number of resulting applicable passwords (Γ), which is more desirable in our attack. However, we consider a more secure version of Gong's method, by assuming that $k_i \in \mathcal{K}$, $i = 1, \ldots, n$. With this assumption, Theorem 1 results in:

Corollary 2. If $k_i \in \mathcal{K}$, $i = 1, \ldots, n$, then $\Delta \subseteq \Gamma$, and therefore, $m \geq n$.

The case ($m = n$) falls into the basic scheme, which is already considered. Now we examine the case when ($m > n$).

3.2 Attacking the extended scheme ($m > n$)

Gong [6, Page 170] extends his method by calculating the checksum as,

$$w_1 \| w_2 \| \cdots \| w_n \| g(K \bmod q, M), \qquad (5)$$

for a suitable $q \in \mathbb{N}$. Employing modular reduction increases m, the size of Γ, if $q < |\mathcal{K}|$. Gong does not disclose the state of n $(= |\Delta|)$, and therefore, we consider two cases in which n is either fixed (always the same n being used) or it is an arbitrary integer (which may vary every time). [1]

---

[1] For a given message M, Alice may fix the number of password collisions (n) and publicize it such that the corresponding hash value will be accepted only if it is verified by n password collisions. This is not discussed in the original paper, however.

If n is not fixed, E can construct a fraudulent checksum by solving the following m equations in $GF(p)$, for $a_1, \ldots, a_m$,

$$
\begin{cases}
K' + a_1 \cdot g(k_{r_1}, M') + \cdots + a_m \cdot g(k_{r_1}, M')^m = k_{r_1} \\
\vdots \qquad\qquad \vdots \qquad\qquad\qquad \vdots \\
K' + a_1 \cdot g(k_{r_m}, M') + \cdots + a_m \cdot g(k_{r_m}, M')^m = k_{r_m}
\end{cases}
\tag{6}
$$

where K' is a randomly chosen key and M' is an arbitrary message; and calculating the new checksum as,

$$
w(1) \parallel w(2) \parallel \cdots \parallel w(m) \parallel g(K' \bmod q, M'),
\tag{7}
$$

where $w(x) = K' + a_1 \cdot x + \cdots + a_m \cdot x^m \pmod{p}$.

If n is fixed, the attack will still succeed with significant probability. Let $m = n+1$. Then E can solve the following $(n+1)$ equations for $(n+1)$ variables $a_0, a_1, \ldots, a_n$,

$$
\begin{cases}
a_0 + a_1 \cdot g(k_{r_1}, M') + \cdots + a_n \cdot g(k_{r_1}, M')^n = k_{r_1} \\
\vdots \qquad\qquad \vdots \qquad\qquad\qquad \vdots \\
a_0 + a_1 \cdot g(k_{r_{n+1}}, M') + \cdots + a_n \cdot g(k_{r_{n+1}}, M')^n = k_{r_{n+1}}
\end{cases}
\tag{8}
$$

where M' is an arbitrary message. The valid checksum for M' will be,

$$
w(1) \parallel w(2) \parallel \cdots \parallel w(n) \parallel g(a_0 \bmod q, M'),
\tag{9}
$$

where $w(x) = a_0 + a_1 \cdot x + \cdots + a_n \cdot x^n \pmod{p}$.

If $m > n+1$, since $\Delta \subseteq \Gamma$ (cf. Corollary 1), E can randomly choose $(n+1)$ passwords $\{k_{t_1}, \ldots, k_{t_{n+1}}\}$ from Γ and have A's password (k_1) among them, with the probability of:

$$
\Pr[\, k_1 \in \{k_{t_1}, \ldots, k_{t_{n+1}}\}\,] = \frac{\dbinom{m-1}{n}}{\dbinom{m}{n+1}} = \frac{n+1}{m},
$$

which is a high probability if m is not much larger than n.

Now, E can use $\{k_{t_1}, \ldots, k_{t_{n+1}}\}$ to solve Equation 8 for $a_0, \ldots, a_n$, and calculate a valid checksum for an arbitrary message M' (Equation 9). That is, E can generate a valid checksum with the probability of $\frac{n+1}{m}$.

For example for $n = 9$, A should ensure that the number of the applicable passwords (m) will be at least 10^5 to decrease the probability of attack to 10^{-4}, which is the probability of guessing a 4-digit number — in bank Automatic Teller Machines (ATM).

3.3 Attack by discarding some of the applicable passwords

Another attack on Gong's method is to reduce the size of Γ, the set of all applicable passwords, by discarding the inappropriate passwords.

We have already proved that if n is not fixed or $m \leq n+1$, there are attacks in which E succeeds with the probability of 1 (100%). Now, assume n is fixed and $m > n+1$, and denote by $\Lambda = \{K_{k_{r_1}}, \ldots, K_{k_{r_m}}\}$ the collection of the resulting keys that correspond to the passwords in $\Gamma = \{k_{r_1}, \ldots, k_{r_m}\}$ (cf. Section 3). Note that, Λ has some repeated elements. This is true because $K_{k_i} = K_{k_j}$, $\forall k_i, k_j \in \Delta \subseteq \Gamma$. We partition Λ into l sub-collections $\Lambda_1, \ldots, \Lambda_l$ corresponding to the distinct values of Λ. That is, $K_\alpha = K_\beta, \forall K_\alpha, K_\beta \in \Lambda_t, t = 1, \ldots, l$.

On one hand, it is obvious that there exists a Λ_t such that $K_{k_i} \in \Lambda_t, \forall k_i \in \Delta$. On the other hand, to derive $K_{k_\alpha} \in \Lambda$ from $k_\alpha \in \Gamma$, we used Equation 3 which maps the small password space $\mathcal{K}$ into the large key space $GF(p)$. Therefore, one cannot expect to come up with many (more than n) distinct passwords ($\notin \Delta$) that are mapped to the same key $K \in \Lambda$. Hence, the above Λ_t, where $\{K_{k_1}, \ldots, K_{k_n}\} \subseteq \Lambda_t$, can be easily distinguished among the other portions, and the claim is emphasized when n, the number of password collisions, is large (cf. the experimental results in Table 2).

Consequently, we can select Λ_t as the portion that includes $K_{k_1}, \ldots, K_{k_n}$, and use the techniques in the previous sections to attack the method. In particular, even if n is fixed and $|\Lambda_t| > n+1$, the probability of successful attack is $\frac{n+1}{|\Lambda_t|}$ ($\geq \frac{n+1}{m}$).

To decrease the probability of an attack, Alice should use a collisionful hash function in Equation 1, not a keyed hash function $g()$. Also, if she can find one more key collision set of at least n elements which result in a different key K', she can halve the success probability. However, this is a very difficult task due to the difficulty of solving Equation 3 for $b_1, \ldots, b_n$, and k_1, when b_0 is a given fixed key. In other words, since $g()$ is generally not invertible, it is hard to find s ($\geq n$) key collisions $k_{t_1}, \ldots, k_{t_s}$ that result in a fixed b_0 ($\neq K$), using Equation 3.

3.4 Attack by using t pairs of [message, checksum]

As suggested by Gong, when the same password is used to calculate several checksums, A should reuse some of the password collisions no : to decrease the number of applicable passwords [6, Section 4]. Suppose t pairs of [message, checksum] are available and enemy has found the corresponding t sets of acceptable passwords. If the size of the intersection of these sets is less than $n+2$, the techniques given in Section 3.2 can break the system (100%). Otherwise, this intersection can be partitioned, similar to the way described in the previous section, to select the appropriate set with a high chance. This probability will significantly increase when t, the number of the given pairs, increases. In fact, the chance of reducing the number of guessed passwords to $k_1, \ldots, k_n$ will significantly increase.

$\|\mathcal{K}\|$	$\|\Gamma\|$
2^{10}	5
2^{12}	5
2^{14}	5
2^{16}	5
2^{18}	5
2^{20}	5

Table 1. *Basic scheme, where the checksum is* $w_1 \| \cdots \| w_n \| g(K, M)$. *For* $n = 5$, *the number of resulting applicable passwords* $(\|\Gamma\|)$ *was exactly equal to 5, in all cases. Therefore,* $\Gamma = \Delta$ $(m = n)$.

$\|\mathcal{K}\|$	q	$\|\Gamma\|$	Partition of Λ											
			$\|\Lambda_1\|$	$\|\Lambda_2\|$	$\|\Lambda_3\|$	$\|\Lambda_4\|$	$\|\Lambda_5\|$	$\|\Lambda_6\|$	$\|\Lambda_7\|$	$\|\Lambda_8\|$	$\|\Lambda_9\|$	$\|\Lambda_{10}\|$	$\|\Lambda_{11}\|$	$\|\Lambda_{12}\|$
2^{10}	127	14	5	1	1	1	1	1	1	1	1	1		
2^{12}	511	16	5	1	1	1	1	1	1	1	1	1	1	1
2^{14}	2047	14	5	1	1	1	1	1	1	1	1	1		
2^{16}	8191	13	5	1	1	1	1	1	1	1	1			
2^{18}	32767	12	5	1	1	1	1	1	1	1				
2^{20}	131071	13	5	1	1	1	1	1	1	1	1			
2^{10}	511	7	5	1	1									
2^{12}	2047	8	5	1	1	1								
2^{14}	8191	7	5	1	1									
2^{16}	32767	5	5											
2^{18}	131071	8	5	1	1	1								
2^{20}	524287	8	5	1	1	1								

Table 2. *Extended scheme, where the checksum is* $w_1 \| w_2 \| \cdots \| w_n \| g(K \bmod q, M)$. *For* $n = 5$, *the number of resulting applicable passwords* $(\|\Gamma\|)$ *was usually larger than 5, but after partitioning* Λ, *in each case, there was only one partition* (Λ_1) *with 5 elements* $(\Lambda_1 = \Delta)$. *This table is part of our extensive experimental results.*

4 Practical Results of the Attack

Authors have implemented Gong's method and the corresponding attacks on a SUN SPARC station ELC. The experiments completely coincide with the previously mentioned theories and support our claims about the weaknesses of the proposed selectable collisionful hash function.

Table 1 illustrates the results of our attack on the basic scheme. In all cases, the number of password collisions (n) was chosen to be 5. It shows that in all cases we could exactly find the five password collisions and forge the checksum based on Section 3.1.

Table 2 is the results of our attack on the extended scheme. Different modulo reductions are examined, where in all cases we could select the exact valid password collisions based on Section 3.3. It is important to notice that Section 3.4 gives even a more powerful attack when several checksums are available. However, we could break the scheme without assuming multiple available checksums.

The results show that with Gong's assumptions (especially the small password space), it is possible to attack his method. In the next section, we present methods which are secure if additional assumptions are met.

5 Securing the Method

In this section we show that the security of Gong's method under certain restricting assumptions is related to the security of *Sibling Intractable Function Families (SIFF)* [11]. This ensures the security of the scheme for a large password space. However we note that assuming a large password space might not be realistic in practice and hence propose alternative methods that reduce the probability of a successful attack.

5.1 Gong's Construction and SIFF

Suppose a message M, a randomly chosen key $K \in GF(p)$, and n password collisions $k_1, \ldots, k_n$ are given. Define $h(x) = g(x, M)$, where $g()$ is a secure keyed hash function. We note that $h()$ is one-way, because $g()$ is one-way on both parameters. We further assume that $h()$ is collision resistant. An example of $h()$ which satisfies these assumptions can be obtained if we start from a collision resistant hash function $H()$, and define $g()$ as $g(k, M) = H(k \parallel M)$. It can be seen that $g(k, M)$ is collision resistant on both parameters and hence $h(x) = g(x, M)$ will be collision resistant.

Now calculate $x_i = h(k_i)$, $i = 1, \ldots, n$ and solve the n equations,

$$
\begin{cases}
a_1 \cdot x_1 + \cdots + a_n \cdot x_1^n = k_1 - K \\
a_1 \cdot x_2 + \cdots + a_n \cdot x_2^n = k_2 - K \\
\vdots \qquad\qquad\qquad\qquad \vdots \\
a_1 \cdot x_n + \cdots + a_n \cdot x_n^n = k_n - K
\end{cases}
$$

for $a_1, \ldots, a_n$. We form $u()$ as,

$$
u(y) = f(y) - a_1 \cdot y - \cdots - a_n \cdot y^n,
$$

where $f(x_i) = k_i$, $i = 1, \ldots, n$.

We note that $(u \circ h)$, which is Gong's construction with extra requirements on $h()$, can be turned into an n-SIFF if $h()$ is chosen to be a 1-SIFF, which is a one-to-one and one-way function family (cf. [11]). An example of such a function family can be obtained by using exponentiation over finite fields. The reader is referred to [11] for a more detailed description of SIFF.

This ensures the security of Gong's construction if $g()$ is properly chosen and, in practice, implies that for a large password space the method resists all possible attacks.

5.2   Small Password Space

As noted in section 5.1, the security of Gong's construction can only be guaranteed for large password spaces. Also, if A could memorize a long password, she could directly use a secure keyed hash function to calculate a secure checksum. These assumptions might not be realistic in practical cases. In the following, we propose alternative solutions by relying on more reasonable assumptions which can provide smaller chance of success for an intruder.

1. Suppose A always uses n passwords $k_1, \ldots, k_n$ to calculate the checksums (they can be words chosen from a phrase). She can solve,

$$\begin{cases} a_0 + a_1 \cdot g(k_1, M) + \cdots + a_{n-1} \cdot g(k_1, M)^{n-1} = k_1 \\ \quad\vdots \qquad\qquad\qquad\qquad\qquad\qquad\qquad \vdots \\ a_0 + a_1 \cdot g(k_n, M) + \cdots + a_{n-1} \cdot g(k_n, M)^{n-1} = k_n \end{cases}$$

for $a_0, \ldots, a_{n-1}$, and calculate the checksum as $g(a_0, M)$. This will be appended to the message M and can be verified only by solving the above equations.

An adversary (E) should guess n passwords from $\mathcal{K}$ and check whether the resulting a_0 satisfies the checksum (there are $\binom{|\mathcal{K}|}{n}$ possible selections). A proper choice of n will prevent E to find the correct selection which results in the genuine $g(a_0, M)$.

Moreover, if $g()$ is collisionful on the first input parameter (cf. [3]), E will not be sure that she has found the right passwords. In fact, E may find an a_0' such that $g(a_0', M) = g(a_0, M)$, but it does not necessarily result in $g(a_0', M') = g(a_0, M')$, for another message M'.

However, disadvantage of this method is the difficulty of memorizing n passwords, when n is large.

2. Let c be the least integer such that 2^c computations are infeasible. Further assume a user password has on average d bits of information. (Clearly, $d < c$ and 2^d computations are feasible.) The checksum of a given message M is calculated as $h(k_1 \parallel R \parallel h(M))$, where k_1 is A's password, R is a randomly chosen $(c-d)$-bit number, and $h()$ is a collision resistant hash function [1, 8]. To verify the checksum, A exhaustively tests 2^{c-d} possible values of R and calculates $h(k_1 \parallel R' \parallel h(M))$ for each candidate $R' \in GF(2^{c-d})$. A match indicates that the checksum is valid, because $h()$ is collision resistant. Since both k_1 and R, which have in total $d + (c - d) = c$ bits of uncertainty, should be guessed by an enemy to verify the checksum, a random guessing attack is thwarted.

Note that, this verification has a maximum overhead of 2^{c-d} computations, but instead, selectable password collisions are not demanded. Furthermore, one may use $h((h(k_1) \bmod 2^b) \parallel R \parallel h(M))$, for a suitable integer b ($\leq d$), to provide password collisions. In this case, R should be $(c - b)$ bits.

For example, assume $c = 64$, $d = 50$, and $h()$ results in 128-bit digests. A can verify the checksum $h(k_1 \parallel R \parallel h(M))$ by computing $h()$ for at most 2^{14}

candidate R's. This takes about 2 seconds on a SUN SPARC station ELC, when $h()$ is MD5 [9]. Verification time is almost independent of the message length, since $h(M)$ needs to be calculated only once (not 2^{14} times).

Disadvantage of this method is the difficulty of finding a constant c which suits all users. In practice, different computing powers result in different values of c. Therefore, the largest amount should be chosen, which is not desirable on slow machines, because 2^{c-d} computations may become time consuming.

6 Conclusion

We showed that Gong's collision-selectable method of providing integrity is not secure, and an attacker with reasonable computing power can forge a checksum of an arbitrary message (or binary code). Assuming extra properties for the underlying hash function, it is possible to prove the security of Gong's construction under all attacks, when the password space is large.

Finally we have proposed alternative methods that require additional assumptions and meanwhile provide higher security (smaller chance of success for the enemy).

References

1. S. Bakhtiari, R. Safavi-Naini, and J. Pieprzyk, "Cryptographic Hash Functions: A Survey," Tech. Rep. 95-09, Department of Computer Science, University of Wollongong, July 1995.

2. S. Bakhtiari, R. Safavi-Naini, and J. Pieprzyk, "Password-Based Authenticated Key Exchange using Collisionful Hash Functions," in *the Astralian Conference on Information Security and Privacy*, 1996. (To Appear).

3. T. A. Berson, L. Gong, and T. M. A. Lomas, "Secure, Keyed, and Collisionful Hash Functions," Tech. Rep. (included in) SRI-CSL-94-08, SRI International Laboratory, Menlo Park, California, Dec. 1993. The revised version (September 2, 1994).

4. J. L. Carter and M. N. Wegman, "Universal Class of Hash Functions," *Journal of Computer and System Sciences*, vol. 18, no. 2, pp. 143–154, 1979.

5. I. B. Damgård, "A Design Principle for Hash Functions," in *Advances in Cryptology, Proceedings of CRYPTO '89*, pp. 416–427, Oct. 1989.

6. L. Gong, "Collisionful Keyed Hash Functions with Selectable Collisions," *Information Processing Letters*, vol. 55, pp. 167–170, 1995.

7. M. Naor and M. Yung, "Universal One-Way Hash Functions and Their Cryptographic Applications," in *Proceedings of the 21st ACM Symposium on Theory of Computing*, pp. 33–43, 1989.

8. B. Preneel, *Analysis and Design of Cryptographic Hash Functions*. PhD thesis, Katholieke University Leuven, Jan. 1993.

9. R. L. Rivest, "The MD5 Message-Digest Algorithm." RFC 1321, Apr. 1992. Network Working Group, MIT Laboratory for Computer Science and RSA Data Security, Inc.

10. M. N. Wegman and J. L. Carter, "New Hash Functions and Their Use in Authentication and Set Equality," *Journal of Computer and System Sciences*, vol. 22, pp. 265–279, 1981.
11. Y. Zheng, T. Hardjono, and J. Pieprzyk, "The Sibling Intractable Function Family (SIFF): Notion, Construction and Applications," *IEICE Trans. Fundamentals*, vol. E76-A, Jan. 1993.

On Password-Based Authenticated Key Exchange Using Collisionful Hash Functions

S. Bakhtiari, R. Safavi-Naini, J. Pieprzyk

Centre for Computer Security Research
Department of Computer Science
University of Wollongong, Wollongong
NSW 2522, Australia

Abstract. This paper presents an attack on Anderson and Lomas's proposed password-based authenticated key exchange protocol that uses collisionful hash functions. The weaknesses of the protocol when an old session key is compromised are studied and alternative solutions are given.

1 Introduction

Cryptographic hash functions are used for providing security in a wide range of applications [2, 8]. A collision-free hash function uniformly maps an arbitrary length message into a fixed length message digest, so that, finding two distinct messages that produce the same digest is computationally infeasible. This property of hash functions is used to provide data integrity.

A typical application of hash functions is to generate a checksum of a message whose integrity needs to be protected. Additionally, one can incorporate a secret key in the hashing process function to provide protection against an active intruder who wishes to modify the message or impersonate the message originator. Such functions are referred to as *keyed hash functions* [2, 5]. Berson *et al.* [5] introduced *collisionful hash functions* as an extension of keyed hash functions. Such functions are especially useful in a situation where the secret key is poorly chosen (e.g., password) and is therefore susceptible to guessing attacks. For a given pair of *[message , digest]*, a collisionful hash function guarantees that multiple keys satisfy the above pair. Hence, even if a guessed key produces a correct digest for a known message, an attacker cannot be certain that the guess is correct.

Anderson and Lomas [1] used collisionful hash functions to propose an augmentation of the well-known Diffie-Hellman key exchange [6] that provides protection against middle-person attack. Their method verifies the initial key exchange by means of a subsequent authentication stage based on communicant's passwords, which are assumed to be poorly chosen. The authentication stage uses a collisionful hash function in order to provide a safeguard against password guessing attacks. This paper shows that if an old session key is compromised, then an attacker can successfully guess a secret password, without being detected, and hence, compromise the whole system.

The rest of this paper is organized as follows. An overview of the Diffie-Hellman key exchange protocol is given in Section 2. Section 3 describes Anderson and Lomas's scheme. Section 4 analyzes the possibilities of guessing a shared password when an old session key is exposed. Alternative solutions are given in Section 5. We conclude the paper in Section 6.

2 Diffie-Hellman Key Exchange

In a seminal paper [6], Diffie and Hellman proposed a method to securely establish a shared secret between two users who do not share any prior secrets. Assume that g is the publicly known generator of a large cyclic group $\mathcal{G}$, where $|\mathcal{G}|$ and $\star$ denote the order of the group and the group operation, respectively. Let g^r denotes $g \star g \star \cdots \star g$ (r times) — this is in fact the exponentiation in $\mathcal{G}$ (e.g. if $\mathcal{G}$ is the multiplicative group of $GF(p)$, then g^r denotes $g^r \bmod p$). Suppose U and H are interested in establishing a secret session key, using the Diffie-Hellman key exchange protocol. U randomly chooses a positive integer $r_U < |\mathcal{G}|$, and sends g^{r_U} to H. Similarly, H randomly chooses $r_H < |\mathcal{G}|$ and sends g^{r_H} to U. That is,

$$1.\ U \to H : g^{r_U},$$
$$2.\ H \to U : g^{r_H}.$$

Both U and H can calculate the desired session key as $k = g^{r_U \cdot r_H}$. Assuming that the discrete logarithm problem is hard, an enemy cannot obtain k from the available public information $\mathcal{G}$, g, g^{r_U}, and g^{r_H}.

A well-known problem with the Diffie-Hellman key exchange is the middle-person attack — an enemy Eve (E) can interpose herself in the middle and send g^{r_E} to both U and H, where r_E is a positive integer chosen by E, such that $r_E < |\mathcal{G}|$. As a result, U and H inadvertently end up with two different keys $k_U = g^{r_E \cdot r_U}$ and $k_H = g^{r_E \cdot r_H}$, respectively, which are both known to E. Such an attack can be detected if U and H mutually authenticate the calculated session keys. Indeed, there are many protocols aimed at extending the basic Diffie-Hellman key exchange to provide the required authentication [4, 10]. When authentication is based on user passwords, the protocols must be resistant to guessing attack. It is also advantageous to devise alternative protocols that do not employ conventional encryption algorithms for this purpose [1].

3 Anderson and Lomas' Scheme

Motivated by the above considerations, Anderson and Lomas suggested a scheme (referred to as the AL protocol) where the basic Diffie-Hellman key exchange is augmented by means of a subsequent password-based authentication protocol. The scheme uses a collisionful hash function $q()$,

$$q(K, M) = h((h(K \parallel M) \bmod 2^m) \parallel M),$$

where $h()$ is a collision-free hash function and '$\|$' denotes concatenation. The function $q()$ has the property that if the first input (K) is given, it is hard to find collisions on the second input; however if the second input (M) is given, it is not hard to find collisions on the first input (depending on the choice of m). It is also assumed that both U and H share a secret password P, where *a password has n bits of information and eavesdropper E can perform 2^n computations*. The authentication protocol runs as follows:

$$3.\ U \rightarrow H : q(P, k_U),$$
$$4.\ H \rightarrow U : q(h(P), k_H).$$

U sends $q(P, k_U)$ to H and H computes $q(P, k_H)$ and compares it with the received value. Since $q()$ is collision-free on the second input, a successful match indicates that $k_U = k_H$. Consequently, H obtains assurance that k_H is shared with U since it is infeasible for outsiders to calculate $q(P, k_U)$. On the other hand, an unsuccessful match indicates that an attacker is at work.

Suppose that E, who this time knows both distinct keys k_U and k_H, intercepts $q(P, k_U)$ and finds a candidate password P' such that $q(P, k_U) = q(P', k_U)$.

Anderson and Lomas claim that since there are, on average, 2^{n-m} possible passwords that satisfy the equation, and all are equally likely, she cannot verify P' without further information. Although she can send $q(P', k_H)$ to H and her guess is verified if H replies, this can happen only at the risk of exposing herself in the event of a wrong guess. The probability of a successful guessing attack of this kind is minimized when $m = \frac{n}{2}$, and is equal to $2^{1-\frac{n}{2}}$.

In this paper we show that an attacker can find the password if an old session key is somehow compromised. (For instance, the session key might have been used in a weak cryptographic algorithm which leaks the key information.) This is a serious security threat, as all further communications between U and H will be compromised. An attack under similar conditions is considered by the authors of the well-known EKE protocol [11].

We also show that the probability of attack is slightly different from that measured in the original paper, and calculate m when that probability is minimized.

4   Attacking Anderson and Lomas' Scheme

Suppose U and H have successfully established a session key using the AL protocol. In other words, U and H have a verified session key k $(= k_U = k_H)$ by using the above protocol. Further assume that k is somehow compromised and that E has recorded $q(P, k)$ and $q(h(P), k)$ from the corresponding run of the protocol between U and H. We also assume, as in the original paper [1], that E can perform 2^n computations to exhaustively search the password space. The attack proceeds as follows.

		Candidate passwords		
n	m	S1	S2	S3
14	7	130	2	1
16	8	260	2	1
18	9	525	2	1
20	10	1023	2	1
22	11	2046	2	1
24	12	4087	2	1

The result of implementing our attack on Anderson and Lomas's proposed protocol, where $m = \frac{n}{2}$. The hash function used in this implementation is MD5, and the password is uniquely determined in the third step. Each result (the number of candidate passwords) is the outcome of the average made on 100 trials.

Table 1.

S1. E exhaustively searches the password space and uses the known pair $[k, q(P, k)]$ to find all the candidate passwords P_i, $i = 1, \ldots, t$, for some integer t, such that $q(P_i, k) = q(P, k)$.

S2. E then calculates $q(h(P_i), k)$ for every candidate password P_i, and reduces the set of likely candidates by retaining only those P_i's that $q(h(P_i), k)$ matches the known value $q(h(P), k)$.

S3. E masquerades as H and sets up a key k_U with U using the Diffie-Hellman key exchange. She then intercepts message 3 from U to H of the subsequent authentication protocol and uses the known pair $[k_U, q(P, k_U)]$ to uniquely determine the password from the reduced set obtained in Step 2.

Knowing the correct password P, E can successfully impersonate as either U or H. E can also successfully complete the protocol and come up with two distinct keys k_U and k_H, which are shared with U and H, respectively.

Table 1 illustrates the results of the implementation of our attack on Anderson and Lomas's protocol, where $m = \frac{n}{2}$, MD5 [9] is used as the hash function $h()$, and each result is the average of 100 trials. The results of Step 1 show that the number of candidate passwords is almost equal to 2^m. In Step 2, the number of candidate passwords is reduced to 2. Finally, a unique password is found in Step 3.

In the following, these results are formalized and theoretical justification of the experiments are given.

4.1 Probability of Success

Figure 1 shows the schematic flowchart of the evaluation of $q(P, k)$, where

$$\begin{cases} u(P, k) = h(P \parallel k), \\ w(u) = u \bmod 2^m. \end{cases}$$

Let k be the shared key between U and H, obtained from a previous run of the protocol, and which is somehow compromised. Denote by $\mathcal{P}$ the set of 2^n possible passwords, and assume 2^n verifications of passwords is a feasible computation. Note that, collisionful hash functions are motivated from the fact that poorly

Schematic flowchart of the collisionful hash function $q()$. In this figure, $P \in \mathcal{P}$ (password space), $k \in \mathcal{G}$ (key space), $u(x, k) = h(x \parallel k)$, and $w(u) = u \bmod 2^m$. The resulting $q \in \mathcal{D}$ (digest space) is $h((h(P \parallel k) \bmod 2^m) \parallel k)$. Typically, $|\mathcal{P}| = 2^n$, $|\mathcal{G}| = 2^{64}$, and $|\mathcal{D}| = 2^{128}$.

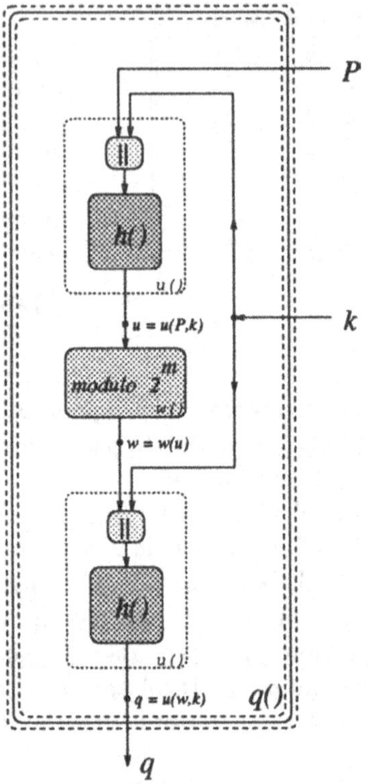

Fig. 1.

chosen passwords reduce the size of password space, and therefore, make them vulnerable to guessing attack. It is recommended not to use passwords in an ordinary keyed hash function, since a password that satisfies the checksum is the correct password with high probability. Anderson and Lomas's proposed scheme is based on this assumption:

> *"It is well known that humans cannot in general remember good keys, and that the passwords which they are able to remember are likely to succumb to guessing attacks. We shall therefore assume that U and H share a password P with n bits of entropy, while the eavesdropper E can perform 2^n computation."* [1, Page 1040 (Paragraph 5)]

Proposition 1. $u(P, k) \neq u(P', k), \forall P \neq P' \in \mathcal{P}$.

Proof: Suppose there exists $P \neq P' \in \mathcal{P}$ such that $u(P, k) = u(P', k)$. Defining $x = (P \parallel k)$ and $x' = (P' \parallel k)$, we have,

$$h(x) = h(P \parallel k) = u(P, k) = u(P', k) = h(P' \parallel k) = h(x').$$

Since $x \neq x'$, the above is a contradiction to the fact that $h()$ is collision-free. $\square$

The above proposition indicates that if the password space is exhaustively tested, the corresponding $h(P \parallel k)$'s are distinct.

Proposition 2. *If $q(P, k) = q(P', k)$, then $h(P \parallel k) \bmod 2^m = h(P' \parallel k) \bmod 2^m$.*

Proof: Let $x = h(P \parallel k) \bmod 2^m$ and $x' = h(P' \parallel k) \bmod 2^m$. If $x \neq x'$, then $h(x \parallel k) \neq h(x' \parallel k)$, because $h()$ is collision-free. This results in $q(P, k) \neq q(P', k)$ which is a contradiction. □

The above proposition ensures that whenever there is a collision on the first parameter of $q()$, it has been occurred in the previous step ($h(P \parallel k) \bmod 2^m$).

Theorem 3. *If the output of the hash function $h()$ is random, where every bit is considered to be independent of others, and $\mathcal{P}$ is exhaustively searched, the expected number of passwords which collide with a given password $P \in \mathcal{P}$ is $\frac{2^n - 1}{2^m} + 1$.*

Proof: Let $w = w(u(P, k)) \in GF(2^m)$. With the above assumptions, for every $P' \in \mathcal{P}$, the corresponding w' ($= w(u(P', k))$) is a random m-bit number and will be equal to w with the probability of $p = \frac{1}{2^m}$. Excluding P itself, one should test $|\mathcal{P}| - 1$ ($= 2^n - 1$) remaining passwords. Now, considering the binomial probability distribution (cf. [7, Chapter 5]) with success probability $p = \frac{1}{2^m}$ and $r = 2^n - 1$ trials, the expected number of successes (occurrence of w) is $r \cdot p = \frac{2^n - 1}{2^m}$. Including P, which collides with itself, results in $\frac{2^n - 1}{2^m} + 1$. □

Using the above theorem, we expect to reduce the number of candidate passwords from 2^n to $\frac{2^n - 1}{2^m} + 1$, in the first step (S1) of our attack (compare the S1 results in tables 1 and 2). The correct password can be guessed with the probability of $p = (\frac{2^n - 1}{2^m} + 1)^{-1}$.

Corollary 4. *Let Q denote the reduced set of the candidate passwords obtained in the above theorem. If $w = w(u(h(P), k))$, then the expected number of passwords in Q which collide with P is $\frac{2^n - 1}{2^{2m}} + 1$.*

Proof: Similar to the previous proof, the probability of success in each trial is $p = \frac{1}{2^m}$ and the number of trials is $r = \frac{2^n - 1}{2^m}$ (excluding P itself). Therefore, the expected number of the occurrence of w will be $r \cdot p = \frac{(2^n - 1)/2^m}{2^m} = \frac{2^n - 1}{2^{2m}}$. Including P will result in $\frac{2^n - 1}{2^{2m}} + 1$. □

Similar kind of argument can be used to prove the following:

Corollary 5. *By repeating password reduction procedure t times, the expected number of candidate passwords will be equal to $\frac{2^n - 1}{2^{tm}} + 1$.*

Table 2 illustrates the numerical values of the above results, when $m = \frac{n}{2}$ (Anderson and Lomas's ideal case). A comparison between Table 1 and Table 2 confirms that our practical results conform with theoretical ones, and the opponent can uniquely determine the correct password.

	Expected Values		
n m	e_1	e_2	e_3
14 7	129	2	1
16 8	257	2	1
18 9	513	2	1
20 10	1025	2	1
22 11	2049	2	1
24 12	4097	2	1

The expected values $e_1 = (\frac{2^n-1}{2^m} + 1)$, $e_2 = (\frac{2^n-1}{2^{2m}} + 1)$, and $e_3 = (\frac{2^n-1}{2^{3m}} + 1)$ corresponding to the three steps (S1, S2, and S3) of our attack, in which $m = \frac{n}{2}$. (Compare with Table 1.)

Table 2.

4.2 Optimum choice of m

We show that if m is properly chosen, the probability of our attack will be minimized. Assume s old session keys are compromised. Therefore, s pairs of $[q(P, k), q(h(P), k)]$ are available. Using Corollary 5, the expected number of candidate passwords will be reduced to $\frac{2^n-1}{2^{2sm}} + 1$. Now if E masquerades as H and set up a key k_U with U, she receives $q(P, k_U)$ and can follow one of the following attacks:

A1. *Randomly choosing a password from the reduced set of candidate passwords:* E can use the extra piece of information, $q(P, k_U)$, and further reduce the expected number of candidate passwords to $\frac{2^n-1}{2^{(2s+1)m}} + 1$ (cf. Corollary 5). Now she can correctly guess the password with the probability of,

$$p_1 = \frac{1}{\frac{2^n-1}{2^{(2s+1)m}} + 1} = \frac{2^{(2s+1)m}}{2^{(2s+1)m} + 2^n - 1}.$$

A2. *Randomly choosing a password after correctly guessing $w(u(P, k_H))$:* Suppose E randomly selects $w = w(u(P, k_H)) \in GF(2^m)$ and sends $q(P, k_H)$ ($= u(w, k_H)$) to H. The probability that w is correctly chosen is 2^{-m}.
If w is guessed correctly, H will send back $q(h(P), k_H)$. Now E can take the three extra pieces of information ($q(P, k_U)$, $q(P, k_H)$, and $q(h(P), k_H)$) and reduce the expected number of candidate passwords to $\frac{2^n-1}{2^{(2s+3)m}} + 1$. Accordingly, she can guess the right password with the probability of $(\frac{2^n-1}{2^{(2s+3)m}} + 1)^{-1}$, and therefore, the total probability of success is measured as,

$$p_2 = 2^{-m}\frac{1}{\frac{2^n-1}{2^{(2s+3)m}} + 1} = \frac{2^{2(s+1)m}}{2^{(2s+3)m} + 2^n - 1}.$$

A3. *Guessing $w(u(P, k_H))$ and $w(u(h(P), k_U))$:* Similar to the previous attack, E randomly selects $w = w(u(P, k_H)) \in GF(2^m)$ and sends the corresponding $q(P, k_H)$ to H. If H responds with $q(h(P), k_H)$, E will again randomly select $w(u(h(P), k_U))$ and send the corresponding $q(h(P), k_U)$ to U. The probability of success in this attack is measured as,

$$p_3 = 2^{-m}2^{-m} = \frac{1}{2^{2m}}.$$

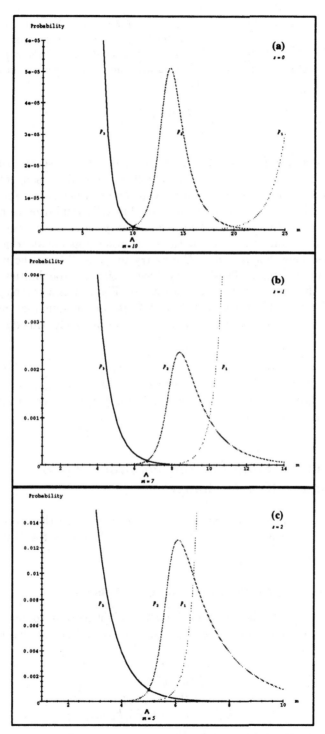

Fig. 2. Comparing p_1, p_2, and p_3, when $n = 40$ and $s = 0$, 1, and 2.

		Candidate passwords		
n	m	S1	S2	S3
14	2	4098	1026	257
18	3	32763	4096	512
22	4	262100	16404	1024

The result of three steps of our attack on the extended protocol (A1). In this case, E cannot uniquely determine the correct password with the knowledge of an old session key.

Table 3.

We note that the first two attacks allow E to find the secret password P, while in A3, she is only able to establish herself as a middle person. If the goal of the attack is defined as the latter one, E can choose either of the above attacks and will succeed with the corresponding probability. To minimize the probability of success, m should be computed such that $\max(p_1, p_2, p_3)$ is minimized.

It is also important to note that there are other possible attacks which are not mentioned here. For instance, E may try to guess P from the information obtained in the A3 attack. These attacks have negligible probabilities of success, compared to the three attacks discussed above. Figure 2 is a comparison of p_1, p_2, and p_3 for $n = 40$ and $s = 0, 1,$ and 2. It can be seen that $m = 10, 7,$ and 5 are good choices when $s = 0, 1,$ and 2, respectively. ($\max(p_1, p_2, p_3)$ is minimized.)

Example 1. Suppose one old session key is compromised ($s = 1$). Table 3 illustrates the practical results of the first attack (A1 with $p = (\frac{2^n - 1}{2^{3m}} + 1)^{-1}$), where best m's are chosen for several choices of n. Table 4 is the expected values which fairly coincides with Table 3.

5 Alternative Solutions for Password-Based Authenticated Key Exchange

In this section we present alternative solutions to decrease the probability of a successful attack, when a password is used to authenticate the Diffie-Hellman key exchange protocol.

5.1 Augmenting Anderson and Lomas's Scheme

To prevent the attack when an old session key might be compromised, U and H can use Anderson and Lomas's scheme but agree on a different shared key. The protocol runs as,

$$1. U \rightarrow H : g^{r_U},$$
$$2. H \rightarrow U : g^{r_H},$$
$$3. U \rightarrow H : q(P, k_U),$$
$$4. H \rightarrow U : q(h(P), k_H),$$

where $k_U = (g^{r_H})^{r_U}$ and $k_H = (g^{r_U})^{r_H}$. The new session key k will be calculated as $f(k_H) (= f(k_U))$, where $f()$ is a one-way function (it can be the same as $h()$).

		Expected Values		
n	m	e_1	e_2	e_3
14	2	4097	1025	257
18	3	32769	4097	513
22	4	262145	16385	1025

The expected values for three steps (extended scheme), where $e_1 = (\frac{2^n-1}{2^m} + 1)$, $e_2 = (\frac{2^n-1}{2^{2m}} + 1)$, and $e_3 = (\frac{2^n-1}{2^{3m}} + 1)$.

Table 4.

Therefore, even though a session key k might be compromised, E cannot guess P from the recorded values $q(P, g^{r_U \cdot r_H})$ and $q(h(P), g^{r_H \cdot r_U})$, because $g^{r_U \cdot r_H} = g^{r_H \cdot r_U} = f^{-1}(k)$ and $f()$ cannot be inverted.

5.2 Other authenticated key exchange protocols

In the following, we present two key exchange protocols that are augmented Diffie-Hellman protocols, and use a password to protect the shared key against an active spoofer.

1. Suppose $\mathcal{P}$ is the set of positive integers smaller than 2^n, and g is the generator of a large finite cyclic group $\mathcal{G}$ of order $|\mathcal{G}|$, and $P' \in \mathcal{P}$ is the smallest password greater than P such that $gcd(P', |\mathcal{G}|) = 1$ (i.e. P' and $|\mathcal{G}|$ are coprime). When U needs to establish a shared session key with H, U randomly chooses a positive integer r_U ($< |\mathcal{G}|$) and sends $g^{P' \cdot r_U}$ to H. Similarly, H randomly chooses a positive integer r_H ($< |\mathcal{G}|$) and sends $g^{P' \cdot r_H}$ to U. That is,

$$1. \, U \rightarrow H : g^{P' \cdot r_U},$$
$$2. \, H \rightarrow U : g^{P' \cdot r_H}.$$

Now, both U and H can calculate the desired session key as $k = g^{P' \cdot r_U \cdot r_H}$, by raising the received value to their randomly chosen values. Assuming that the discrete logarithm problem is hard, an active spoofer E cannot gain anything by modifying the transmitted values $g^{P' \cdot r_U}$ and $g^{P' \cdot r_H}$. Note that the above method simulates the Diffie-Hellman key exchange with a hidden generator $g^{P'}$. P' should be chosen co-prime to $|\mathcal{G}|$ to ensure that $g^{P'}$ is itself a generator of $\mathcal{G}$.

E can however force the protocol to result in two different keys shared with U and H (middle-person attack), or replay some of the old transmitted values (replay attack). *Random challenge* numbers can be used for a handshake between U and H to thwart any modification from an active spoofer or possible channel noises. Let $E_k(x)$ denote the encryption of x using the key k, and k_U and k_H be the keys calculated by U and H, respectively. The protocol is completed as follows.

$$1.\ U \to H : g^{P' \cdot r_U},$$
$$2.\ H \to U : g^{P' \cdot r_H},\ E_{k_H}(c_H),$$
$$3.\ U \to H : E_{k_U}(c_U ; c_H),$$
$$4.\ H \to U : E_{k_H}(c_U),$$

where c_U and c_H are randomly chosen challenge numbers by U and H, respectively. The protocol finishes successfully only if k_U and k_H are equal.

2. Similar to the above protocol, U and H can protect the transmitting values by XORing (bitwise exclusive-or) them to secret masks, generated from the password P:

$$1.\ U \to H : g^{r_U} \oplus h_1(P),$$
$$2.\ H \to U : g^{r_H} \oplus h_2(P),$$

where $h_1()$ and $h_2()$ are one-way hash functions (it is assumed that the size of message digests for $h_1()$ and $h_2()$ are equal to $\log_2 |\mathcal{G}|$). The session key will be calculated by both parties as $k = g^{r_U \cdot r_H}$.

Again, E cannot attack the protocol, except middle-person attack or replay attack. However, these two attacks can be prevented by random challenge numbers. In the following we give a hand-shake protocol in which the encryption function is replaced by a keyed hash function $v()$:

$$1.\ U \to H : g^{r_U} \oplus h_1(P),$$
$$2.\ H \to U : g^{r_H} \oplus h_2(P),\ v(k_H, c_H),\ c_H,$$
$$3.\ U \to H : v(k_U, (c_U \parallel c_H)),\ c_U,$$
$$4.\ H \to U : v(k_H, c_U).$$

The protocol finishes successfully only if k_U and k_H are equal.

Security of the first protocol is based on the discrete logarithm problem. It needs multiplications $P' \cdot r_U$ and $P' \cdot r_H$ in a large group which can be efficiently implemented. The second protocol is more efficient and only needs XOR operations. Note that, if $h_1() = h_2()$, then E can calculate $x = g^{r_U} \oplus g^{r_H}$ by XORing the first two transmitted messages. Since the session key is $g^{r_U \cdot r_H}$ and discrete logarithm is hard, E will gain no information about the session key nor the password.

Also, because of the low speed and export restriction of encryption functions, it is recommended to use hash functions for the hand-shaking, in the first protocol.

6 Conclusion

We showed that Anderson and Lomas's password-based authenticated Diffie-Hellman key exchange protocol is insecure if an old session key is compromised. We analyzed the probability of success of various attacks and suggested the best choice for m which minimizes the probability of success of our attack. Alternative protocols that achieve the same goal in a secure manner were also proposed. They let the users securely establish a shared session key, using a shared password.

Acknowledgment: We are grateful to Anish Mathuria for helpful discussions.

References

1. R. J. Anderson and T. M. A. Lomas, "Fortifying Key Negotiation Schemes with Poorly Chosen Passwords," *Electronics Letters*, vol. 30, pp. 1040–1041, June 1994.
2. S. Bakhtiari, R. Safavi-Naini, and J. Pieprzyk, "Cryptographic Hash Functions: A Survey," Tech. Rep. 95-09, Department of Computer Science, University of Wollongong, July 1995.
3. S. Bakhtiari, R. Safavi-Naini, and J. Pieprzyk, "On Selectable Collisionful Hash Functions," in *the Astralian Conference on Information Security and Privacy*, 1996. (To Appear).
4. S. M. Bellovin and M. Merritt, "Augmented Encrypted Key Exchange: a Password-based Protocol Secure Against Dictionary Attacks and Password File Compromise," tech. rep., AT&T, Nov. 1993.
5. T. A. Berson, L. Gong, and T. M. A. Lomas, "Secure, Keyed, and Collisionful Hash Functions," Tech. Rep. (included in) SRI-CSL-94-08, SRI International Laboratory, Menlo Park, California, Dec. 1993. The revised version (September 2, 1994).
6. W. Diffie and M. Hellman, "New Directions in Cryptography," *IEEE Transactions on Information Theory*, vol. IT-22, pp. 644–654, Nov. 1976.
7. S. R. Harrison and H. U. Tamaschke, *Applied Statistical Analysis*. Prentice-Hall, 1984.
8. B. Preneel, *Analysis and Design of Cryptographic Hash Functions*. PhD thesis, Katholieke University Leuven, Jan. 1993.
9. R. L. Rivest, "The MD5 Message-Digest Algorithm." RFC 1321, Apr. 1992. Network Working Group, MIT Laboratory for Computer Science and RSA Data Security, Inc.
10. R. L. Rivest and A. Shamir, "How to Expose an Eavesdropper," *Communications of the ACM*, vol. 27, 1984.
11. J. G. Steiner, B. C. Neuman, and J. I. Schiller, "Kerberos: An Authentication Service for Open Network Systems," in *Winter 1988 USENIX Conference*, (Dallas, TX), pp. 191–201, USENIX Association, 1988.

A New Universal Test for Bit Strings

Babak Sadeghiyan and Javad Mohajeri *

Amir-Kabir University of Technology
Sharif University of Technology
Tehran, Iran

Abstract. A new universal statistical test for random bit generators is presented. The test is an attempt to implement the concepts of the next-bit test theory. Given the statistics of a binary string and a small portion of it, the parameter of the test is related to the amount of next bits which can be guessed. The test detects local as well as overall non-random behaviour of the string. It can be applied for all the strings, from rather small to the large. It is easy to implement, and well suited for cryptographic applications.

1 Introduction

A pseudorandom bit generator is a bit generator which generates bit strings deterministically but they appear as if they are generated randomly. In other words, they appear to be statistically independent and uniformly distributed.

The next-bit test was shown by Yao [1] to be a universal test for bit generators. The idea of the test, informally saying, is that given any i bits of a string generated by the generator, to predict the next bit of the string with a probability of success significantly greater than 1/2. When it is impossible to predict the next bit efficiently the generator is pseudorandom. This test is universal, in the sense that when a generator passes such a test it passes any test for generators. However, the test has a theoretical value, and there is no such a test in practise.

In practise, many different tests are carried on the strings generated by a bit generator to evaluate its performance. These practical tests are divided into two groups, i.e., complexity tests and statistical tests. Complexity tests evaluate how long of a generated string is required to reconstruct the whole string [2],[3],[4]. Statistical tests evaluate whether the generator performs according to a specific probabilistic model. If it does, it is evaluated as a good generator. We refer the interested readers to [5] and [6] for a thorough discussion on these models.

This paper is an attempt to design and implement a practical test based on the idea of the *next bit* test, but not in its general case. Contrary to Maurer's universal test which needs a large amount of bits to apply the test on, the new test can be applied on a string with any length.

---

* This project has been supported by the Electronics Research Center of Sharif University of Technology.

2 Preliminaries

We use the following notions and definitions throughout the paper, following the notations used by Schrift and Shamir appearing in [7]. Let s_1^n denotes a binary string of length n in $\{0,1\}^n$. Let the i-th bit of the string is denoted by s_i. A substring starting with the j-th bit and ending with the k-th bit is denoted s_j^k. The notation $O(\nu(n))$ is used for any function $f(n)$ that vanishes faster than any polynomial.

Theoretical results are stated in terms of probabilistic polynomial-time algorithms. The notions of probability and independence are the standard notions from probability theory.

Definition 1: An ensemble S is a sequence $\{S_n\}$, where S_n is a probability distribution on $\{0,1\}^n$.

Definition 2: An ensemble S is called *uniform*, if S_n is the uniform probability distribution for every n, that is:

$$Prob\{s_1^n = \alpha\} = \frac{1}{2^n}$$

for every $\alpha \in \{0,1\}^n$. We denote the uniform ensemble by U.

A *source* is a string generator. A generator may be represented by the ensemble presenting the probability distributions of the generation of the bit strings. The *source* is used for the *ensemble* when it is more convenient. We denote by $Prob_S(E)$ the probability of an event taking place when the probability distribution is defined by the source ensemble S.

3 Theoretical Universal Tests

Next Bit Test : A source S passes the next bit test, if for every $1 < i \le n$ and for every probabilistic polynomial-time algorithm $A : \{0,1\}^{i-1} \to \{0,1\}$,

$$| Prob_S\{A(s_1^{i-1}) = s_i\} - \frac{1}{2} | \le O(\nu(n))$$

Schrift and Shamir introduced a test called *the predict or pass test* for biased sources in [7], where biased sources are defined as follows:

Definition 3: A source S is *biased* toward 1 with a fixed bias b , $\frac{1}{2} \le b < 1$, if for every i:

$$Prob_S\{s_i = 1\} = b$$

The source S is called *independent biased*, if all the bits are independent.

The *Predict or Pass* (POP) test is stated as follows:

POP Test : A biased source S passes the POP test, if for every i, $1 < i \le n$, for every fixed c, and every probabilistic polynomial-time algorithm $A : \{0,1\}^{i-1} \to \{0,1,?\}$, if $Prob\{A(s_1^{i-1}) \ne ?\} \ge \frac{1}{n^c}$ then

$$| Prob_S\{A(s_1^{i-1}) = s_i | A(s_1^{i-1}) \ne ?\} - b | \le O(\nu(n))$$

A outputs 1 when it predicts the next bit to be 1, and outputs 0 when it predicts the next bit to be 0. If A can not give a prediction, it outputs a ?. Schrift and Shamir also proved the following theorem based on the POP test.

Theorem 1: A source is a perfect independent biased source if and only if it passes the POP test.

The definitions for perfect independence and indistinguishability are given as follows:

Definition 4: Two sources S_1 and S_2 are polynomially indistinguishable, if for every probabilistic polynomial-time (distinguishing) algorithm $D : \{0,1\}^n \rightarrow \{0,1\}$,

$$| Prob(D(S_1) = 1) - Prob(D(S_2) = 1) | \leq O(\nu(n))$$

Definition 5: A source S is a perfect independence biased source, with some fixed bias b, if it is polynomially indistinguishable from the independent biased source, B, with the same bias b.

4 Practical Statistical Tests

In practise, many different tests are performed on a string to evaluate whether it is generated by a random string generator, i.e., to evaluate the bits are independent and unbiased. Standard statistical tests compare the overall behaviour of the string with some particular probabilistic models. A very common practical statistical test is frequency test, which is used to determine whether a generator is biased. The test is based on the binary memory less source (BMS) model where the bit generator outputs statistically independent and identically distributed random variables. The test is characterized by a parameter $p = \frac{1}{2}$, the probability of emitting ones. In this test, first the number of zeroes and ones of the string is counted, and are called n_0 and n_1 respectively. Then a value called χ^2 is computed as follows:

$$\chi^2 = \frac{(n_1 - n_0)^2}{n}$$

If $n_0 = n_1 = \frac{n}{2}$ then $\chi^2 = 0$. The greater the discrepancy between the observed and the expected frequencies the larger the value of χ^2 is. If the value is no greater than 3.84 the string is accepted to be a random string, for a 5% significance level. This value is extracted from a chi-squared table with 1 degree of freedom. When $\chi^2 = 3.84$, it means that $\frac{(n_1 - n_0)^2}{n} = 3.84$. Without loss of generality, consider $n_1 \geq n_0$. As $n_1 = n - n_0$, then:

$$\frac{n_1}{n} = \frac{1 + \sqrt{\frac{\chi^2}{n}}}{2}$$

For instance, consider $n = 1000$ bits, then it can be computed that maximum $\frac{n_1}{n} = 0.53$. It is worth noting that any 1000-bit binary string with a bias $0.47 \leq b \leq 0.53$ passes this test.

Another common test is *serial* test, where dibits are considered. A dibit can be either a "00" or a "01" or a "10" or a "11". In this test, first the number of occurrences of these patterns on the string are counted, and are called n_{00}, n_{01}, n_{10} and n_{11} respectively. Then, χ^2 is computed as follows:

$$\chi^2 = \frac{4}{n-1}\sum_{i=0}^{1}\sum_{j=0}^{1}(n_{ij})^2 - \frac{2}{n}\sum_{i=0}^{1}(n_i)^2 + 1$$

If $n_{00} = n_{01} = n_{10} = n_{11} = \frac{n}{4}$ then $\chi^2 = 0$. The greater the discrepancy between the observed and the expected frequencies the larger the value of χ^2 is. If the value is no greater than 5.99 the string is accepted to be a random string, for a 5% significance level.

Without loss of generality, consider $n_{11} \geq n_{01}$. When $\chi^2 = 5.99$, then it is worth noting that at most $\frac{n_{11}}{n} = \frac{\sqrt{4.49n - 4.99}}{2n} + \frac{1}{4}$. For instance, consider $n = 1000$ bits, then it can be computed that a string with the maximum $\frac{n_{11}}{n} = 0.2736$ may pass the test.

Another common test is *poker* test, where patterns of m bits on the string are considered. There are 2^m different patterns. For a pseudorandom string we expect each pattern happens with a probability around $\frac{1}{2^m}$. Again, some bias is allowed. Notice that all the above tests do not indicate deviations from the bias locally on the string.

The countings of the above tests can be drawn in the following tree, where each node represent the number of occurrences of each pattern.

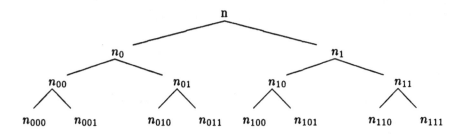

Note that, a number at each node can be computed from the numbers appearing in the two nodes directly under it in the next layer. For instance, $n_0 = n_{00} + n_{01}$ or $n_{00} + n_{01} + 1$. Hence, if there is a difference between the numbers of two nodes, the difference would remain in the underlying edges of the nodes.

Knuth mentions in [8] that, in order to have a valid result of statistical tests, the chosen length of the string should be such that to allow each pattern to appear at least 5 times, as a simple rule of thumb.

Example 1: consider the following string and the related tree, where $n = 70$:
1100110101100101000110100010000011011011010101000011100111000001001011

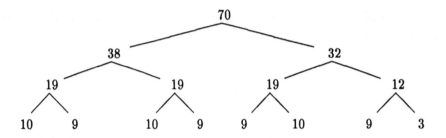

5 New Test

Now we present a new practical universal test, while the idea behind the new practical test can be related to the ideas of statistical tests and the *predict or pass* test.

We arrange the above tree from a different point of view. Consider a weighted tree where in the nodes we write down the patterns of each layer, and the ratio of the number of the patterns in the next layer to the previous layer on the edges (or branches) connecting the nodes. To avoid extra complexities in our computations, we append $m-1$ bits of the beginning of the string to its tail for the countings of layer m. So, the number of each pattern can be computed from the addition of the numbers of patterns related to the two nodes directly under it in the next layer.

For a large enough random string, we expect all the ratios be near $\frac{1}{2}$. However, for a pseudorandom one always some deviation is allowed. For the string of the Example 1, the first four layers of this tree is shown below.

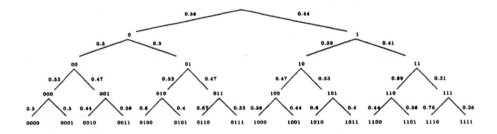

When the bits are generated independently, given any part of the string the next bit would be a 1 with a constant probability. When forming the above tree for such a string, we expect that we get a constant bias for all the nodes. In fact, when the bits are generated independently with some bias, obtaining a constant bias for all the nodes is a necessary condition. Note that the converse is not true.

Notice that the probabilities deviates from $\frac{1}{2}$ towards 0 and 1, as we move down the above tree. This is a general property for any string, as the number of occurrences of the patterns become less and less, by moving down to lower layers on a tree. This property makes the result of performing statistical tests, such as poker test, with large m invalid. However, this observation indicates us that we can predict the next bit of many patterns with large m, given the statistics on that lower layer. For instance, when pattern 00111 appears on the string of the Example 1, the next bit is 0 with a probability of 1.

For a 70-bit string, it can be computed that if the bias is $0.3829 \leq b \leq 0.6171$, the string passes the frequency test, otherwise it would be rejected to be a pseudorandom string . We can extend this decision to any other string with a different length. As this is a necessary condition we propose the following definition:

Definition 6: Let s_1^n be a string with length n. The *decision threshold* is defined as:

$$\alpha = \frac{1 + \sqrt{\frac{\chi^2}{n}}}{2}$$

where χ^2 is the value corresponding to the significance level required in the test.

For random sequences with high complexities, such as de bruijn sequences, we expect that there is one occurrence of each pattern at layer $\ell = \log_2(n)$, when $n = 2^\ell$. For less complex sequences, some patterns do not occur at layer ℓ and some other patterns occur once or more than once. This makes the probabilities which appear on some of the edges from layer $\ell - 1$ to layer ℓ nodes to be 1 (or 0). Having such probabilities and considering that this probabilities is related to the one (or near one) occurrence of a pattern, we would be able to reconstruct part of the string. The larger reconstructed part means predicting a larger part of the string, having statistics of the layer ℓ.

To define a threshold length of reconstruction, we consider the following definition and theorem, which is a natural extension of the *POP* test.

Extended POP Test : A biased source S passes the Extended-POP test, if for every i, ℓ, $1 < i, \ell \leq n$, for every fixed c, and every probabilistic polynomial-time algorithm $A : \{0,1\}^{i-1} \rightarrow \{\{0,1\}^\ell, ?\}$, if $Prob\{A(s_1^{i-1}) \neq ?\} \geq \frac{1}{n^c}$ then,

$$| \ Prob_S\{A(s_1^{i-1}) = s_i^{i-1+\ell} | A(s_1^{i-1}) \neq ?\} - b \ | \leq O(\nu(n))$$

Theorem 2: The following conditions are equivalent:
i) A biased source is a perfect independent biased source.
ii) A biased source passes the extended POP test.

The proof of the above theorem can be obtained from theorem 1 easily. If a source can not pass the extended POP test, it means there is a probabilistic polynomial-time test which given a block of the string can guess the next block efficiently. Therefore, the test A can predict s_i given the previous block, and the source is not a perfect independent biased source. The converse can also be detailed easily.

As we mentioned earlier, the probability of the next bit on the string should not exceed the calculated bias, for a pseudo-randomly generated string to pass the frequency test. As this is a necessary condition, we may define the calculated bias to be the threshold of decision for the probabilistic polynomial-time algorithm A, such that when the next bit on the string appears with a probability more than the bias, i.e. $b \leq p$ or $p \leq 1 - b$, the algorithm predicts the bit on the string. When the next bit on the string appears with a probability less than the bias, i.e. $1 - b \leq p \leq b$, the algorithm outputs a ?.

Having the above introduction in mind, we give a precise description of the steps of the new test, on a string of length n, as follows:

1. Calculate the threshold of decision, α, according to the formula:

$$\alpha = \frac{1 + \sqrt{\frac{\chi^2}{n}}}{2}$$

2. Calculate $\ell = round(\log_2(n))$
3. Append $\ell - 1$ bits of the beginning of the string to its tail, and divide the string into overlapping ℓ-bit sections.
4. Count the number of occurrence of each pattern of length ℓ.
5. Form part of the tree at layers $\ell - 1$ and ℓ, and write down the corresponding probabilities on each edge.
6. For each node at layer $\ell - 1$, if the next bit (either 1 or 0) appears with a probability more than α then the next bit is predicted accordingly, otherwise the next bit can not be decided.
7. For each node at layer $\ell - 1$, calculate the length of the string which can be predicted after it.

Notice that in some cases, an infinite length string may be predicted after some nodes in the last step. As the countings of each node of layer ℓ is given, we restrict the length of the predicted string according to the occurrences of the patterns.

We seek for local non-random behaviour and overall non-random behaviour of the string. Evaluation of the results is done in the following ways:

- For local non-random behaviour: If there is any node at layer $\ell - 1$ where more than $\ell + 1$ bits can be predicted after it, this means that there is a block of length ℓ where its next block could be guessed. Thus, there is some local non-randomness on the string, and the string may be rejected as being generated with a generator with required properties.
  The above test is a universal statistical test, as any of deviation of the probabilities appearing on the branches of upper layers from $\frac{1}{2}$, will cause a more deviation on the lower branches probabilities. In the new test, the deviations will manifests itself as the local non-random behavior measured on the string, while each kind of statistical tests are measurings of the deviations from $\frac{1}{2}$ on the upper layer branches of the tree. In other words, when a string passes the new practical test, it must pass standard tests, such as frequency test, serial test and poker test. Note that there may be strings which may pass some standard statistical tests, while they show local non-random behaviour.

– For overall non-random behaviour: We can take advantage of the results of the new test to obtain an evaluation on the overall behaviour of the strings. This is done by forming a histogram, giving the number of the nodes with respect to the number of predicted bits after them. Then measuring whether the shape of the resulted curve conforms to a curve, obtained from the following computations.

6   Curve of the Histogram

For a random string, the probabilities which appear on the branches between layer $\ell - 1$ and layer ℓ is a random variable between $(0, 1)$. If these probabilities are out of $(1 - \alpha, \alpha)$ then the next bit is predicted. But, if the probabilities are within the above range we may not decide about the next bit, and it can not be predicted. We call this range β to be equal to $(2\alpha - 1)$. For each node, we draw the path of the nodes that it meets when a next bit is predicted. For instance, from the node 00111 the next bit which can be guessed is 0, bringing us to the node 01110. We call the nodes at this stage *stage-1 nodes*. Again for the node 01110 the next node is 11100. We call the nodes at this stage *stage-2 nodes*. We can count the number of nodes at different stages in this way.

For a complex sequence, such as de bruijn sequences, the number of stage-0 nodes is: $N_0 = 2^{round(\log_2(n)-1)}$. From many nodes, we can not predict the next bit. The number of such nodes is equal to $\aleph_0 = \beta N_0$. From any remaining node, we can advance one stage further and guess the next bit. The number of the remaining nodes is equal to $(1-\beta)N_0$ nodes. Not all of these nodes reach to a different node of stage 1. The number of stage-1 nodes is: $N_1 = N_0(1-\beta)(1-\frac{\gamma}{2})$, where γ is the probability that two nodes reach to the same node of the next stage. (Notice that: $N_1 = N_0(1 - \beta)[(1-\gamma) + \frac{\gamma}{2}]$.) On the string of the Example 1, after both the node 01110 and the node 11110 a bit 0 can be guessed. So both the nodes reach to the node 11100.

From many nodes at stage 1, we can not advance further. The number of such nodes is equal to $\aleph_1 = \beta N_1$. The number of stage-0 nodes reaching to such stage-1 nodes is equal to: $NA_1 = \frac{\aleph_1}{1-\frac{\gamma}{2}}$. Notice that, NA_1 is the number of layer $\ell - 1$ nodes, where one bit after them can be predicted.

The number of stage-2 nodes is equal to: $N_2 = N_1(1-\beta)(1-\frac{\gamma}{2})$. The number of stage-2 nodes from which we can not advance one step further is: $\aleph_2 = \beta N_2$, and the number of corresponding stage-0 nodes is equal to $NA_2 = \frac{\aleph_2}{(1-\frac{\gamma}{2})^2}$. In general, the number of stage-i nodes is equal to: $N_i = N_{i-1}(1-\beta)(1-\frac{\gamma}{2})$. The number of stage-i nodes from which we can not advance one step further is: $\aleph_i = \beta N_i$, and the number of corresponding stage-0 nodes is equal to $NA_i = \frac{\aleph_i}{(1-\frac{\gamma}{2})^i}$. Again, notice that NA_i is the number of layer $\ell - 1$ nodes where i bits after them can be predicted.

For the above computations, a lower bound on γ can be calculated from α, as follows. On the tree of an independent source, two nodes with a particular $m - 2$ bit interim pattern will merge in the next stage, if the head of the two is different and the tail is the same. For every node there is only one such a different node. We can compute γ as the probability that the tails of two nodes are the same, given a particular $m-1$ bit pattern. Assuming that all the nodes appear with the same probability at layer $\ell - 1$, and assuming that the probability of collision is the same at all the stages, then: $\gamma = 1 + 2\alpha^2 - 2\alpha$. As the number of stage-0 nodes is equal to $2^{\ell-1}$ and we defined the number of advancement to be $\ell + 1$, then we can draw a curve to be the shape of the histogram, enabling us to evaluate the result of the experiment. Notice that for a random string we expect the measured $\aleph_0$ be bigger than or equal to what the curve shows. In contrast, we expect the measured $\aleph_i$, when $i > 1$, be less than or equal to what the curve shows.

7 Conclusions and Open Problems

In this paper, we have presented a new test which enables us to detect both local and overall non-random behaviours on a string. Contrary to Maurer's universal test which needs a large amount of bits to apply the test on, it can be applied on the string with any length. We set the limit on predicting the next bits to be as big as one block. Depending on the application, one may choose other limits, thus permitting to distinguish local non-randomness with other lengths. In the examples given in the paper, we have considered the significance level to be 5%, while other levels can also be employed depending on the application. To evaluate the overall behaviour of the string, we compare the histogram of the result with a curve. As we have taken some approximations in obtaining the required one, we believe this computations can be worked on to get a more precise and improved result.

8 References

[1] A. Yao, "Theory and application of trapdoor functions", Proc. 23rd FOCS, 1982, pp. 80-91.

[2] L. Brown et al., "A generalized test-bed for analyzing block and stream ciphers", Proc. of IFIP 1991/Information Security.

[3] H. Gustafson et al., "Comparison of block ciphers", in Lecture Notes on Computer Science, Vol. 453, Proceedings of Auscrypt '90, pp.153-165, 1990.

[4] J. Ziv, "Compression tests for randomness and estimating the statistical model of an individual sequences", Sequences, 1990, pp.366-373.

[5] H.Beker and F. Piper, "Cipher Systems", Northwood Books,1982.

[6] U. Maurer, "A universal statistical test for random bit generators", Journal of Cryptology, Vol. 5, No. 2, 1992, pp. 89-105.

[7] A. Schrift and A. Shamir, "Universal tests for nonuniform distributions", Journal of Cryptology, Vol. 6, No. 3, 1993, pp. 119-113.

[8] D. Knuth, "The Art of Computer Programming", Vol. 2, Addison-Wesley, 1973.

An Alternative Model of Quantum Key Agreement via Photon Coupling

Yi Mu[1] and Yuliang Zheng[2]

[1] Department of Computing, University of Western Sydney,
Nepean, Kingswood, NSW 2747, Australia
[2] School of Computing & Information Technology, Monash University,
McMahons Road, Frankston, VIC 3199, Australia

Abstract. It has recently been shown that shared cryptographic quantum bits are achievable through the use of an optical coupler, instead of polarised photons. We show that such shared cryptographic bits can also be produced by using a different optical apparatus - a beam-splitter. An important advantage of such a system is that it could be experimentally more feasible than an optical coupler.

1 Introduction

The central idea behind quantum cryptography is that an eavesdropper cannot monitor transmission based on quantum mechanics without being noticed by participants. This feature is based upon quantum mechanical phenomena such as Heisenberg's uncertainty principle and quantum correlation. The later is represented by the EPR or Einstein-Podolsky-Rosen-Bohm *gedankenexperiment* [9, 1]. A well-known protocol was suggested by Bennett, Brassard and co-workers in Refs. [6, 5]. This protocol is now called BB protocol. The BB protocol shows that information can be enclosed in one of four nonorthogonal quantum states (based on photon polarisation) on two bases in such a way that any attempt to extract the information by an eavesdropper will randomise and hence destroy the information. In other words, the eavesdropper's acts will definitely cause a change in the signal between the legitimate users, which therefore reveals the presence of the eavesdropper. On the other hand it has been demonstrated that EPR and Bell's theorem or inequality [3] are also useful in quantum cryptography. Protocols based EPR and Bell's theorem exploit the properties of quantum-correlated particles [10]. A further simplified protocol which does not use Bell's inequality has been proposed by Bennett *et al*[8]. Although there are some other interesting protocols, for instance, by photon interferometry[4], teleporting [7], rejected-data[2], and so on, the BB protocol and Ekert's protocol are the most typical models in quantum cryptography.

Recently, it has been shown that without using polarised photons one can also achieve a secure quantum cryptographic protocol [11]. This system is based on an optical apparatus - optical coupler. In practice, however, there may exist certain difficulties to achieve efficient photon coupling largely due to the fact a signal beam in the system is calculatedly chosen to be very weak in order to avoid potential beamsplitting attacks.

In this paper, using a beam-splitter, we develop a new quantum cryptosystem which also allows a cryptographic key bit to be encoded using four *nonorthogonal* quantum states described by non-commuting *quadrature phase amplitudes* (not photon polarisations !). We suggest that the proposed new system present a more promising solution from the experimental point of view.

Similarly to the system in Ref. [11], in the present system the nonorthogonal states are designed to have a large multi-overlap, hence it is impossible to obtain a certain result if a measurement is performed on only one of these states. This property forms the basis of security against any potential eavesdropping.

2   Background on quantum states and uncertainty

In this section, we briefly introduce some basic knowledge of quantum states, including coherent states and squeezed states, which will later be used to describe our system.

For a quantum field mode c, we can write it in the form of $c = c_1 + ic_2$, where c_1 and c_2 are quadrature phase amplitudes. The inequality of uncertainty for the quadrature phase amplitudes is given by

$$\langle \Delta c_1^2 \rangle \langle \Delta c_2^2 \rangle \geq 1/16. \tag{1}$$

where $\langle \Delta c_1^2 \rangle$ $(\langle \Delta c_2^2 \rangle)$ denotes the variance of c_1 (c_2). Inequality (1) suggests that only one of two quadrature phase amplitudes is certain.

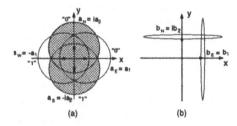

Fig. 1. On planes of quadrature-phase amplitudes, (a) shows Alice's encoding strategy based on four nonorthogonal coherent states; (b) shows Bob's probe modes using squeezed light. Uncertainty of a state is represented by error ellipses for squeezed states and by error circles for coherent states.

For a coherent state, since the photon distribution is Poissonian, the uncertainties for both quadrature-phase amplitudes are equal and the equality in (1) also holds. Hence both variances of the quadrature phase amplitudes are $1/4$. Accordingly, in figure 1 (a) we can see a noise circle for each coherent state, where we have assumed that mode a represents a coherent state with four encoding arrangements $a_E = a_1$, $a_W = -a_1$, $a_N = ia_2$, and $a_S = -ia_2$ (east, west, north, and south states). Under our encoding strategy, overlaps among these states should be as large as possible, thus it is accordingly assumed that the overlap between the east and west states is approximately 65%, so does the overlap between the north and south states. This requires that the mean number of photons for each state should be around 0.1. The absolute magnitude of overlap of two coherent states can be calculated by

$$|\langle\alpha|\beta\rangle|^2 = e^{-|\alpha-\beta|^2}. \tag{2}$$

With the mean number of photons per state being 0.1, it is easy to find that the overlap between the east and west state or the north and south states is 65%, and between the east and north states is around 82% (the same for each other pairs of neighbour states).

When a state is in an overlap between two states, it will not be able to be determined with certainly because it could belong to either of these states. On the other hand, when a state is not in the overlap region, it will be possibly determined without mixing with other states. Since under our arrangement *total* area of overlaps in a state is more than 90% and a large part of area has four overlap layers, it is almost impossible to obtain a certain result when performing a measurement on these states.

Homodyne detection is the most sound scheme for performing a measurement on a quadrature phase amplitude. The value of measurement is actually equal to the projection on the axis of the corresponding detector. We may lock a homodyne detector to an orientation, $x, -x, y$, or $-y$, which suits the measurements for different encodings, and consistently, we define four detection vectors V_x, V_{-x}, V_y, or V_{-y}, which in fact are four noncommuting projection operators.

We first look at homodyne detection performed on a *single* coherent state, the east state or the north state, and ignore the superposition for a while. In order to measure the east state, the homodyne detector must be locked at x direction (i.e., using V_x). This is because it has the largest probability of obtaining the correct result – a value of the mean $\langle a_1 \rangle$, despite the uncertainty $\langle \Delta a_1^2 \rangle =1/4$. When utilising the same projection operator V_x to detect the north state, we will then be unable to obtain a correct value, but have a high probability of obtaining zero (the uncertainty also equals 1/4). On the other hand, if a state does not have any projection on the detection vector, the state will not be able to be determined. For example, using V_x, we cannot determine the west state, since it does not have any useful projection on V_x (except the projection due to noise). It is concluded that for obtaining a correct detection the detection vector must be set accordingly to the direction of the signal state.

Since we are using four nonorthogonal states and each state has a large area of overlap with other states, it is hardly possible to correctly determine one out of these states by using a homodyne detector. This feature presents a promise for us to apply these states to cryptography.

For a squeezed state which is a minimum uncertainty state, the equality of (1) will hold, while the variance of one of the quadrature components is squeezed (to zero for a perfect squeezed state) and the variance of the other quadrature component is enlarged (to infinity for a perfect squeezed state). Assuming that

b is a squeezing mode. The variances of quadrature phase amplitudes can be described by

$$\langle \Delta b_1^2 \rangle = \frac{1}{2}e^{-2r}, \quad \langle \Delta b_2^2 \rangle = \frac{1}{2}e^{2r}. \tag{3}$$

As showed in figure 1 (b), two orthogonal squeezed states are used by Bob as his input to the optical coupler. The mode $b_E = b_1$ corresponds to $r \gg 0$, while the mode $b_N = ib_2$ corresponds to $r \ll 0$. One advantage of using squeezed light is that one of quadrature components can be measured with little influence of quantum noise.

The area of an ellipse for a mode represents uncertainty (or noise). For instance, we can see that, for the squeezed mode $b_E = b_1$ the x component (the projection on x axis) is knowable (small noise. ideally zero), but the y component (the projection on y axis) is uncertain (large noise, ideally infinity). We can explain the other mode similarly.

3 The new system

Our system is constructed using an optical beam-splitter as showed in figure 2, where a cryptographic communication is implemented between Alice and Bob. Alice is the sender who has a signal generator which can produce four nonorthogonal states and Bob is the receiver who measures the signal states by means of a beam-splitter. One feature of the system is that it allows cryptographic signals to be coupled with Bob's squeezed light. The coupling of light pulses provides us with a significant gain in the signal to noise ratio in comparison with that using a conventional coherent light source. This in turn provides us with a more efficient cryptographic key distribution protocol.

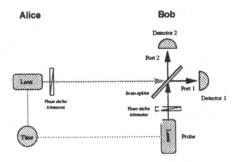

Fig. 2. The schematic diagram of the quantum cryptosystem using an optical beam splitter.

A quantised light field can be represented by a creation operator and an annihilation operator. We assume that Alice's signal mode is expressed by a creation operator $a^\dagger$ or an annihilation operator a and Bob's mode (the probe light) is represented by a creation operator $b^\dagger$ or an annihilation operator b. For a

50-50 beam-splitter with a mirror amplitude refectivity $1/\sqrt{2}$, the output beams obey

$$a' = \frac{1}{\sqrt{2}}[a\exp(i\theta_a) + ib\exp(i\theta_b)] = \frac{i}{\sqrt{2}}(a+b), \qquad (4)$$

$$b' = \frac{1}{\sqrt{2}}[ia\exp(i\theta_a) + b\exp(i\theta_b)] = \frac{1}{\sqrt{2}}(-a+b) \qquad (5)$$

where θ_a and θ_b are the reference phases of mode a and mode b, respectively. It is reasonable to assume $\theta_a = \pi/2$ and $\theta_b = 0$, which results in the second equalities. The above equations can be transferred into equations of quadrature phase amplitudes, with $a = a_1 + ia_2$ and $b = b_1 + ib_2$.

To simplify our discussions, we have employed a symmetric (50-50) beam-splitter. In practice, however, it might be necessary to use an asymmetric one that allows a large portion of Alice's signal to pass through (ideally, all photons). This change would be important, since Alice's signal is very weak.

4 The protocol

The basic intention is to establish a common key between two parties, Alice and Bob, who share no secret information at the beginning of the cryptographic communication. The beam-splitter is controlled by Bob who can independently choose the probe light (squeezed light). Both signal generators are controlled by a time base that guarantees a perfect photon coupling action. The output signal is detected using two homodyne detectors, one for each port. Also, importantly, in order to realise a perfect coupling action in the beam-splitter, Alice and Bob need to choose a phase reference before their communication starts. This can be done by Alice sending a sequence of bright reference pulses to Bob and publicly announcing their phases.

Alice's generator produces a faint coherent light, on the average, 0.1 photon per pulse, i.e., $\langle a^\dagger a \rangle = 0.1$. As we have mentioned, under this assumption the total overlap on a state is over 90%. The probability a signal pulse contains one or more photons is approximately 10%. This figure suggests that 90% of the total pulses are vacuum. Note that it is possible to employ a weaker signal light such that the superposition of the four nonorthogonal states is even larger. However we do not intend to do that, since our assumption is sufficient for our cryptographic protocol. Bob's squeezed light is much brighter and has on average one photon per pulse.

Because of the noise of light, it is very difficult for Bob to identity the correct detection result. In order to resolve this problem, we give the following definition:

Definition screening criterion *An output bit from the beam-splitter is recorded, if and only if Bob finds that two photons are projected on the detector at one port and nothing is projected on the detector at the other port.*

Bob's measurements are based on a homodyne detection scheme, where both detectors are arranged in terms of the probe mode used by Bob himself. Bob

Table 1. The results of the photon coupling. The illustration is based on a quadrature plane. We have assumed equal intensity for both mode a and mode b, the symbol "×" represents "discarded", C represents "Cancelled", E represents "Enhanced", and a sign, character or binary figure in front of "/" has a higher probability of appearance. In other words, those in front of "/" are correct; those behind "/" are associated with the overlap on the corresponding opposite state. The later ones can be corrected eventually.

Alice's mode	Bob's mode	Output from Beamsplitter	Measurement Vector	Status	Result	Final Result
a_E	b_E	$a' = \frac{i}{\sqrt{2}}[(+/-)a_1 + b_1]$	V_y	E/C	0/1	0
		$b' = \frac{1}{\sqrt{2}}[(-/+)a_1 + b_1]$	V_x	C/E		
	b_N	$a' = \frac{i}{\sqrt{2}}[(+/-)a_1 + ib_2]$	V_{-x}	Uncertain	×	
		$b' = \frac{1}{\sqrt{2}}[(-/+)a_1 + ib_2]$	V_y	Uncertain		
a_W	b_E	$a' = \frac{i}{\sqrt{2}}[(-/+)a_1 + b_1]$	V_y	C/E	1/0	1
		$b' = \frac{1}{\sqrt{2}}[(+/-)a_1 + b_1]$	V_x	E/C		
	b_N	$a' = \frac{i}{\sqrt{2}}[(-/+)a_1 + ib_2]$	V_{-x}	Uncertain	×	
		$b' = \frac{1}{\sqrt{2}}[(+/-)a_1 + ib_2]$	V_y	Uncertain		
a_N	b_E	$a' = \frac{i}{\sqrt{2}}[b_1 + (+/-)ia_2]$	V_y	Uncertain	×	
		$b' = \frac{1}{\sqrt{2}}[b_1 - (+/-)ia_2]$	V_x	Uncertain		
	b_N	$a' = \frac{i}{\sqrt{2}}[(+/-)ia_2 + ib_2]$	V_{-x}	E/C	0/1	0
		$b' = \frac{1}{\sqrt{2}}[(-/+)ia_2 + ib_2]$	V_y	C/E		
a_S	b_E	$a' = \frac{i}{\sqrt{2}}[b_1 - (+/-)ia_2]$	V_y	Uncertain	×	
		$b' = \frac{1}{\sqrt{2}}[b_1 + (+/-)ia_2]$	V_x	Uncertain		
	b_N	$a' = \frac{i}{\sqrt{2}}[(-/+)ia_2 + ib_2]$	V_{-x}	C/E	1/0	1
		$b' = \frac{1}{\sqrt{2}}[(+/-)ia_2 + ib_2]$	V_y	E/C		

should have a rule which allows him to determine which detection vector needs to be used.

Definition detection rule *If the probe mode is associated with b_E, the detector at Port 1 is set toward the y direction (using V_y) and the detector at Port 2 to the x direction (using V_x); if the probe mode is associated with b_N, the detector at Port 1 is set to $-x$ direction (using V_{-x}) and the detector at Port 2 to y direction (using V_y).*

Under the detection rule, Bob needs only two sets of detection vectors: $\{V_y, V_x\}$ and $\{V_{-x}, V_y\}$. Each time Bob chooses only one of them.

Our quantum cryptographic key distribution protocol is described as follows:
During the preparation stage, both Alice and Bob need to prepare their data. Assuming that α_i is randomly selected from four quantum states $a = \{a_E, a_W, a_N, a_S\}$, Alice constructs a vector $A = (\alpha_1, \alpha_2, ..., \alpha_n)$ of n random choices, $\alpha_i \in a = \{a_E, a_W, a_N, a_S\}$. a is public information, while A is private

data only known to Alice. Bob independently chooses a vector $B = (\beta_1, \beta_2...., \beta_n)$ of n random choices, $\beta_i \in b = \{b_E, b_N\}$. b is public information, but B is private data only known to Bob.

Phase one: Signal transmission and measurement:
1: Alice sends Bob a $\alpha_i \in A$, while Bob injects a β_i which interacts with α_i in Bob's beam-splitter. All possible outcomes are shown in Table 1. In terms of the subsequent detection and the screening criterion,

$$\text{Bob sets } \beta_i' = \begin{cases} 0 & \text{(a bright flash at Port 1 and nothing at Port 2)}, \\ 1 & \text{(a bright flash at Port 2 and nothing at Port 1)}, \end{cases}$$

Otherwise, Bob deletes the bit. Alice and Bob repeat the process until the whole signal string is sent. "bright flash" means that two photons have been projected on Bob's detection vector.
Bob keeps B and $B' = (\beta_1', \beta_2', ..., \beta_n')$ secret.
2: Bob speaks to Alice publicly for each β_i': *Accept* if Bob "saw" a bright flash at Port 1 (2) and nothing at Port 2 (1) (obeying the screening criterion); *reject* if Bob "saw" flashes at both ports or other instances which do not satisfy the screening criterion.
3: Bob announces to Alice which detection vector has been used for each accepted bit (but nothing about the outcome of the measurement).
4: Alice asks Bob to delete those bits obtained using an incorrect detection vector. For example, Alice may ask him to delete a north-state-related "0" bit which is obtained by using V_x. This step ensures that all flawed bits subject to the overlaps with two closer neighbour states (but not the opposite state) are removed. (We will give more explanation later.)

Phase two: Error correcting:
Up to now, Bob's remaining bits still contain a number of flawed bits subject to overlap with the opposite states. In order to correct (but not remove) the flawed bits, the following steps should be taken:
1: Alice secretly divides all remaining bits related to each state, east, north, west, or south into N groups ($N \geq 100$), where each group contains m bits (in the present case, $m \geq 30$ is appropriate). This requires that the number of original signal bits sent by Alice are sufficient. Each group involves only one signal state, but both binary bits. Amongst these binary bits, one fraction of binary bits ("0" or "1") stem from the correct detections and these bits are the majority; the other fraction of binary bits ("1" or "0") come from the overlap on the opposite state. Note that during the grouping the original positions of the bits were not changed.
2: Alice publicly announces the grouping result, without releasing any encoding information. So nobody knows which group belongs to which state, except Alice herself. Since each Bob's detection vector has been used for two nonorthogonal states, knowing the detection vector of each group releasees no encoding information of the group.

3: Bob calculates the number of "0" or "1" bits in each group. The encoding of the majority bits will represent the encoding of *all bits* in the group. For example, if Bob finds that "0" bits are the majority, he will regard all bits in the group as "0". So far Bob has corrected all mistakes caused by the overlap with the corresponding opposite state and has obtained the encoding information of each group. This step can only be implemented by Bob, because he is the only one who knows the measurement result.

4: Bob tells Alice the positions of all useful bits. Alice knows the full information of these bits.

Upon the completion of communication, Alice and Bob keep the bits which have eventually survived as the secret key.

Our system is summarised in Table 1 and Figure 3. The latter illustrates the protocol.

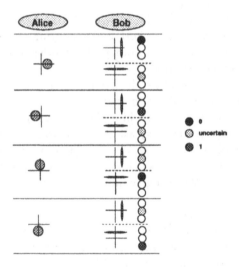

Fig. 3. The summary of the system: light modes and the results are given. The traffic lights are used to illustrate the result of implementing the protocol.

5 Analysis

In comparison with the protocol presented in Ref. [11], the main difference is that in the present protocol, Bob uses different sets of detection vectors. The change is due to the modification of the signal/probe phase.

Table 1 shows all possible detection results obtained by Bob when both light pulses have the same intensity. Instead of explaining all cases in the table, we only focus on the first case, where Alice uses the east state a_E. The explanations for the remaining cases are similar. In the first case, Bob uses b_E (and V_x consistently). According to the coupling equations, there are two possible outcomes: (1) The output at port 1 is enhanced and the output at port 2 is reduced to a vacuum state due to the cancellation. Bob then further checks whether the outputs satisfy the screening criterion. If the answer is yes, a "0" is accordingly recorded. (2) Because of the superposition between the east state and the opposite west state, a large fraction of bits associated with the east state turn out being mixed with the west state, and Bob could then have a false result and a "1" is hence recorded. The latter is obviously wrong, but Bob is aware of his mistake. In order to overcome this problem, Alice divides all accepted bits related to the east state into N (say 100) groups and each group contains m (say 30) bits (see also our analysis to be presented later). By calculating the number of "0" or "1" bits, Bob is able to find the majority bits which will be used to represent the encoding of all bits in the group. The mechanism of this error correcting method is simple: since the overlap between the states is not 100%, there is a larger probability of obtaining the east state rather than the west state. This is obviously true, because only if the superposition is 100%, the probability of obtaining the east state or the west state is $1/2$,

By means of a Q-representation, we can further explain the error correcting method. A coherent state α in a Q-representation is given by

$$Q(\gamma) = \frac{1}{\pi}e^{-|\gamma-\alpha|^2}, \tag{6}$$

which actually represents a quasi-probability of the coherent state. For the east coherent state with an average projection value of 0.33 (an intensity of 0.1 photon) on the x axis (on the quadrature-phase plane), the probability of a projection being around 1 on a small region $(\Delta x) \cdot y$, where $-\infty < y < \infty$, is given by

$$P(projection = 1|\alpha = 0.33) = \frac{1}{\sqrt{\pi}}e^{-0.67^2}\Delta x \approx 0.36\Delta x, \tag{7}$$

while the probability of projection being -1 on the small region is given by

$$P(projection = -1|\alpha = 0.33) = \frac{1}{\sqrt{\pi}}e^{-1.33^2}\Delta x \approx 0.0963\Delta x. \tag{8}$$

It is easy to find that, amongst the total pulses with a value 1 or -1 projection on x axis, the 1-pulses is 79% and the -1-pulses 21%. According to these data, we may roughly calculate the correctness rate of Bob's error correcting: assuming that $m = 30$ and the minimal number of bits m_{min} for Bob to correctly identify the encoding is greater than $m/2 = 15$, we have the correctness rate:

$$P(m_{min} > m/2) = 1 - \sum_{i=1}^{m}\binom{m}{i}(0.79)^i(0.21)^{m-i} \approx 0.9996. \tag{9}$$

This value suggests that Bob is almost 100% correct. Note however that if an eavesdropper wants to measure the signal, she cannot have such a high ratio of 1-pulses to -1-pulses, since her detection is subject to the superposition from other two neighbour states, the north and south states. More serious problem for the eavesdropper is that she does not know which detection vector should be used. Bob does not have this problem, because Alice can ask him to delete all bits owing to the superposition with the two neighbour states and due to using incorrect detection vectors. This case will be further studied in next section.

We now focus on the second case, i.e., Alice still uses $a = a_1$ and Bob uses the other mode b_N (and V_y, consistently). Bob is obviously wrong. Most possibly, the outputs at one or both ports are nonzero, Bob can thus "view" a light flash with a various intensity at one or both ports. These bits are useless and can be removed in terms of the screening criterion. However, since the measurement is subject to the noise or overlaps, we must consider that Bob might occasionally obtain a result which meets the screening criterion. When this happens, Bob will not be able to identify the flaw. In order to get rid of these flawed bits, no matter what measurement result has been obtained, Alice will ask Bob to remove the bit.

We have not explained the influence of overlaps associated with the two neighbour states, the north and south states. These instances actually belong to other two cases where Alice sends the north or south state. The corresponding flawed bits will be handled by Alice and Bob using a similar procedure to that given above.

6 Discussion

We have made clear that the quantum states used in our system are not identifiable due to the superposition. More explanations are provided in this section to detail the various potential eavesdroppings from case to case. Some of discussions here have been given in Ref. [11].

- Intercept/resend:
  An adversary (Eve) would intercept the signal and measure it by using a similar apparatus. If she does so, at least half of her measurements will be random, because she has to randomly select her probe states and detection vectors. Moreover, the remaining half of Bob's measurements are also uncertain due to the superposition with respect to Alice's signal. Therefore, it is impossible for Eve to regenerate and resend the signal to Bob, using her own measurement.
- Direct detection:
  Assume that Eve knows that four projection operators, $\{V_{-x}, V_x, V_{-y},$ and $V_y\}$, can be used to detect Alice's signal and these detection vectors respectively suit detecting a_E, a_W, a_N, and a_S. Eve might then wish to use her detector to measure Alice's signal directly, instead of using an optical coupler. However since she does not know which state has been sent by Alice, she has no better way than to choose a detection vector randomly. The probability of choosing the correct detection vector is obviously 1/4. Fortunately, even if she happens to select the correct detector, her measurement is still uncertain

because of the overlap of the encoding states. If Eve has a correct detection vector and knows that a projection of value 1 is important, it is not hard to find there is a probability of 3/5 for her obtaining a wrong projection belonging to the neighbour states. These bits cannot be identified by Eve. The total success rate of measuring a bit is found to be 1/10. In fact it is impossible for Eve to know whether or not she has used the correct detection vector, since, from Bob's public information, she can only know either V_x or V_y has been used by Bob (V_x or V_y corresponds to two Alice's states). This suggests that even if Eve's success rate is 1/10, she cannot know which detection is successful. Consequently, Eve achieves nothing from such eavesdropping.

– scanning the conversation:

Eve may not do anything but just listens to Alice and Bob's public conversation. After Alice and Bob implement the protocol, Eve is aware which detection vector has been used, which bits were accepted, and which detection vector has been applied to each group chosen by Alice. Because each Bob's detection vector corresponds to two nonorthogonal states, Eve can only guess whether the bits in each group belong to either "0" or "1". Hence, for each individual group, Eve has a 1/2 chance to succeed. However, since the number of groups for each state $N > 100$, Eve's success rate will be less than $1/2^{400}$ or approximately $1/10^{120}$. In practice, it is highly unlikely for Eve to succeed.

– Statistical analysis:

The requirement for the number of bits in each group depends on the superposition of encoding states. As discussed in the previous section, if the average number of photons is 0.1, $m = 30$ is appropriate for Bob to obtain a good success rate. However, if Eve has a little knowledge about the encodings, she could also implement a similar statistical analysis. How can Eve obtain a small piece of information on a group? Eve knows that it will not work, if she intercepts all signal pulses. In order to avoid being detected, Eve may randomly intercept/measure only a small fraction of signal pulses using the four detection vectors, for instance 10% in the total number of pulses, and lets the rest go through without being interfered. Can Eve then have good guesses? In the case $m = 30$, Eve intercepts only 3 pulses (among 30). The measurement on the 3 pulses (based on randomly choosing measuring vector) is not adequate for her to implement a statistical analysis. Moreover, intercepting 10% of total pulses could also result in a substantial influence on Bob's measurement which could reveal Eve's attempt.

However, if the size of m is large, say 1000, with intercepting a small number of bits Eve may then have enough bits used for her statistical analysis. Again, the big problem for her is how to obtain useful encoding information on these bits. The most thinkable way could still be the interception, but according to the discussion in the second paragraph of present section, Eve cannot obtain any useful information even for a single bit. Consequently, even if m is large, Eve is still unable to carry out a statistical analysis. However, there might be some other unseen way such that Eve could obtain a small fraction of information from Alice's signal. A large m will then in principle be useful for Eve. Therefore we should define an upper limit for m. Because the upper limit depends on the superposition of the signal, we can only define a general criterion: the limit on m should be the minimum value where Bob has a satisfied success rate.

7 Conclusion

In this paper, we have shown that using a beam-splitter and four nonorthogonal states is promising for constructing a secure quantum exchange-key system which is not detectable to eavesdroppers. The main contributions of this work are the proof of availability of beam-splitters and the extension of the proceeding system based on an optical coupler [11] to a experimentally more promising model.

References

1. Aspect, A., Grangier, P., Roger, G.: Experimental realization of Einstein - Podolsky - Rosen - Bohm *gedankenexperiment*: A new violation of Bell's inequalities. Phys. Rev. Lett. **49** (1982) 91–94.
2. Barnett, S. M., Huttner, B., Phoenix, S. J. D.: Eavesdropping strategies and rejected-data protocols in quantum cryptography. Journal of Modern Optics **40** (1993) 2501–2513.
3. Bell, J. S.: On the Einstein Podolsky Rosen Paradox. Physics (N.Y.) **1** (1964) 195.
4. Bennett, C. H.: Quantum cryptography using any two nonorthogonal states. Phys. Rev. Lett. **68** (1992) 3121–3124.
5. Bennett, C. H., Bessette, F., Brassard, G., Salvail, L., Smolin, J.: Experimental quantum cryptography. Journal of Cryptology **5** (1992) 3–28.
6. Bennett, C. H., Brassard, G., Breidbard, S., Wiesner, S.:. Quantum cryptography, or unforgeable subway token. In Advanced in Cryptography: Prodeedings of Crypto 82 (1983) Plenum Press pp. 267–275.
7. Bennett, C. H., Brassard, G., Crépeau, C., Jozsa, R., Peres, A., Wootters, W. K.: Teleporting an unknown quantum state via dual classical and Einstein-Podolsky-Rosen channels. Phys. Rev. Lett. **70** (1993) 1895–1899.
8. Bennett, C. H., Brassard, G., Mermin, N. D.: Quantum cryptography without Bell's theorem. Phys. Rev. Lett. **68** (1992) 557–559.
9. Bohm, D. J.:. "Quantum Theory". Prentice-Hall, Englewood Cliffs, N.J. 1951.
10. Ekert, A. K., Rarity, J. G., Tapster, P. R., Palma, G. M.: Practical quantum cryptography based on two-photon interferometry. Phys. Rev. Lett. **69** (1992) 1293–1295.
11. Mu, Y., Seberry, J., Zheng, Y.: Shared cryptographic bits via quantized quandrature phase amplitudes of light. Optics Communications (1996).

AUTHOR INDEX

Springer
and the
environment

At Springer we firmly believe that an
international science publisher has a
special obligation to the environment,
and our corporate policies consistently
reflect this conviction.
We also expect our business partners –
paper mills, printers, packaging
manufacturers, etc. – to commit
themselves to using materials and
production processes that do not harm
the environment. The paper in this
book is made from low- or no-chlorine
pulp and is acid free, in conformance
with international standards for paper
permanency.

Lecture Notes in Computer Science

For information about Vols. 1–1101

please contact your bookseller or Springer-Verlag